Lecture Notes in Statistics

Vol. 1: R.A. Fisher: An Appreciation. Edited by S.E. Fienberg and D.V. Hinkley. XI, 208 pages, 1980.

Vol. 2: Mathematical Statistics and Probability Theory. Proceedings 1978. Edited by W. Klonecki, A. Kozek, and J. Rosiński. XXIV, 373 pages, 1980.

Vol. 3: B.D. Spencer, Benefit-Cost Analysis of Data Used to Allocate Funds. VIII, 296 pages, 1980.

Vol. 4: E.A. van Doorn, Stochastic Monotonicity and Queueing Applications of Birth-Death Processes. VI, 118 pages, 1981.

Vol. 5: T. Rolski, Stationary Random Processes Associated with Point Processes. VI, 139 pages, 1981.

Vol. 6: S.S. Gupta and D.-Y. Huang, Multiple Statistical Decision Theory: Recent Developments. VIII, 104 pages, 1981.

Vol. 7: M. Akahira and K. Takeuchi, Asymptotic Efficiency of Statistical Estimators. VIII, 242 pages, 1981.

Vol. 8: The First Pannonian Symposium on Mathematical Statistics. Edited by P. Révész, L. Schmetterer, and V.M. Zolotarev. VI, 308 pages, 1981.

Vol. 9: B. Jørgensen, Statistical Properties of the Generalized Inverse Gaussian Distribution. VI, 188 pages, 1981.

Vol. 10: A.A. McIntosh, Fitting Linear Models: An Application on Conjugate Gradient Algorithms. VI, 200 pages, 1982.

Vol. 11: D.F. Nicholls and B.G. Quinn, Random Coefficient Autoregressive Models: An Introduction. V, 154 pages, 1982.

Vol. 12: M. Jacobsen, Statistical Analysis of Counting Processes. VII, 226 pages, 1982.

Vol. 13: J. Pfanzagl (with the assistance of W. Wefelmeyer), Contributions to a General Asymptotic Statistical Theory. VII, 315 pages, 1982.

Vol. 14: GLIM 82: Proceedings of the International Conference on Generalised Linear Models. Edited by R. Gilchrist. V, 188 pages, 1982.

Vol. 15: K.R.W. Brewer and M. Hanif, Sampling with Unequal Probabilities. IX, 164 pages, 1983.

Vol. 16: Specifying Statistical Models: From Parametric to Non-Parametric, Using Bayesian or Non-Bayesian Approaches. Edited by J.P. Florens, M. Mouchart, J.P. Raoult, L. Simar, and A.F.M. Smith, XI, 204 pages, 1983.

Vol. 17: I.V. Basawa and D.J. Scott, Asymptotic Optimal Inference for Non-Ergodic Models. IX, 170 pages, 1983.

Vol. 18: W. Britton, Conjugate Duality and the Exponential Fourier Spectrum. V, 226 pages, 1983.

Vol. 19: L. Fernholz, von Mises Calculus For Statistical Functionals. VIII, 124 pages, 1983.

Vol. 20: Mathematical Learning Models – Theory and Algorithms: Proceedings of a Conference. Edited by U. Herkenrath, D. Kalin, W. Vogel. XIV, 226 pages, 1983.

Vol. 21: H. Tong, Threshold Models in Non-linear Time Series Analysis. X, 323 pages, 1983.

Vol. 22: S. Johansen, Functional Relations, Random Coefficients and Nonlinear Regression with Application to Kinetic Data. VIII, 126 pages, 1984.

Vol. 23: D.G. Saphire, Estimation of Victimization Prevalence Using Data from the National Crime Survey. V, 165 pages, 1984.

Vol. 24: T.S. Rao, M.M. Gabr, An Introduction to Bispectral Analysis and Bilinear Time Series Models. VIII, 280 pages, 1984.

Vol. 25: Time Series Analysis of Irregularly Observed Data. Proceedings, 1983. Edited by E. Parzen. VII, 363 pages, 1984.

Vol. 26: Robust and Nonlinear Time Series Analysis. Proceedings, 1983. Edited by J. Franke, W. Härdle and D. Martin. IX, 286 pages, 1984.

Vol. 27: A. Janssen, H. Milbrodt, H. Strasser, Infinitely Divisible Statistical Experiments. VI, 163 pages, 1985.

Vol. 28: S. Amari, Differential-Geometrical Methods in Statistics. V, 290 pages, 1985.

Vol. 29: Statistics in Ornithology. Edited by B.J.T. Morgan and P.M. North. XXV, 418 pages, 1985.

Vol. 30: J. Grandell, Stochastic Models of Air Pollutant Concentration. V, 110 pages, 1985.

Vol. 31: J. Pfanzagl, Asymptotic Expansions for General Statistical Models. VII, 505 pages, 1985.

Vol. 32: Generalized Linear Models. Proceedings, 1985. Edited by R. Gilchrist, B. Francis and J. Whittaker. VI, 178 pages, 1985.

Vol. 33: M. Csörgő, S. Csörgő, L. Horváth, An Asymptotic Theory for Empirical Reliability and Concentration Processes V, 171 pages, 1986.

Vol. 34: D.E. Critchlow, Metric Methods for Analyzing Partially Ranked Data. X, 216 pages, 1985.

Vol. 35: Linear Statistical Inference. Proceedings, 1984. Edited by T. Caliński and W. Klonecki. VI, 318 pages, 1985.

Vol. 36: B. Matérn, Spatial Variation. Second Edition. 151 pages, 1986.

Vol. 37: Advances in Order Restricted Statistical Inference. Proceedings. 1985. Edited by R. Dykstra, T. Robertson and F.T. Wright. VIII, 295 pages, 1986.

Vol. 38: Survey Research Designs: Towards a Better Understanding of Their Costs and Benefits. Edited by R.W. Pearson and R.F. Boruch. V, 129 pages, 1986.

Vol. 39: J.D. Malley, Optimal Unbiased Estimation of Variance Components. IX, 146 pages, 1986.

Vol. 40: H.R. Lerche, Boundary Crossing of Brownian Motion. V, 142 pages, 1986.

Vol. 41: F. Baccelli, P. Brémaud, Palm Probabilities and Stationary Queues. VII, 106 pages, 1987.

Vol. 42: S. Kullback, J.C. Keegel, J.H. Kullback, Topics in Statistical Information Theory. IX, 158 pages, 1987.

Vol. 43: B.C. Arnold, Majorization and the Lorenz Order: A Brief Introduction. VI, 122 pages, 1987.

ctd. on inside back cover

Lecture Notes in Statistics

Edited by J. Berger, S. Fienberg, J. Gani,
K. Krickeberg, I. Olkin, and B. Singer

74

P. Barone A. Frigessi
M. Piccioni (Eds.)

Stochastic Models, Statistical Methods, and Algorithms in Image Analysis

Proceedings of the Special Year on Image Analysis
held in Rome, Italy, 1990

Springer-Verlag
Berlin Heidelberg New York London Paris
Tokyo Hong Kong Barcelona Budapest

Editors

Piero Barone
Arnoldo Frigessi
Istituto per le Applicazioni del Calcolo, C.N.R.
Viale del Policlinico 137, 00161 Rome, Italy

Mauro Piccioni
Dipartimento di Matematica Pura ed Applicata
Università di L'Aquila
L'Aquila, Italy

Mathematical Subject Classification: 62-06, 68U10

ISBN-13:978-0-387-97810-9 e-ISBN-13:978-1-4612-2920-9
DOI: 10.1007/978-1-4612-2920-9

Typesetting: Camera ready by author

47/3140-543210 – Printed on acid-free paper

This volume contains a selection of papers based on the contributions to the 1990 Concentration Year on *Stochastic Models, Statistical Methods and Algorithms in Image Analysis* held at IAC in Rome. The Istituto per le Applicazioni del Calcolo - C.N.R. (IAC) has been organizing since 1988 concentration years on relevant topics in applied mathematics. Scientists in different disciplines are invited to give lectures and stimulate workshops and informal discussions.

Image analysis has known a new development since the introduction of sophisticated statistical and probabilistic methods and the increasing availability of computing facilities. Although motivated by crucial applications, this approach to image processing has also become a source of stimulating mathematical problems and a field in which the boundary between pure and applied research is hard to find.

The papers in this collection range from research notes to state-of-the-art surveys and reviews of relevant applications. The contributions can be roughly classified into three main areas: inference for Markov random fields, methods and algorithms for image restoration and object recognition, applications.

We wish to thank all contributors that made this volume possible, and the participants for their active presence during the Concentration Year. The Scientific Committee included: P. Baldi, D. Geman, U. Grenander, M. Piccioni, E. Presutti, B.Ripley, Y.Vardi. Their advice and suggestions were fundamental for the organization of the Year. Alberto Tesei, director of IAC, followed the whole project and was always of great help. The assistance of the IAC staff, expecially G. Castellini, C. Cecchetti, L. Ellari, A. Guspini, C. Ruggeri, G. Silveri, was always excellent.

Roma, November 15, 1991

Piero Barone Arnoldo Frigessi Mauro Piccioni

TABLE OF CONTENTS

EDGE PRESERVING IMAGE RESTORATION

Mona Abdalla and Jim Kay
Department of Mathematics and Statistics
University of Stirling, Scotland, UK

1 Introduction

Algorithms for the restoration of degraded images often contain unknown parameters which must be selected judiciously by the analyst. This selection process is particularly difficult when there are several parameters and therefore we believe that it is important to develop an automatic means of estimating them. Many real images contain discontinuities and as a result image restoration algorithms that involve smoothing must inevitably produce serious bias unless these edges are modelled explicitly (or implicitly) within the prior model for the unknown image. This article describes a simple approach which is intended to resolve both of these problems.

We develop an algorithm for the simultaneous determination of the unknown image,the edge-strengths and the unknown parameters. Our work has been stimulated by the seminal work of the Gemans, Blake ,Zisserman and Besag [5, 3, 2]. We employ a deterministic approach following on from the work reported in [2, 8], but a stochastic version of our algorithm could be produced by employing the Gibb's sampler. The article proceeds as follows. In section 2 we develop our notation and describe our algorithm in the case where the parameters and edge-strengths are known. Section 3 outlines our approach to modelling edges and estimating their strength. In section 4 we discuss the issue of the estimation of parameters which leads to simple formulae for the estimators. We describe our restoration algorithm in section 5 ; it is deterministic and iterative and successively updates the estimates for the unknown image, the edge-strengths and the unknown parameters in a cyclic process extending the simultaneous estimation-restoration approach of Besag [2]. There have been other approaches to the development and implementation of algorithms involving simultaneous restoration with parameter estimation;see [7] for a brief summary of the relevant literature. In section 6 we describe some practical experiments with the algorithm using real gamma-camera data of a human heart and also a synthetic image.

Finally in section 7 we provide some conclusions and discuss some possibilities for future extensions of this work.

2 Combining the ICM and EM algorithms

We now provide some notation and develop the basis of our algorithm in the case when the unknown parameters are assumed to be known. Our images are two-dimensional but we use one-dimensional notation by stacking the pixel values in a raster scan using lexicographic ordering. Let the vectors x and y denote respectively the true and observed images. The point-spread matrix is denoted by H. We consider two different models for the generation of the data y from the true image x.

$$y_i \sim Po(\sum_{j \in B_i} h_{ij} x_j), \ (i \in D)$$

and

$$y_i = \sum_{j \in B_i} h_{ij} x_j + \epsilon_i, \ (i \in D)$$

where the $\{\epsilon_i\}$ are independent and identically distributed N(0,σ^2) random variables. Here D is the set of pixels in 'data' space, B_i denotes the set of pixels in 'object' space that contribute to y_i. We make the following assumption of conditional independence.

$$p(y \mid x) = \prod_{i \in D} p(y_i \mid x_{B_i})$$

We take as a prior for x the pairwise-interaction markov random field model

$$p(x) \propto \exp(-V(x)),$$

where

$$V(x) = \sum_{i \sim j} G(x_i, x_j).$$

Here i $\sim j$ means that pixels i and j are neighbours. The ICM algorithm [2] proceeds as follows. Given a current estimate of x, x^{old}, a new estimate is obtained via

$$x_r^{new} = argmax \ p(x_r \mid x_{\partial r}^{old}, y), \ (\forall r \in D)$$

That is, we require to calculate local conditional modes. This involves the simultaneous solution of the usually highly non-local and non-linear equations

$$\partial/\partial x_r\, p(x_r \mid x_{\partial r}^{old}, y) = 0, \;\; (\forall r \in D),$$

which destroy the simplicity of the ICM algorithm. In order to ease the computational burden we employ the EM algorithm [4] within ICM to perform these local optimisations. This leads to a set of simultaneous non-linear equations involving only co-ordinatewise nonlinearities. In the important case of the Poisson additive regression model we will obtain an explicit solution which will guarantee positivity and is ,we believe, a new algorithm. We now develop the algorithm in the Poisson case but the analysis holds more generally.

Following Green [8], we define as the complete data $\{z_{ij}\}$, where these random variables are assumed to be mutually independent with

$$z_{ij} \sim Po(h_{ij}x_j), \;\; (j \in B_i; i \in D)$$

Then

$$y_i \sim Po(\sum_{j \in B_i} h_{ij}x_j), \;\; (i \in D)$$

and so the data model is preserved. We now consider the E and M steps in this particular context.

E step

We 'estimate' z_{ij} by

$$E(z_{ij} \mid y_i, x^{old})$$

M step

We choose x_r, for each r, to maximise the expectation of

$$\prod_{i \in D} \prod_{j \in B_i} p(z_{ij} \mid x_j) p(x)$$

In this expression note that in the calculation of x_r all the other components of x are fixed at their 'old' values. We take

$$V(x) = \sum_{i \sim j} (x_i - x_j)^2.$$

In the Poisson case , the E-step gives

$$\hat{z}_{ij} = E(z_{ij} \mid y_i, x^{old}) = y_i h_{ij} x_j / \sum_{j \in B_i} h_{ij} x_j.$$

The M-step involves the solution of the quadratic equation

$$2\beta n_r x_r^2 - (2\beta n_r \bar{x}_{\partial i}^{old} - \sum_{i \in D_r} h_{ir}) x_r - \sum_{i \in D_r} \hat{z}_{ir} = 0$$

for x_r at the rth pixel for each r. Here n_r denotes the number of pixels in the neighbourhood of pixel r and $D_r = \{i : r \in B_i\}$. This solution may be shown to yield a maximum and to be positive. It is possible to prove a local convergence result for this particular prior which holds for every positive value of the smoothing parameter β. This represents an improvement over the OSL algorithm [8] in this particular case. See [11] for further details.

3 Preservation of Edges

Real images often contain discontinuities which may be incorporated into the prior by taking $V(x)$ to be

$$V(x) = \sum_{i \sim j} (x_i - x_j)^2 (1 - e_{ij}),$$

where e_{ij} lies in [0,1] ; the extremes $e_{ij} = 0, 1$ indicate respectively the absence or presence of an edge between pixels i and j ; see [3, 5]. An alternative interpretation of this prior is that it allows the restoration process to adapt to the underlying smoothness of the unknown image by smoothing to different degrees in different parts of the image. We require to 'estimate' the $\{e_{ij}\}$ from the data and use the simple idea of directional split-smoothers for this purpose.

Consider the direction through pixels i and j. Calculate an average using the pixels on the same side of the direction as pixel i and similarly for the pixels on the same side as pixel j, denoting them respectively by $m(P_i)$ and $m(P_j)$. We then estimate the strength of the edge by

$$c_{ij} = \mid m(P_i) - m(P_j) \mid .$$

We then map then edge-strengths into the interval [0,1] using the following filter. Consider the empirical distribution of the edge-strengths $\{c_{ij}\}$. Define q_1 and q_2 to be the p_1th and p_2th quantiles,respectively, of the $\{c_{ij}\}$. In practice p_1 and p_2 must be supplied by the user. Now estimate e_{ij} by $F(c_{ij})$, where

$$\mathrm{F(z)} = \begin{cases} 0 & z \leq q_1 \\ (z-q_1)/(q_2-q_1) & q_1 \leq z \leq q_2 \\ 1 & z \geq q_2 \end{cases}$$

In experiments we found that the performance of the algorithm depends critically on the supplied value of p_2 but was robust against the selection of the p_1 parameter. Essentially this means that the user must supply a value for p_2, with $1 - p_2$ being the percentage of definite edges in the unknown image. Hence the attempt at automation has failed. It is interesting to note that Blake and Zisserman have an additional parameter in their prior which they interpret as the cost incurred by breaking the continuity. However in their approach it is not easy to set this parameter. In contrast the p_2 parameter has the advantage of having a clear interpretation in terms of the restoration task and so is easier to determine from previous experience.

4 Parameter Estimation

In order that our algorithm be as automatic as possible we require to estimate the unknown parameters in the prior model and data model, respectively. Given a current estimate of x, x^{old}, we estimate the parameters of the data model by maximising the estimated likelihood

$$\prod_{i \in D} p(y_i \mid x_{B_i}^{old}, \theta)$$

In the Poisson data model there are no additional (nuisance) parameters but in the Gaussian case we obtain an explicit formula for the parameter estimates. This is given by

$$\hat{\sigma}^2 = \frac{\sum_{i \in D}(y_i - \sum_{j \in B_i} h_{ij} x_j^{old})^2}{n}.$$

Here n denotes the number of pixels in D. We estimate unknown parameters in the prior model by maximising the 'plug-in' pseudo-likelihood

$$\prod_{i \in D} p(x_i^{old} \mid x_{\partial i}^{old}, \beta)$$

Note that our primary purpose is that of restoration and this particular approach would not be recommended for parameter estimation per se as it is likely to suffer from appreciable bias because the current estimate for x is being taken as the truth. We will discuss this further in section 7 below. Note also that in this maximisation we plug in the current estimates of the edge-strengths as if they were known. This leads to an explicit, simple formula for the estimate of β. This is given by

$$\hat{\beta}^{-1} = \frac{2}{n} \sum_{i \in D} (x_i^{old} w_i - v_i)^2 / w_i),$$

where

$$w_i = \sum_{j \in B_i} (1 - e_{ij})$$

and

$$v_i = \sum_{j \in B_i} x_j^{old} (1 - e_{ij}).$$

The theoretical properties of this estimator are unknown. However in the case of undegraded images the pseudo-likelihood estimator is consistent ; see [6].

5 An iterative edge-preserving restoration algorithm

The algorithm has three basic features, namely the estimation of x, the estimation of the edge-strengths e and the estimation of the unknown parameters θ and β. In the Gaussian case the iterative formula for updating x given the current parameter estimates and estimated edge-strengths is

$$x_r^{new} = \frac{\sum_{i \in D_r} h_{ir}(y_i - \sum_{j \neq r} h_{ij} x_j^{old}) + 2\beta\sigma^2 v_r}{\sum_{i \in D_r} h_{ir}^2 + 2\beta\sigma^2 w_r}.$$

In the Poisson case we take x_r^{new} to be the positive solution of the quadratic equation

$$2\beta w_r x_r^2 - (2\beta v_r - \sum_{i \in D_r} h_{ir}) x_r - \sum_{i \in D_r} \hat{z}_{ir} = 0.$$

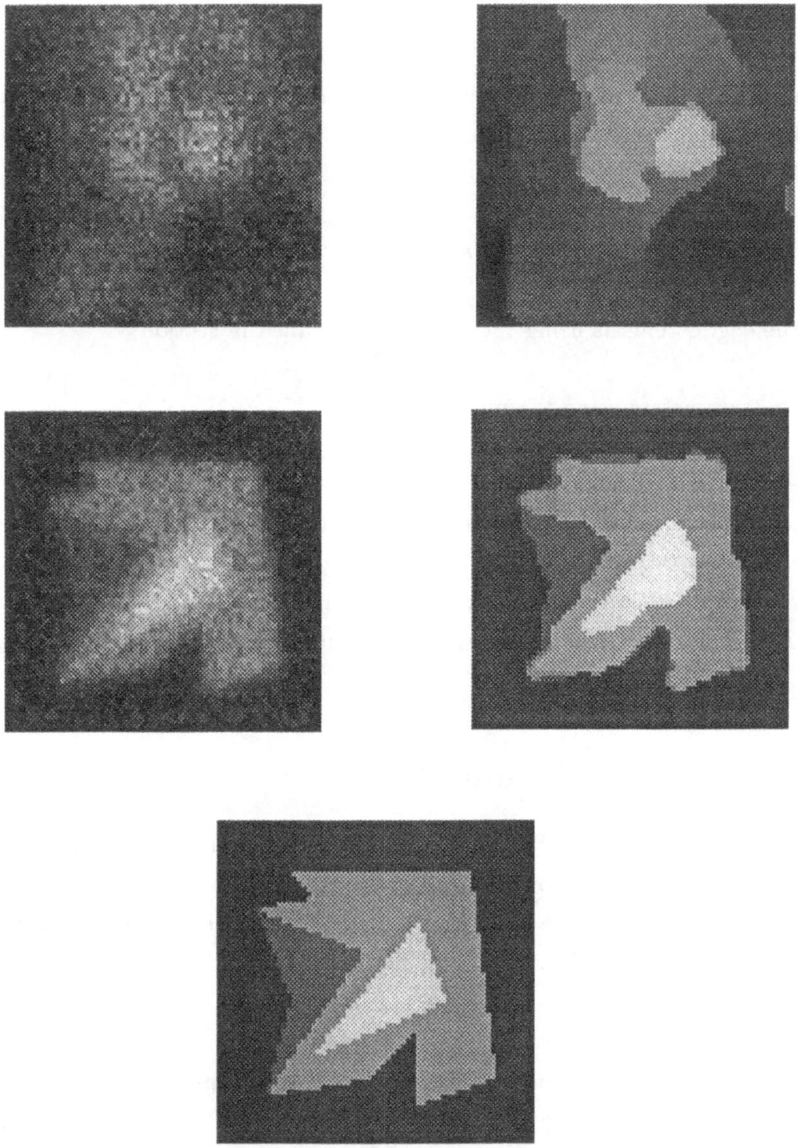

Figure 1 : Reading from top to bottom and from left to right we have the HEART data with restoration and the degraded,restored and true versions of CONIM.

The algorithm proceeds as follows.

Step 0

Choose an initial estimate for x. This could be the data or the uniform distribution or a random image. Use guesses for the unknown parameters θ and β and take e to be zero , that is, switch off all the edges.

Step 1

Perform one iteration of the ICM-EM algorithm to produce a new estimate of x.

Step 2

Estimate the edge-strengths using the procedure outlined in section 3.

Step 3

Estimate the parameters using the methods described in section 4.

Step 4

Repeat step 1 until convergence.

In our practical experience this algorithm has converged with real gamma-camera images of the human heart and also with synthetic images. While it is possible to prove a local convergence result when the parameters are known the convergence is not guaranteed theoretically when the parameters are changed at each iteration. The theoretical behaviour is therefore a challenging open problem.

6 Practical Experiments

We report here on experiments conducted with two images CONIM and HEART. CONIM is a synthetic 64 by 64 piecewise-constant image having grey-levels 100,120,150 and 180. We subjected CONIM to a 5 by 5 geometric blur with parameter 0.5 and then added independent Gaussian noise with a standard deviation of 10. The degraded version of CONIM , the restoration and the true image are displayed in Figure 1. We were relatively pleased with the quality of the restoration although clearly further work could be performed to sharpen up the edges. Our results with this algorithm compare favourably with our previous work [1, 10] on these images.

HEART is a real image obtained from the Department of Nuclear Medicine at Stobhill Hospital,Glasgow. The physicists at the hospital measured the point-spread function of the gamma-camera and these estimated values were used in the computations; the blur was thought to be of size 27 by 27. The HEART data and the restoration obtained using the ICM-EM algorithm appear in Figure 1. Clearly we do not know the truth here but

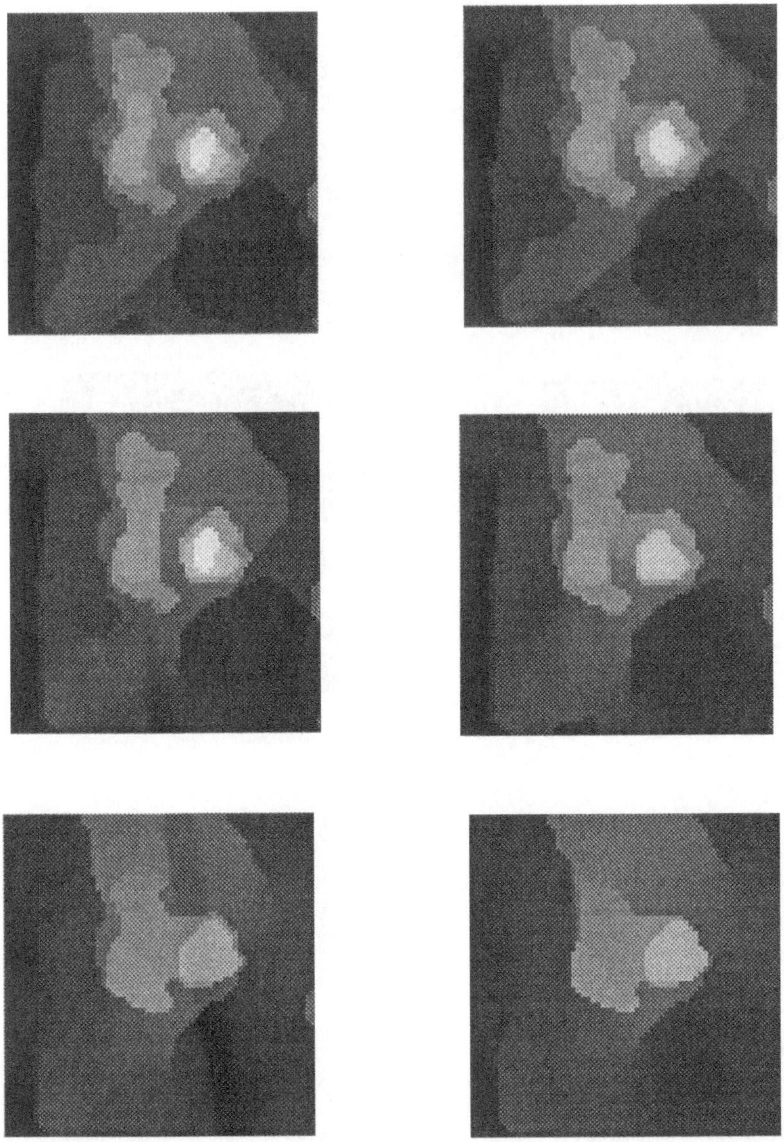

Figure 2 : Reading from top to bottom and from left to right we have a p2 movie with values .75 ,.8 ,.83 ,.85 ,.9 ,.93 and with p1=0.1 throughout

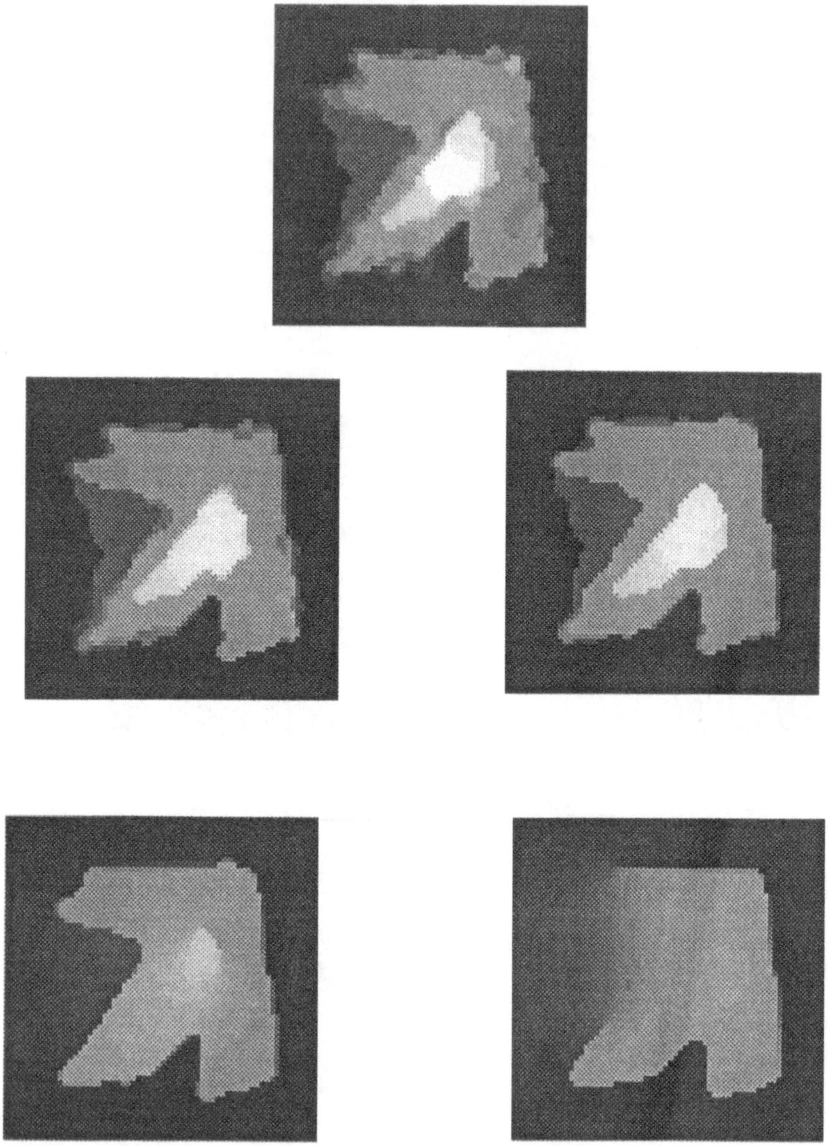

Figure 3 : Reading from top to bottom and from left to right we have a p2 movie with values .7 ,.8 ,.85 ,.9 ,.95 and with p1=0.1 throughout the physicists at the hospital thought that this result was promising in that a reasonably good estimate of the left ventricle (the white region) was possible. The measurement of the area of this ventricle forms part of the assessment of the efficiency with the heart is functioning.

We found that the quality of the restoration was relatively insensitive to the choice of p_1 but depended rather critically on the value supplied for p_2. Figures 2 and 3 show p_2-movies for the HEART and CONIM images,respectively. For these images it is important for p_2 to be in the range [0.8,0.9].

7 Discussion and Future Work

We have reported on some simple experiments designed with the ultimate aim of designing an automatic algorithm for the restoration of images that contain discontinuities by outlining procedures for the determination of continuous edge-strengths and the on-line estimation of parameters . We have worked at the pixel level without attempting to incorporate any global information about the unknown scene into the restoration process. This would clearly improve the results by providing greater connectivity of the edge-elements and continuity of the extracted boundaries.

However we are encouraged by the results. Work is currently in progress to develop better methods for the estimation of the unknown prior parameters. One idea is to use a local empirical Bayes approach within the penalised EM algorithm. Such an approach can be used even when the degradation process contains blurring. This can be compared with a global empirical Bayes approach and also with a hyper-parametric approach: see [9]. We also are extending our approach to deal with the restoration of sequences of images and three-dimensional scenes. In the case of sequences the discontnuities in the time dimension will be indicative of motion of the underlying objects in the scene and so there are possibilities for the detection of motion in the presence of appreciable noise and with non-Gaussian degradation processes. We are also developing multi-resolution approaches using the penalised EM algorithm.

8 Acknowledgements

We wish to thank the SERC for financial support under their Initiative in Complex Stochastic Systems. We also wish to express our gratitude to David Sumner and Andy Kelman of the Department of Nuclear Medicine in Glasgow's Stobhill hospital for providing the heart data.

References

[1] ABDALLA, M. (1990). Contribution to the discussion of Silverman et al. *J. R. Statist. Soc. B 306-308.*

[2] BESAG, J. E. (1986). On the statistical analysis of dirty pictures(with discussion). *J. R. Statist. Soc. B 259-302.*

[3] BLAKE, A. and ZISSERMAN, A. (1987). Visual Reconstruction. *MIT Press, Cambridge Mass.*

[4] DEMPSTER, A.P., LAIRD, N.M. and RUBIN, D.B. (1977). Maximum likelihood from incomplete data via the EM algorithm (with discussion). *J. R. Statist. Soc. B 1-38.*

[5] GEMAN, S. and GEMAN, D. (1984). Stochastic relaxation, Gibb's distributions and the Bayesian restoration of images. *IEEE Trans. Pattern Analysis and Machine Intelligence. PAMI-6, 721-741.*

[6] GEMAN, S. and GRAFFIGNE, C. (1987). Markov random field models and their applications to computer vision. *In A.M.Gleason (Ed.) Proc. Int. Congress of Mathematicians 1986. Washington DC. American Mathematical Society.*

[7] GRAY, A.J., KAY ,J.W. and TITTERINGTON, D.M. (1991). On the estimation of noisy binary Markov random fields. *Pattern Recognition. To appear.*

[8] GREEN, P. J. (1990). On the use of the EM algorithm for penalised likelihood estimation. *J. R. Statist. Soc. B 443-452.*

[9] GULL, S. F. (1989). Developments in maximum entropy data analysis. *In Maximum Entropy and Bayesian Methods, J.Skilling(Ed.), Kluwer, Dordrecht.*

[10] KAY, J. W. (1990). Contribution to the discussion of Silverman et al. *J. R. Statist. Soc. B 304-306.*

[11] KAY, J. W. (1991). Some convergence properties of the ICM-EM algorithm. *In preparation.*

BOLTZMANN MACHINES : HIGH-ORDER INTERACTIONS AND SYNCHRONOUS LEARNING

Robert AZENCOTT(*)

École Normale Supérieure (DIAM)

and

Université Paris-Sud

1. INTRODUCTION

A now classical innovative paper [H.S.A.] by Hinton-Sejnowski-Ackley introduced a class of formal neural networks, the Boltzmann machines, governed by *asynchronous* stochastic dynamics, *quadratic* energy functions, and *pairwise* interactions defined by synaptic weights. One of the exciting aspects of [H.S.A.] was the derivation of a locally implementable learning rule linked to a scheme of decreasing (artificial) temperatures, in the spirit of simulated annealing.

However actual simulations of these machines for pattern classification problems have run into practical difficulties, the main one being the heavy load of computing time involved. Thus, the "neural network community" has often had a tendancy to consider the Boltzmann machines as useless slow learners.

We feel that these speed problems are enhanced by two facts, inherent to the

(*) CNRS labs : [LMENS] and [Stat. App. Univ. Paris-Sud]
Address : R. Azencott, ENS, 45 rue d'Ulm - F-75230 Paris Cedex 05
Tel. 33-1-47702465 - FAX : 33-1-48010754.

original [H.S.A.] presentation : at low temperature T, the stabilization times of these stochastic networks are *extremely long* (they grow exponentially with $1/T$), and moreover the *sequential update* of neurons, where only one neuron fires at a time is an obvious crucial cause of lengthy computations.

Hence we have been suggesting the use of learning at *constant temperatures* (suitably selected) and the implementation of highly *synchronous* neural updates, which in view of the availability of parallel hardware is quite natural in this context.

This raised the mathematical problem of devising suitable *synchronous learning rules*, which we have described in a recent paper [A], for the pairwise interactions situation. An interesting new feature was the natural emergence of *one-step delayed correlations* between the activities of pairs of neurons, as crucial indicators for weights updates.

However, in many low-level vision applications, the use of Markov field approaches instigated by D. and S. Geman [G.G], can be *reinterpreted in terms of sequential* Boltzmann machines for which *the energies are far more general than quadratic functions and for which cliques of interacting neurons have cardinals higher than three.* We have sketched the reinterpretation of Markov field approaches in the forthcoming paper [A]. Of course this requires the derivation of *new learning rules for sequential dynamics, general energies and high-order interactions.* The proper formal setup is described here and involves *a pair of dual networks* : the neural network S and the clique indicators network K. The synapses link *only* individual neurons to individual cliques, and synaptic weights w_c are indexed by cliques $C \in K$.

Our *sequential learning rules* for these generalized Boltzmann machines are still *locally implementable* ; they involve the *correlation between current clique activity* $J_C(X^n)$ *and the current score* $\lambda(X^n_R)$ of the machine response. We compute these scores through the introduction of fairly arbitrary *loss functions* which compare desired and current responses of the machine.

For *very specific and simple choices of loss function*, these multiple interaction, general energies, sequential learning rules include in particular a weight update Δw_c proportional to the difference in average clique activity $E(J_c)$ between two regimes (clamped output and unclamped output) which generalizes by a quite different route the sequential learning rule derived in [H.S.A.] for quadratic energy functions and pairwise interactions.

Of course, it is quite tempting and natural to study these general Boltzmann machines in the synchronous case too, in order to gain important speed factors. It turns out that *synchronous dynamics with high-order interactions* involve a serious mathematical difficulty : *the equilibrium probability distribution M on the set of global neurons configurations cannot be computed explicitly* in general. Hence in the present paper, we use more sophisticated probabilistic tools to compute the gradient $\frac{\partial M}{\partial w_c}$, and to interpret the results at the empirical level.

We have thus obtained here *quite new learning rules for synchronous dynamics in presence of high-order interactions and general energies.* These rules are still *implementable locally.* For each clique C they involve *the sum of strings delayed correlations* between the *past clique unexpected transition activity* $u_c(X^{n-k-1}, X^{n-k})$ and *the current score* $\lambda(X_R^n)$ of the machine response. Here again scores involve general loss functions, computed at the level of response units and fed back to all clique indicators. For a clique C, the *transition activity* between successive configurations x and y is computed by $[\sum_{s \in C} J_c(x_{c-s}, y_s)]$, a quantity which appears naturally in our gradient computations. Finally we sketch a few results on the *choice of optimal temperatures* T, and obtain interesting physical and probabilistic interpretations of the gradient, with respect to T, of the expected score.

In [A], we have deduced from these synchronous learning rules a set of *algorithms for parameter estimation in synchronous Markov field model fitting*, an interesting new twist in Markov field approaches for low-level vision models.

The learning rules derived here are currently being tested experimentally in our research group DIAM at ENS (Paris), using simulations on computers offering high degrees of parallelization (Cray, connection machine). The possibilities of implementing generalized Boltzmann machines on specialized hardware are being currently evaluated in collaboration with P. Garda and other researchers at ENS Paris and I.E.F. Orsay. The vision application of synchronous Markov field models are currently studied in collaboration with other researchers at ENS Paris (*cf.* recent work in collaboration with A. Doutrieux, L. Younès, J. Lacaille).

2. NEURAL NETWORKS WITH MULTIPLE INTERACTIONS

Consider an arbitrary finite set S of "formal neurons". The *state* x_s of neuron s will be a variable with values in an arbitrary finite set A. The *configuration* $x = (x_s)_{s \in S}$ of the network S is an arbitrary element of $\Omega = A^S$.

Fix an arbitrary family K of subsets of S, which will be called the set of *cliques* in the network S. The *activity* of any clique $C = \{s_1 \cdots s_p\} \in K$ will be measured by an *interaction potential*

$$J_C(x) = J_C(x_{x_1}, \cdots, x_{s_p}).$$

These interaction potentials are *arbitrary numerical functions of the clique configuration*

$$x_C = (x_{s_1}, \cdots, x_{s_p}).$$

For each clique $C \in K$, introduce a numerical parameter $w_c \in \mathbf{R}$ which will be called the *clique weight.* The *weight vector* $w = (w_c)_{c \in K}$ in $\mathbf{R}^K$ will parametrize the network architecture defined by $\{A, S, K, (J_C)_{C \in K}$.

Recall that widely used standard models of format neurons tend to consider only pairwise interactions, that is to say situations where all cliques have cardinal ≤ 2. On the other hand, Markov field theory and particularly its application to image analysis often involve cliques of cardinal > 3.

3. SEQUENTIAL STOCHASTIC DYNAMICS

Introduce now the *sequential energy function*

$$U(x) = \sum_{C \in K} w_C \, J_C(x) \tag{3.1}$$

which measures the *weighted activity of the cliques.* A classical Glauber type of sequential stochastic dynamics can now be defined to ensure that the sequential network has at equilibrium (in the long run) a Gibbs probability distribution on the set Ω of configurations, given by

(3.2)
$$G(x) = \frac{1}{Z} \exp\left[-\frac{U(x)}{T}\right]$$

(3.3)
$$Z = \sum_{y \in \Omega} \exp\left[-\frac{U(y)}{T}\right],$$

where Z is the *partition function*,and T is a positive fixed parameter called the *temperature.*

In the sequential dynamics, at each instant $n \in N$, only one of the neurons attempts to modify its state. Call s_n its index, which is generally preassigned by an arbitrary deterministic sequence $(s_1 \cdots s_n)$ visiting *periodically* all neurons $s \in S$. Such a sequence can also be random, provided *it is ergodic and uniformly distributed on S*. In either case, whenever the current random configuration $X^n = x$, and the neuron $s_n = s \in S$ is preassigned as seen above for a possible change of state at time n, then the new configuration X^{n+1} coincides with X^n at all neurons in $[S - s_n]$, and the conditional distribution of $X_{s_n}^{n+1}$ is given by :

(3.4)
$$P\left(X_{s_n}^{n+1} = a \,|\, X^n = x, X^{n-1}, \cdots, X^0\right) = G_s(a \,|\, x_{S-s})$$

where $G_s(a \,|\, z)$ is the *conditional probability* $G(X_s = a \,|\, X_{S-s} = z)$ under the Gibbs distribution G defined in (3.2).

Define the set Ns of *neighbours of neurons* s as the set of all $t \in S$ such as $t \neq s$ and there is some clique $C \in K$ containing both s and t. Then the *local action potential* of configuration x at site s is defined by :

(3.5)
$$V_s(x) = V_s(x_s, x_{Ns}) = - \sum_{\{\text{cliques } C \text{ containing} s\}} w_C \, J_C(x)$$

and (3.1) classically yields :

(3.6)
$$G_s(a \,|\, x_{S-s}) = \frac{e^{V_s(a, x_{Ns})/T}}{\sum_{\alpha \in A} e^{V_s(\alpha, x_{Ns})/T}}$$

which shows that the random updating of neuron s is purely "*local*" since it involves only the configuration x restricted to s and its set of neighbours Ns.

As is well known, whatever the initial configuration X^0 of the network, the limit distribution is given by :

(3.7) $$\lim_{n\to+\infty} P(X^n = x) = G(x)$$

where $G(x)$ is the Gibbs distribution (3.1). We point out that *this sequential dynamics takes place at fixed temperature.*

4. SYNCHRONOUS STOCHASTIC DYNAMICS

In view of the purely local computations involved in asynchronous random updates, it is quite tempting to *parallelize completely* the random updating, and hence to define *synchronous stochastic dynamics*

$$P_{\text{syn}}\,(X^{x+1} = y \,|\, X^n, X^{n-1}, \cdots, X^0) = \prod_{s\in S} G_s(y_s \,|\, X^n_{S-s})$$

where G is the Gibbs distribution (3.1) above. This simply means that all neurons $s \in S$ update their states with *simultaneous independent random choices*, each one of the individual random choices being governed by a conditional law computed exactly as in the sequential case, by (3.5).

Since X^n is obviously again an ergodic Markov chain on the state space Ω, *the limit distribution*

(4.2) $$M(x) = \lim_{n\to+\infty} P_{\text{syn}}(X^n = x))$$

exists and does not depend on the initial configurations X^0. However, *in general* M *does not coincide at all with the Gibbs distribution* G.

Of course M is the invariant distribution for P_{syn}, and hence calling $Q(x,y)$, $x \in \Omega$, $y \in \Omega$ the one step transition of P_{syn}, we have :

(4.3) $$\sum_{x\in\Omega} M(x)\,Q(x,y) = M(y)$$

which will be as usual noted in matrix form

$$(4.4) \qquad M\,Q = M.$$

Note also that formula (4.1) implies

$$(4.5) \qquad Q(x,y) = \prod_{s \in S} G_s(y_s \,|\, x_{S-s})$$

with $G_s(.\,|\,.)$ given by (3.6).

The main difficulty below will be the fact that in the generic case M is only known *implicitly* through equations (4.4) (4.5). In fact the only cases where M has been computed explicitly are those where all cliques have cardinal ≤ 2. We refer to our paper [A] for a detailed treatment of this important particular case, which of course includes the case of Boltzmann machines with quadratic energy introduced by Hinton-Sejnowski-Ackley. Let us recall one of our results from [A] :

4.6. THEOREM.— *Assume that all cliques $C \in K$ are of cardinal ≤ 2. Then the synchronous stationary measure M is given by*

$$(4.7) \qquad M(x) = m \prod_{s \in S} \Big[\sum_{a \in A} \exp \frac{1}{T} V_s\,(a, x_{N_s})\Big]$$

where the constant m is determined by $\sum_{x \in \Omega} M(x) = 1$. Moreover M verifies the microbalance equations

$$(4.8) \qquad M(x)\,Q(x,y) = M(y)\,Q(y,x).$$

In particular the doubly infinite stationary synchronous Markov chain $(Y^n)_{n \in \mathbf{Z}}$ is then *reversible* in the sense that

$$(4.9) \qquad P_{\text{syn}}(Y^{n+1} = y \,|\, Y^n = x) = P_{\text{syn}}(Y^n = y \,|\, Y^{n+1} = x) = Q(x,y).$$

5. LOCAL IMPLEMENTATION OF THE STOCHASTIC DYNAMICS

5.1. A pair of dual networks

A convenient architecture is based on a pair (S, K) of dual networks, where S is the network of neurons, and K the network of *clique indicators*. To each clique $C \in K$ is associated *a cell called the indicator of clique C, whose state y_C* belongs to a finite subset of R, and is given by

(5.2) $$y_C = J_C(x).$$

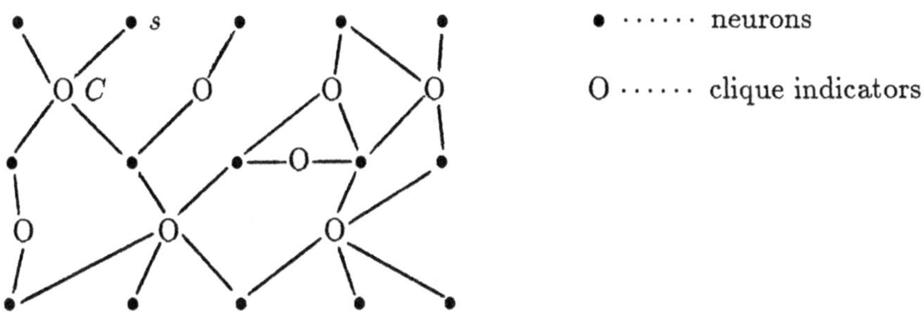

Example of a small network with cliques of various cardinals

5.2. Dynamics of the dual networks

The only *connections* in the pair (S, K) of networks are *links between one neuron s and one clique indicator C*. Namely *such a link exists if and only if s belongs to C*. At this stage such a connection is simply a transmissions channel. From s to C, this channel transmits the state x_s to the clique indicator C, which once it has gotten hold of all the $(x_s)_{s \in S}$ computes deterministically its state $y_C = J_C(x_C)$.

Conversely from C to s, the connection transmits the state y_C to the neuron s, which can then compute its action potential

$$U_s(x) = - \sum_{\text{C containing s}} w_C \, y_C$$

by a simple sum over all cliques connected to s. Then the neuron s can use $U_s(x)$ for its own random updating. Detailed setups for the effective computation of such an update will be proposed elsewhere in a more realistic context, and are easily imagined (see §9.12).

Hence on the pair (S, K) of dual networks, *the sequential and the synchronous stochastic dynamics are purely local.*

6. GENERALIZED BOLTZMANN MACHINES AS PATTERN CLASSIFIERS

Consider a stochastic network S with the above structure, defined by

$$\{A, S, K, (J_C)_{C \in K}\, ,\, w = (w_C)_{C \in K}\}.$$

Select and fix *two arbitrary disjoint subsets* D *and* R *in* S, *the data units* D and *the response units* R. Their union $D \cup R$ will be the set of *visible units*, and the other neurons $H = S - (D \cup R)$ constitute the set of *hidden units*, to come back to a terminology introduced by Hinton-Sejnowski-Ackley [H.S.A.].

The environment provides on the data set D a family of inputs which are random configurations $d \in A^D$. To each input configuration $d \in A^D$, we want to associate a preassigned output configuration $r = F(d) \in A^R$ of the response units. The map $F : A^D \to A^R$ is for the moment assumed known to the supervisor of the learning process, at least on a "*training set*" also called a *set of examples*, which is simply a subset Γ of A^D.

The stochastic network will now be used to emulate F, assuming the weight vector w properly adjusted during a previous training period which will be studied further down.

To compute the response of the network to the input $d \in A^D$, the data units are *clamped* on the initial stimulus $X_D^0 = d$ while the rest of the network starts with an arbitrary initial configuration, and evolves freely according to one of the two stochastic dynamics defined above (sequential or synchronous), *at fixed temperature* T. When stochastic equilibrium is (approximately) reached, the response units still have a *random configuration* $X_R^n \in A^R$, but the asymptotic distribution of X_R^n is well defined. When the data units D are clamped on the input $d \in A^D$, the network configuration x remains in the set

$$\Omega^d = \{x \in \Omega = A^S \,|\, x_D = d\} \tag{6.1}$$

which is of course isomorphic to the set A^{S-D} of configurations for the *reduced network* $(S - D)$. Then the *sequential equilibrium distribution for the (sequential) Markov chain* (X^n) *with clamped inputs* $\{X_D^n \equiv d$ for all $n \geq 0\}$ is the Gibbs measure on Ω^d

$$\begin{aligned} G^d(x) &= \frac{1}{Z^d} \exp\left[-\frac{1}{T} U(x)\right] \qquad \text{for all } x \in \Omega^d \\ Z^d &= \sum_{x \in \Omega^d} G^d(x). \end{aligned} \tag{6.2}$$

Clearly, $G^d(x)$ coincides with the conditional probability $G[X = x \,|\, X_D = d]$ for all $x \in \Omega^d$.

The *synchronous Markov chain* (X^n) *with clamped inputs* $X_D^n \equiv d$ for all $n \geq 0$, has *a stationary distribution* M^d *on* Ω^d, which is the unique solution of the matrix equation

$$M^d \, Q_d = M^d.$$

Here $Q_d(x, y)$ denotes, for $x, y \in \Omega^d$, the transition matrix of the synchronous chain (X^n) with clamped inputs $X_D^n \equiv d$. Hence we obviously have

$$Q_d(x, y) = \prod_{s \in S-D} G_s(y_s \,|\, x_{S-s}) \qquad \text{for all } x, y \in \Omega^d \tag{6.4}$$

which in view of (3.6) can be written explicitly as

$$Q_d(x, y) = \frac{1}{Z(x)} \exp \frac{1}{T} \left[\sum_{s \in S-D} V_s(y_s, x_{N_s}\right] \qquad \text{for } x, y \in \Omega^d$$

with

$$Z(x) = \prod_{s \in S-D} \left[\sum_{a \in A} \exp \frac{1}{T} V_s(a, x_{N_s}\right].$$

The asymptotic marginal distribution of the output X_R^n for clamped inputs $X_D^n \equiv D \in A^d$ is then obtained by summing $M^d(x)$ (or $G^d(x)$ in the asymptotic case) over hidden configurations $x_H \in A^H$.

The response $\widehat{F}_n(d)$ of the machine to the input $D \in A^d$ will be the random configuration X_R^n of response units *with n large enough to ensure that stochastic equilibrium has practically been reached.* The pattern classifier $\widehat{F}_n : A^D \to A^R$ thus emulated by the machine is a random mapping. Note that again these classifiers $\widehat{F}_n$ are emulated at fixed temperature T. The choice of T will be evoked further below in §10.

The main problem in practical applications is to select the weight vector $w \in \mathbf{R}^K$ so that $\widehat{F}_n$ is *as close as possible to a preassigned classifier* $F : A^D \to A^R$.

To evaluate the performance of the machine as an emulator of F, we introduce *a loss function* $L(x_R, x'_R)$ for pairs of output configurations. Namely
$L : A^R \times A^R \to \mathbf{R}^+$ *is an arbitrary positive function, equal to zero whenever* $x_R = x'_R$.

Now for each configuration $x \in \Omega$, we can compute *the score* $\lambda(x)$ *of* x by

$$\lambda(x) = L[F(x_D), x_R)]. \tag{6.5}$$

A natural Bayesian point of view is to assume that *the environment provides inputs with a fixed a priori probability distribution p on A^D.* Les us point out that the actual knowledge of p will not be necessary below.

We introduce a random configuration Y with values in Ω having the distribution

$$P(Y = y) = p(y_D)\, G^{y_D}(y) \qquad \text{in the sequential case} \tag{6.6}$$

$$P_{\text{syn}}(Y = y) = p(y_D)\, M^{y_D}(y) \qquad \text{in the synchronous case} \tag{6.7}$$

To simulate Y, we use the *Markov chain X^n with clamped inputs* defined by the constraints

$$\begin{cases} X_D^n \equiv X_D^0 \\ \text{distribution } (X_D^0) \equiv p \end{cases} \tag{6.8}$$

and free stochastic dynamics on $S - D$; if the dynamics on $S - D$ remains sequential (resp. synchronous), the limit distribution of X^n is the corresponding distribution (6.6) [resp. (6.7)] of Y.

In this context, *the expected score*

$$\ell = E[(y)] = \lim_{n\to\infty} E[\lambda(X^n)] \tag{6.9}$$

is a natural quantity to *minimize* with respect to the weight vector $w \in \mathbf{R}^R$ which parametrizes the machine. If F_n is the classifier obtained by reading the response X_R^n of the machine at time n, one has obviously

$$\ell = \lim_{n\to\infty} E\, L[F(X_D^0), \widehat{F}_n(X_D^0)]. \tag{6.10}$$

With our previous notations, we have in *the synchronous case*

$$\ell = \sum_{x\in\Omega} p(x_D)\, M^{x_D}(x)\, \lambda(x) \tag{6.11}$$

while *in the sequential case* M^{x_D} is replaced by G^{x_D}.

The *learning rule* will simply be a gradient descent on the score function, of the following type : the kth update of the weights will be given by

$$w^{k+1} - w^k = -\eta_k \frac{\partial \ell}{\partial w}(w^k) \tag{6.12}$$

where $\eta_k > 0$ is a *slowly decreasing* gain coefficient. We shall suggest here the choice $\eta_k = \frac{\alpha}{\beta+k}$ where *the coefficient* $\alpha > 0$, $\beta > 0$ *remain fixed during learning at fixed given temperature* T, and of course should be T-dependent if several temperatures are used. This slow decrease at speed (ℓ/k) for the gain is classical for gradient descent algorithms, and has the advantage of at least forcing convergence of the sequence (w^k) $k = 1, 2, \cdots$ whenever the sequence comes back infinitely often in the neighborhood of an isolated local minimum of the expected score ℓ.

Thus *for all practical purposes*, either the sequence w^k is unbounded, or the sequence $\ell(w^k)$ converges almost surely to a *local* minimum of ℓ.

The main problem is of course to compute $\frac{\partial \ell}{\partial w}$ through a *fast and parallel algorithm implementable on a dual neuronal architecture* of the type described in §5 above. In particular $\frac{\partial \ell}{\partial w_C}$ *should only involve computations confined to the neighborhood* NC *of clique* C.

Let us tackle first the much easier case of sequential dynamics.

7. LEARNING RULES FOR THE SEQUENTIAL CASE

The explicit formulas (3.1) (3.2) (3.3) for the energy U, the Gibbs distribution G, and the partition function Z immediately give

$$\frac{1}{Z}\frac{\partial Z}{\partial w_C} = -\frac{1}{T}\,E[J_C(x)] \qquad \text{for all cliques } C \in K \tag{7.1}$$

where X is a random configuration on S having the Gibbs distribution G, and then

$$-\frac{\partial G}{\partial w_C}(x) = \frac{1}{T}\,G(x)\,[J_C(x) - \overline{J}_C] \qquad \text{for all } x \in \Omega \tag{7.2}$$

where

$$\overline{J}_C = E[J_C(x)] = \sum_{x\in\Omega} J_C(x)\,G(x). \tag{7.3}$$

Let $d \in A^D$ be an arbitrary input, and call Ω^d the set of configurations x with clamped input $x_D = d$. Let G^d be the sequential Gibbs measure on Ω^d [*cf.* (6.1)), (6.2)]. We may now apply formulas (7.2) (7.3) to G^d. Hence we set for all $d \in A^D$

$$\overline{J}_C(d) = \sum_{\{x\in\Omega \mid x_D \equiv d\}} J_C(x)\,G^d(x). \tag{7.4}$$

and we define *the centered activity of clique* C by

$$j_C(x) = J_C(x) - \overline{J}_C(x_D) \qquad \text{for all } x \in \Omega. \tag{7.5}$$

Then formula (7.3) yields for all $d \in A^D$

$$-\frac{\partial G^d}{\partial w_C}(x) = \frac{1}{T}\,G^d(x)\,j_C(x) \text{ for } x \in \Omega \text{ and } x_D \equiv d. \tag{7.6}$$

From (6.11) we immediately get in the sequential case

$$\frac{\partial \ell}{\partial w_C} = \sum_{x\in\Omega} p(x_D)\,\lambda(x)\,\frac{\partial}{\partial w_C}\left[G^{x_D}(x)\right]. \tag{7.7}$$

and hence in view of (7.6)

$$-\frac{\partial \ell}{\partial w_C} = \frac{1}{T}\sum_{x\in\Omega} p(x_D)\,G^{x_D}(x)\,j_C(x)\,\lambda(x) \tag{7.8}$$

which is immediately interpretable as

$$(7.9)\qquad -\frac{\partial \ell}{\partial w_C} = \frac{1}{T}\, E[j_C(Y)\,\lambda(Y)]$$

where the random configuration Y is defined by (6.6).

Since by construction

$$E\,[J_C(Y)\,|\,Y_D] = \overline{J}(Y_D)$$

we see that (7.9) has an interpretation in terms of *correlations*, by

$$(7.10)\qquad -\frac{\partial \ell}{\partial w_C} = \frac{1}{T}\,\mathrm{cor}\,[j_C(Y), \lambda(Y)] = \frac{1}{T}\, E\{\mathrm{cor}_{Y_D}\,[J_C(Y), \lambda(Y)]\}$$

where for *all inputs* $d \in A^D$ *one defines* cor_d *as the correlation with respect to the conditional distribution of* Y *given* $\{Y_D = d\}$.

Using the sequential Markov chain X^n with clamped inputs introduced in (6.8), we have obviously

$$(7.11)\qquad \mathrm{cor}_d\,[J_C(Y), \lambda(Y)] = \lim_{n\to\infty}\, \mathrm{cor}\,\{[J_C(X^n), \lambda(X^n)]\,|\,X_D^0 = d\}$$

and hence using the ergodicity of the chain X^n conditioned by $X_D^0 = d \in A^D$, we get

$$(7.12)\qquad \mathrm{cor}_d\,[J_C(Y), \lambda(Y)] = \lim_{n\to\infty}\left[\frac{1}{n}\sum_{k=1}^{n} J_C(Z^n)\,\lambda(Z^n) - \overline{J}_{C,n}\,\overline{\lambda}_n\right]$$

where

$$\overline{J}_{C,n} = \frac{1}{n}\sum_{k=1}^{n} J_C(Z^n)$$

$$\overline{\lambda}_n = \overline{\lambda}_n = \frac{1}{n}\sum_{k=1}^{n} \lambda(Z^n)$$

(Z^n) is the sequential Markov chain on $(S-D)$ with clamped input $Z_D^n \equiv d$ for all $n \geq 0$.

Call $\mathrm{cor}_d^n\,[J_C, \lambda)$ *the empirical correlation at time* n *between clique activity* J_C *and score* λ, *conditioned by* $Z_D^k \equiv d$, $k = 0, 1, \cdots, n$. Fix a random sample $\Gamma \subset A^D$

of inputs, *selected with the a priori distribution* p, which is equivalent to saying that *the training set* Γ *of inputs is generated by the environment*, and let $N = \text{card}(\Gamma)$. Then (7.10) (7.12) yield the practical approximation

$$-\frac{\partial \ell}{\partial w_C} = \frac{1}{T} \lim_{N,n\to\infty} \frac{1}{N} \sum_{d\in\Gamma} \text{cor}_d^n (J_C, \lambda). \tag{7.13}$$

This approximation is of course easy to implement locally through a *feedback from the response units to the clique indicator* C, which provides the locally computed score $\lambda(X^k) = L[F(X_D^k), X_R^k]$. This is summarized by the sketch below.

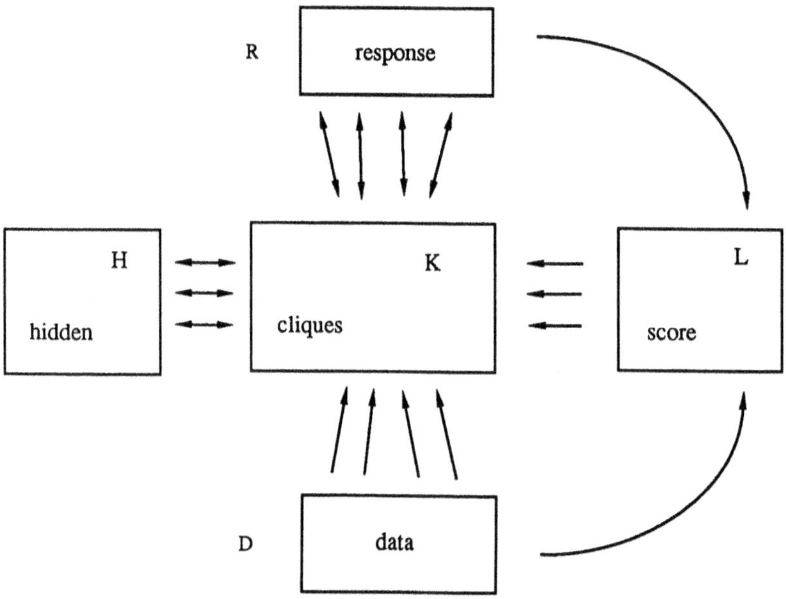

7.14. Communications between various computing blocks in the dual network with score feedback

In sketch 7.14, the block L computes the instantaneous score between the desired response $F(X_D)$ to the input X_D and the current machine response X_R, and then feeds back the score $\lambda(X) = L[F(X_D), X_R]$ to all cliques $C \in K$.

7.15. An important example of score function

Let us now apply the preceding result to a particular score function L_{lik} defined by

(7.16)
$$L_{lik}(r, r') = \begin{cases} 0 \text{ if } r = r' \\ 1 \text{ if } r \neq r' \end{cases} \qquad r, r' \in A^R.$$

Then the associated expected score ℓ_{lik} is given by

(7.17)
$$\ell_{lik} = P[Y_R \neq F(Y_D)]$$

where Y is the random configuration defined previously in §6, and hence minimizing ℓ_{lik} is equivalent to solving in $w \in \mathbf{R}^K$

$$\max_{w \in \mathbf{R}^K} P[Y_R = F(Y_D)].$$

In this case (7.9) becomes

(7.19)
$$-\frac{\partial \ell_{lik}}{\partial w_C} = \frac{1}{T} E\left[j_C(Y) 1_{Y_R} = F(y_D)\right].$$

But *minimizing* ℓ_{lik} *is equivalent to maximizing*

(7.20)
$$\log(1 - \ell_{lik}) = \log P[Y_R = R(Y_D)]$$

and (7.19) implies

(7.21)
$$\frac{\partial}{\partial w_C} [\log(1 - \ell_{lik})] = \frac{1}{T} E\left[j_C(Y) \,|\, Y_R = F(Y_D)\right].$$

On the other hand, elementary manipulations and the definition $j_C(Y) = J_C(Y) - J_C(Y_D)$ yield

(7.22)
$$\begin{aligned} E\left[j_C(Y) \,|\, Y_R = F(Y_D)\right] &= E\left\{E[j_C(Y) \,|\, Y_D \,;\, Y_R = F(Y_D)]\right\} \\ &= E\left\{E[j_C(Y) \,|\, Y_D \,;\, Y_R = F(Y_D)] - \overline{J}_C(Y_D)\right] \end{aligned}$$

Recall that $\overline{J}_C(Y_D) = E[J_C(Y) \,|\, Y_D]$ to conclude that

(7.23)
$$\frac{\partial}{\partial w_C} \log P[Y_R = F(Y_D)] = \frac{1}{T} E\left\{J_C^1(Y_D) - J_C^2(Y_D)\right\}.$$

Here for each input $d \in A^D$,

(7.24) $J_C^1(d)$ is *the expected activity of clique* C *at stochastic equilibrium when the input* Y_D *remains clamped on* d and *the output* Y_R *remains clamped on* $F(d)$*,while the hidden neurons* $s \in S-D-R$ *evolve according to the sequential dynamics.* **(7.25)** $J_C^2(d)$ is *the expected activity of clique* C *at stochastic equilibrium when the input* Y_D *remains clamped on* d *while all the remaining neurons* $s \in S-D-R$ *evolve freely according to the sequential stochastic dynamics.*

Since the weight update Δw_C is proportional to $\frac{\partial}{\partial w_C} \log(1 - \ell_{lik})$, we see that **(7.26)** Δw_C *is propotional to the difference in average activity (for clique* C*) between two regimes : clamped output and free output.* In both regimes the data units remain clamped on the initial input $d \in A^D$, which should run through *a random training set* $\Gamma \subset A^D$ *"generated" by the environment*, in the sense specified above.

In this setup the expectations in (7.23) can correctly be estimated by the ordinary average on $d \in \Gamma$ and for each d and each regime by empirical averages of clique activity over time.

Of course the learning rule defined by (7.23) (7.24) (7.25) (7.26) generalizes directly the learning rule proposed by Hinton-Sejnovski-Ackley [H.S.A.] for sequential Boltzmann machines with quadratic energies, in which all cliques C contain only two neurons and $J_C(x) = -x_s x_t$ for $C = \{s, t\}$. However we see that the learning rule (7.26) is linked to a particularly rigid choice of loss function, and is only one example within the much wider family of learning rules which we have just derived here.

8. THE GRADIENT OF THE STATIONARY MEASURE IN THE SYNCHRONOUS CASE

In the particular case where all cliques in K have cardinal ≤ 2 a direct computation of the gradient M' of the synchronous stationary measure M, using the explicit expression (4.7), is possible and we have carried it out in our paper [A], deriving from it several natural learning rules.

However we want to handle here the case of general stochastic networks with multiple interactions, for which M cannot be computed, and is known only implicitly as the unique solution of $MQ = M$. From this equation we get

$$M' = MQ' + M'Q. \tag{8.1}$$

An iterative use of (8.1) immediately yields

$$M' - M'Q^{n+1} = MQ'[I + Q + Q^2 + \cdots + Q^n]. \tag{8.2}$$

Now we have for all $x, y \in A^S = \Omega$

$$\lim_{n \to +\infty} Q^n(x, y) = M(y)$$

and hence

$$\lim_{n \to +\infty} M'Q^n(y) = \left[\sum_{x \in \Omega} M'(x)\right] M(y).$$

Since $\sum_{x \in \Omega} M(x) \equiv 1$, we get

$$\lim_{n \to +\infty} M'Q^n = 0 \tag{8.3}$$

and the following result

8.4 THEOREM.— *If the ergodic transition matrix Q depends smoothly on a parameter $w \in \mathbf{R}^K$, then the invariant probability measure M of Q also depends smoothly on w and we have*

$$M' = \lim_{n \to +\infty} MQ'(I + Q + \cdots + Q^n). \tag{8.5}$$

Call K_s *the set of cliques in K containing a given neuron $S \in S$* and let

$$g(x_{N_s}) = \sum_{a \in A} \exp \frac{1}{T} \left[- \sum_{C \in K_s} w_C \, J_C(x_{C-s}, a) \right]. \tag{8.6}$$

From (3.5) (3.6), we get

$$G_s(y_s \mid x_{S-s}) = \frac{1}{g(x_{N_s})} \exp \frac{1}{T} \left[- \sum_{C \in K_s} w_C \, J_C(x_{C-s}, y_s) \right]. \tag{8.7}$$

And hence if C is an arbitrary fixed clique, *we have for* $s \in C$

$$T \frac{\partial \log G_s}{\partial w_C}(y_s \,|\, x_{S-s}) = -J_C(x_{C-s}, y_s) + \sum_{a \in A} J_C(x_{C-s}, a)\, G_s(a \,|\, x_{S-s}) \tag{8.8}$$

where the notation $J_C(x_{C-s}, a)$ stands short for $J_C(z)$ with $z_{C-s} = x_{C-s}$ and $z_s = a$. Of course we also have

$$\frac{\partial \log G_s}{\partial w_C}(y_s \,|\, x_{S-s}) = 0 \qquad \text{for } s \notin C. \tag{8.9}$$

On the other hand, the relation (4.5) says that, in the synchronous dynamics,

$$Q(x, y) = \prod_{s \in S} G_s(y_s \,|\, x_{S-s}) \tag{8.10}$$

and hence we get in the synchronous case

$$\frac{\partial \log Q}{\partial w_C}(x, y) = \sum_{s \in S} \frac{\partial \log G_s}{\partial w_C}(y_s \,|\, x_{S-s}). \tag{8.11}$$

In view of (8.8) the right-hand side of (8.11) depends only on x_{NC} and y_C where

$$NC = \cup_{s \in C} Ns \tag{8.12}$$

is the neighborhood of clique C.

To give a probabilistic interpretation of the right-hand sides of (8.8) (8.11), call X^n, $n = 1, 2, \cdots$ the successive random configurations of the network S in the synchronous dynamics. Then clearly

$$G_s(a \,|\, x_{S-s}) = P\left(X_s^{n+1} = a \,|\, X_{S-s}^n\right) \tag{8.13}$$

and hence

$$\sum_{a \in A} J_C(x_{C-s}, a)\, G_s(a \,|\, x_{S-s}) = E\left[J_C\left(X_{C-s}^n, X_s^{n+1}\right) \,|\, X_{S-s}^n\right]. \tag{8.14}$$

Introduce then the notion of *transition activity* $a_C(x, y)$ *for the clique* C, defined by

$$a_C(x, y) = \sum_{s \in C} J_C(x_{C-s}, y_s). \tag{8.15}$$

Note that $a_C(x,y)$ depends only on x_C, y_C. The *expected transition activity at time n, for clique C, given the configuration* X^n, is then

(8.16)
$$\overline{a}_C(X^n) = E\left[a_C(X^n, X^{n+1}) \,|\, X^n\right]$$

where $\overline{a}_C(x)$ depends only on x_{NC} and is given by

(8.17)
$$\overline{a}_C(x) = \sum_{s\in C}\sum_{\beta\in A} J_(x_{C-s}, \beta)\, G_s(\beta \,|\, x_{S-s}).$$

Introduce now *centered transition activity* $[a_C(x,y) - \overline{a}_C(x)]$, which we prefer to call the *unexpected transition activity of clique C.*

(8.18)
$$u_C(x,y) = a_C(x,y) - \overline{a}_C(x).$$

Clearly $u_C(x,y)$ depends only on (x_{NC}) and hence is still a *local* notion.

We may now interpret (8.8) (8.11) to get

(8.19)
$$\frac{\partial Q(x,y)}{\partial w_C} = -\frac{1}{T}\, Q(x,y)\, u_C(x,y).$$

Using (8.5) (8.19), we now obtain for all $z \in \Omega$

(8.20)
$$-T\,\frac{\partial M}{\partial w_C}(z) = \sum_{k=0}^{+\infty}\ \sum_{x\in\Omega, y\in\Omega} M(x)\, u_C(x,y)\, Q(x,y)\, Q^k(y,z).$$

The following result gives a crucial probabilistic interpretation of (8.20).

8.21. THEOREM.— *Call* $(Y^n)_{n\in\mathbf{Z}}$ *the* doubly infinite stationary synchronous Markov chain *of network configurations, having the synchronous transition matrix* Q, *and such that every* Y^n *has the stationary synchronous distribution* M. *For every function* $f : \Omega \to \mathbf{R}$, *every clique* $C \in K$, *every* $n \in \mathbf{Z}$, *we have :*

(8.22)
$$-\sum_{z\in\Omega} f(x)\,\frac{\partial M}{\partial w_C}(z) = \frac{1}{T}\sum_{k=1}^{+\infty} \operatorname{cor}\left[f(Y^n), u_C(Y^{n-k}, Y^{n-k+1})\right]$$

where $u_C(x,y)$ *is the* unexpected transition activity *of clique* C *defined by* (8.18) *above.*

Proof. — Using (8.20), one gets immediately

$$(8.23)\qquad -T\sum_{z\in\Omega} f(z)\frac{\partial M}{\partial w_C(z)} = \sum_{k=0}^{+\infty} E\left[u_C(Y^0,Y^1)\,f(Y^{k+1})\right].$$

By construction, we clearly have

$$(8.24)\qquad E\left[u_C(Y^0,Y^1)\right]=0$$

and hence

$$E\left[u_C(Y^0,Y^1)\,f(Y^{k+1})\right] = \mathrm{cor}\left[f(Y^{k+1}),u_C(Y^0,Y^1)\right]$$

a correlation which, in view of the stationary of (Y^n), coincides with

$$\mathrm{cor}\left[f(Y^n),u_C(Y^{n-k-1},Y^{n-k})\right]$$

for arbitrary integers $n \in \mathbf{Z}$. This proves the equivalence between (8.23) and the announced formula (8.22).

From a practical point of view, $(Y^n)_{n\in\mathbf{Z}}$ cannot be simulated in general since M is unknown. However, *large segments of that chain can easily be approximately simulated to an arbitrary degree of accuracy.* Indeed, let $(X^n)_{n=0,1,2,\cdots}$ be the (easily simulated) sequence of random configurations obtained by synchronous stochastic dynamics with arbitrary initial configuration X^0. For N large, the sequence $(X^{N+n})_{n=0,1,2,\cdots}$ has finite joint distributions arbitrarily close to those of $(Y^{j+n})_{n=0,1,2,\cdots}$ for arbitrary fixed $j \in \mathbf{Z}$. This suggests several practical approximations of (8.22) based on (X^n).

Define first the *cumulative transition activity of clique C, between instants i and j, with $i \le j+1$, by*

$$(8.25)\qquad \mathrm{cum}_C(i,j)=\sum_{k=i}^{j-1} u_C(X^k,X^{k+1}).$$

Several practical approximations are summarized in the following theorem :

8.26. THEOREM.— *Let (X^n) $n = 0,1,2,\cdots$ be the chain of configurations obtained by synchronous dynamics with arbitrary initial configuration X^0. Let C be* *an arbitrary clique in* K, w_C *its weight, and* cum_C *its cumulative transition activity* [*cf.* (8.25)]. *For every function* $f : \Omega \to \mathbf{R}$, *the following three limits exist*

$$\lim_{n,k \to +\infty} \operatorname{cor}\left[f(X^{n+k}), \mathrm{cum}_C(n, n+k)\right] \tag{8.27}$$

$$\lim_{n \to +\infty} \operatorname{cor}\left[f(X^{2n}), \mathrm{cum}_C(n, 2n)\right] \tag{8.28}$$

$$\lim_{n \to +\infty} \operatorname{cor}\left[f(X^{n}), \mathrm{cum}_C(0, n)\right] \tag{8.29}$$

and they coincide with $\left[-T \sum_{z \in \Omega} f(z) \frac{\partial M}{\partial w_C}(z)\right]$, *where* M *is the synchronous stationary distribution on* Ω.

Proof.— The proof of Th. 8.2 is an easy technical variation on the crucial formula (8.22) and details will be given elsewhere. In fact, it can also be shown that all limits (8.27) (8.28) (8.29) are *uniform* with respect to the clique $C \in K$, the function f, and the weight vector w provide $\|f\|_\infty$ and $\|w\|$ remain *bounded*. Speeds of convergence for those three limits can also be computed explicitly.

Of course in view of (8.24), correlations in (8.27) (8.28) (8.29) may be replaced by expectations. Also, using the ergodicity of X^n, *these correlations can be correctly estimated by long time averages* such as, for (8.27),

$$\lim_{q \to \infty} \frac{1}{q} \sum_{j=1}^{q} f(X^{n+j+k})\, \mathrm{cum}_C(n+j, n+j+k) \tag{8.30}$$

with a similar expression for (8.28).

9. LEARNING RULES IN THE SYNCHRONOUS GENERAL BOLTZMANN MACHINE

We place ourselves in the general case of an energy function involving interactions of arbitrary orders and with synchronous dynamics. For the simpler case of synchronous machines with only pairwise interaction, we refer to [A].

Formula (6.11) for the expected score implies

(9.1)
$$\frac{\partial \ell}{\partial w_C} = \sum_{x \in \Omega} p(x_D) \left[\frac{\partial}{\partial w_C} M^{x_D}(x) \right] \lambda(x).$$

To compute $\left[\frac{\partial}{\partial w_C} M^{x_D}(x)\right]$ for $x_D = d \in A^D$, we can apply the results of §8 to the reduced network $S - D$, with energy function $U^d(x_{S-D}) \equiv U(x)$ for all x such that $x_D = d$, and synchronous stochastic dynamics on $S - D$.

Call X^n the *synchronous Markov chain with clamped inputs* defined in (6.8) by

(9.2)
$$X_D^n \equiv X_D^0 \qquad \text{and} \qquad \left\{\text{distribution } X_D^0 \equiv p\right\}.$$

To the notion of transition activity a_C of clique C, we have associated its expected, unexpected, and cumulative versions $\overline{a}_C$, u_C, cum_C. *Since the basic network is here reduced to $S - D$, we introduce the reduced versions of a_C, $\overline{a}_C$, u_C, cum_C, defined by*

(9.3)
$$a_C(x, y) = \sum_{s \in C \cap (S-D)} J_C(x_{C-s}, y_s)$$

(9.4)
$$\overline{a}_C(x) = \sum_{s \in C \cap (S-D)} \sum_{\alpha \in A} J_C(x_{C-s}, \alpha)\, G_s(\alpha \,|\, x_{S-s})$$

(9.5)
$$u_C(x, y) = a_C(x, y) - \overline{a}_C(x)$$

(9.6)
$$\text{cum}_C(i, j) = \sum_{k=i}^{j-1} u_C(X^k, X^{k+1}).$$

The direct application of (8.27) yields for each input $d \in A^D$

(9.7)
$$-\sum_{\left[x_{S-D} \in A^{S-D},\, x_D \equiv d\right]} \lambda(x) \frac{\partial M^d}{\partial w_C}(x) = \frac{1}{T} \lim_{n,k \to \infty} \text{cor}_d \left[\lambda(X^{n+k}), \text{cum}_C(n, n+k)\right]$$

where as in (7.10), $\text{cor}_d(V, W)$ denotes *the correlation of V and W with respect to the conditional distribution given $X_D^0 = d$.*

Using (9.1) we then obtain

$$(9.8) \qquad -\frac{\partial \ell}{\partial w_C} = \frac{1}{T} \lim_{n,k\to\infty} E\left\{\mathrm{cor}_{X_D^0}\left[\lambda(X^{n+k}); \mathrm{cum}_C(n, n+k)\right]\right\}$$

and as in §8, *the correlations given* X_D^0 *may be replaced in* (9.8) *by expectations given* X_D^0 *as well as empirical correlations given* X_D^0. From (9.8), we now deduce a family of synchronous learning rules.

9.9. Approximate learning rules for the synchronous case

Consider a general Boltzmann machine with multiple interactions and synchronous dynamics. Fix a loss function L on pairs of outputs and a desired input-output mapping $F : A^D \to A^R$. Let $\lambda(x) = L[F(x_D), x_R]$ be the associated score of configuration $x \in \Omega$.

Fix a *training set* Γ *of inputs, generated by the environment*, so that Γ is a finite random sample of the real life *a priori* distribution p of inputs. *Fix the temperature* T. Choose then two "large" integers n, k and two positive parameters α, β. To n, k, Γ, α, β, we now associate *an approximate learning rule* LR for which the qth update of weight w_C, $C \in K$, is defined by

$$(9.10) \qquad \Delta w_C = \frac{1}{\mathrm{Card}\,\Gamma} \sum_{d\in\Gamma} \Delta w_C(d)$$

$$(9.11) \qquad \Delta w_C(d) = \frac{\alpha}{\beta + q}\left[\frac{1}{n}\sum_{j=1+k}^{n+k} \lambda(X^j)\,\mathrm{cum}_C(j-k, j)\right]$$

where for all j in (9.11) the Markov chain X^j has clamped inputs $X_D \equiv d$ and synchronous dynamics on $(S - D)$. As in 9.6, cum_C is the (reduced) cumulative transitional activity of clique C over a past of length k.

Then for large n, k, and card(Γ), the learning rule LR will tend to achieve local minimization of the expected score ℓ.

We now give another interpretation of the synchronous learning rules LR. Indeed the empirical correlation between score and past cumulative transition activity can be rewritten as

$$\sum_{i=1}^{k}\left[\frac{1}{n}\sum_{j=1+k}^{n+k}\lambda(X^j)\,u_C(X^{j-i},X^{j-i+1})\right]$$

which is *the sum of k empirical correlations between score and delayed unexpected transition activity*. This points out the importance of *delays* in automatic learning processes for synchronous networks. As easily seen, *for i large, the correlation between* $\lambda(X^j)$ *and* $u_C(X^{j-i},X^{j-i+1})$ *tends to zero at exponential speed*, and hence in many situations, only a fairly moderate number k of delays are significant.

9.12. Implementation of synchronous learning rules

Just as in the asynchronous case, the learning rule (9.10) (9.11) requires a feedback of the score $\lambda(X^n)$ to all clique indicators $C \in K$. Thus the communication scheme is similar to the sketch 7.15 described above.

Consider the pair (S, K) of dual networks introduced in Section 5, *fix an input* $d \in A^D$, and call X^n the configuration of neurons at time n. The flow of parallel computations — communications between configurations n and $n+1$ can be roughly described as follows, *the data units D remaining clamped on the input* $d \in A^D$.

(9.13) At the end of period n, the memory of each clique C has stored precisely

- the local configuration X_C^n
- the past unexpected transitional activities

$$u_C^j = u_C(X^{j-1},X^j) \qquad j = n, n-1, \cdots, n-k+1$$

- the cumulative transitional activity

$$\mathrm{cum}_C^n = \sum_{j=n-k+1}^{n} u_C^j$$

- the empirical correlation

$$n\,\mathrm{cor}_C^n = \sum_{j=1}^{n}\lambda(X^j)\,\mathrm{cum}_C^j.$$

(9.14) For each possible individual neural state $a \in A$, execute the following loop of *parallel* computations

9.14.1. each clique C computes for each $s \in C$ the lateral activity

$$\gamma_C(s,a) = J_C(X^n_{C-s}, a)$$

and transmits $\gamma_C(s,a)$ to the neuron $s \in C$.

9.14.2. every neuron $s \in S - D$ computes

$$\pi_s(a) = \exp -\frac{1}{T} \sum_{C \ni s} w_C \, \gamma_C(s,a)$$

and transmits $\pi_s(a)$ to all cliques C containing s.

9.14.3. - if A is not exhausted, go back to the beginning of loop (9.14)

- if A is exhausted, go to (9.15).

(9.15) Every neuron $s \in S - D$ computes a (random) state X^{n+1}_s with distribution $P(X^{n+1}_s = a) = \frac{\pi_s(a)}{\pi_s}$, where $\pi_s = \sum_{a \in A} \pi_s(a)$, and transmits X^{n+1}_s to all cliques C containing s.

(9.16) The score $\lambda^{n+1} = \lambda(X^{n+1})$ is computed in the L-block (*cf.* 7.15) and fed back to all cliques C.

(9.17) Every clique C computes successively

- $\pi_s = \sum_{a \in A} \pi_s(a)$ for all $s \in C \cap (S - D)$
- $\overline{a}_C(X^n) = \sum_{s \in C \cap (S-D)} \sum_{a \in A} \gamma_C(s,a) \frac{\pi_s(a)}{\pi_s}$
- $a_C(X^n, X^{n+1}) = \sum_{s \in C \cap (S-D)} J_C(X^n_{C-s}, X^{n+1}_s)$
- $u^{n+1}_C = a_C(X^n_{C-s}, X^{n+1}) - \overline{a}_C(X^n)$
- $\mathrm{cum}^{n+1}_C = \mathrm{cum}^n_C + u^{n+1}_C - u^{n+1-k}_C$
- $(n+1)\,\mathrm{cor}^{n+1}_C = n\,\mathrm{cor}^n_C + \lambda^{n+1}\,\mathrm{cum}^{n+1}_C$

During the learning phase, for each input d in the training set Γ, the algorithm (9.13) $\cdots$ (9.17) is iterated with clamped input,until the correlations cor^n_C stabilize (with an upper bound on n of course). Call $\mathrm{cor}_{d,C}$ this limit correlation.

The weight vector w will be updated after each complete pass of the training set Γ. If the current pass corresponds to the qth-update of w, this update Δw_C is given by

$$\Delta w_C = \frac{\alpha}{\beta + q} \frac{1}{\text{Card}(\Gamma)} \sum_{d \in \Gamma} \text{cor}_{d,C}. \tag{9.18}$$

Recall that *we assume the training set Γ of inputs to be a finite random sample "generated" by the environment.* This last point is crucial for the validity of (9.18) and has often been overlooked in the neural network literature. As pointed out by Bourlard [BO] in another context, *the empirical a priori distribution exhibited by the training set should be close to the a priori distribution of real-life inputs.*

10. OPTIMAL CHOICE OF THE TEMPERATURE

As was pointed out several times, the general Boltzmann machines considered here are meant to operate at *fixed temperature* T. However T is a natural and important parameter, and hence we may try to optimize the choice of T in the learning phase. Note however *a crucial point* : since the machine is actually parametrized by $\left[\frac{w}{T}\right]$, *no generality is lost in principle* if $T \equiv 1$.

Introduce as before the desired input-output mapping $F : A^D \to A^R$ and a loss function $L : A^R \times A^R \to \mathbf{R}^+$; call $\ell = \lim_{n\to\infty} E[\lambda(X^n)]$ the expected score. To choose an optimal T, it is natural to use gradient descent to *minimize* ℓ in $T \in \mathbf{R}^+$. Thus after each complete pass of the training set $\{d^1 \cdots d^n\}$ of random inputs (provided by the environment), we may introduce a temperature update

$$\Delta T \quad \text{proportional to} \quad -\frac{\partial \ell}{\partial T} \tag{10.1}$$

with small decreasing gain as for Δw.

The computation of $\frac{\partial \ell}{\partial T}$ is an easy consequence of the computation of $\frac{\partial \ell}{\partial w}$.

10.2 PROPOSITION.— *Consider a general Boltzmann machine with either synchronous or sequential dynamics. Select a learning rule as in §7 and 9 above for the* *weight vector w, and call Δw the corresponding weight update after one complete pass of the training set. The gradient descent in temperature is then given by*

$$(10.3) \qquad \Delta T = -\frac{1}{T}\Big[\sum_{C\in K} w_C\,\Delta w_C\Big] = -\frac{1}{2T}\,\Delta(\|w\|^2).$$

Indeed one has for the expected score ℓ

$$(10.4) \qquad \frac{\partial \ell}{\partial T} = -\frac{1}{T}\Big[\sum_{C\in K} w_C\,\frac{\partial \ell}{\partial w_C}\Big].$$

Proof.— For any function $g : \mathbf{R}^K \to \mathbf{R}$, one has trivially, if $f(w,T) = g(\frac{w}{T})$

$$\frac{\partial v}{\partial w} = \frac{1}{T}\,g'\left(\frac{w}{T}\right) \quad \text{et} \quad \frac{\partial f}{\partial T} = -\frac{1}{T^2}\,\langle w\,.\,g'\left(\frac{w}{T}\right)\rangle$$

which proves immediately (10.4) and hence (10.3).

10.5 PROPOSITION.— *Consider a general* sequential *Boltzmann machine with energy function U. Let (X^n) $n = 0,1,\cdots$ be the sequential Markov chain for which $X^n_D \equiv X^0_D$ for all n, and X^0_D has the a priori distribution of inputs generated by the environment. Let $\ell = \lim_{n\to\infty} E[\lambda(X^n)]$ be the expected score. We then have*

$$(10.6) \qquad \frac{\partial \ell}{\partial T} = \frac{1}{T^2}\lim_{n\to\infty} E\left\{\mathrm{cor}_{X^0_D}\,[U(X^n),\lambda(X^n)]\right\}$$

where as in 7.10, $\mathrm{cor}_{X^0_D}$ denotes the conditional correlation given $X^n_D \equiv \cdots \equiv X^0_D$.

Proof.— Call Y the random configuration defined by (6.6). From formula (7.10) and (10.4), we get immediately

$$(10.7) \qquad \frac{\partial \ell}{\partial T} = \frac{1}{T^2}\,E\left\{\mathrm{cor}_{Y_D}\,[U(Y),\lambda(Y)]\right\}$$

since $U(y) \equiv \sum_{C\in K} w_C\,J_C(y)$. Then (10.6) is a direct consequence of (10.7).

10.8. Another probabilistic interpretation

Let G be the Gibbs distribution on Ω associated to energy U at temperature T. Since $\log G(y) = -\frac{U(y)}{T}$

$$\log G(y) = -\frac{1}{T}\,U(y) - \log Z$$

where log Z is a constant, we get the interesting interpretation

(10.9) $$\frac{\partial \ell}{\partial T} = -\frac{1}{T} E\left\{\mathrm{cor}_{Y_D}\left[\log G(Y), \lambda(Y)\right]\right\}$$

so that the sequential temperature update ΔT should be proportional to the average correlation between the log likelihood and a random configuration and its score.

10.10. Global transition energy for synchronous networks

Consider now a synchronous Boltzmann machine with energy $U = \sum_{C \in K} w_C J_C$ and data units D. In (9.3) (9.4) (9.5), we have defined the notions (reduced to $S - D$) of clique transition activities $a_C(x, y)$ as well as their expected and unexpected versions $a_C(x)$, $u_C(x, y)$.

We now introduce three global quantities, the *transition energies* $a(x, y)$, the *expected transition energy* $a(x)$, and the *unexpected transition energy* $u(x, y)$ defined for pairs $x, y \in \Omega$ of global configurations, by

(10.11) $$a(x, y) = \sum_{C \in K} w_C \, a_C(x, y)$$

(10.12) $$\overline{a}(x) = \sum_{C \in K} w_C \, \overline{a}_C(x)$$

(10.13) $$u(x, y) = \sum_{C \in K} w_C \, u_C(x, y)$$

with a_C, $\overline{a}_C$, u_C given by (9.3) (9.4) (9.5).

It turns out that a, $\overline{a}$, u have interesting global interpretations in terms of the *synchronous Markov chain* (X^n) *on* S *with clamped inputs on* D which verifies $X_D^n \equiv X_D^0$ for all $n \geq 0$. For each input $d \in A^D$, let

(10.14) $$\Omega^d = \{x \in \Omega = A^X \mid x_D = d\}$$

and for $x, y \in \Omega$, call $Q_d(x, y)$ the transition matrix of X^n given $X_D^0 \equiv d$.

From definitions (9.3) (9.10) (3.5), we get for $x, y \in \Omega$

$$a(x,y) = -\sum_{s \in S-D} V_s(y_s, x_{N_s}) \tag{10.15}$$

where the V_s are the local action potentials (*cf.* (3.5)). In view of the explicit expression (6.4) of $Q_d(x,y)$, from (10.15) we deduce *for* $x, y \in \Omega_d$

$$-\frac{1}{T}\, a(x,y) = \log Q_d(x,y) + \log Z_d(x) \tag{10.16}$$

where $Z_d(x)$ has been defined in (6.2).

On the other hand, (9.4) (10.12) and (6.4) give the interpretation of $a(x)$ as a conditional expectation

$$\overline{a}(x) = \sum_{y \in \Omega_d} Q_d(x,y)\, a(x,y) \qquad \text{for } x \in \Omega_d \tag{10.17}$$

Now a comparison of (10.13) (10.16) (10.17) provides trivially a new interpretation of $u(x,y)$ for $x, y \in \Omega_d$

$$-\frac{1}{T}\, u(x,y) = \log Q_d(x,y) - \sum_{z \in \Omega_d} Q_d(x,z) \log Q_d(x,z). \tag{10.18}$$

Thus we now introduce the *entropy* $\mathrm{ent}_d(x)$ *of the transition distribution* $Q_d(x,.)$ classically defined by

$$\mathrm{ent}_d(x) = -\sum_{z \in \Omega_d} Q_d(x,y) \log Q_d(x,y) \tag{10.19}$$

and interpret (10.18) in the following statement

10.20 PROPOSITION.— *For an arbitrary input* $d \in A^D$*, let* $Q_d(x,y)$*,* $x, y \in \Omega_d$ *be the transition matrix of the synchronous chain* (X^n) *with clamped inputs* $X^n_D \equiv d$*. Let* $\mathrm{ent}_d(x)$ *be the entropy of* $Q_d(x,.)$*. Then the global unexpected transition energy* $u = \sum_{C \in K} w_C\, u_C$ *of the network is given by*

$$\frac{1}{T}\, u(x,y) = -\log Q_d(x,y) - \mathrm{ent}_d(x) \qquad \text{for all } x, y \in \Omega_d. \tag{10.21}$$

We now introduce the *cumulative transition in energy* of the network along a sequence $(X^i X^{i+1} \cdots X^j)$ random configurations, defined by

$$\text{(10.22)} \qquad \operatorname{cum}(i,j) = \sum_{k=i}^{j-1} u((X^k X^{k+1}) = \sum_{C \in K} w_C \operatorname{cum}_C(i,j).$$

In view of (10.21), *we see that* $\left[\frac{1}{T}\operatorname{cum}(i,j)\right]$ *can roughly be interpreted as a centered version of the joint log likelihood of the sequence* $(X^i X^{i+1} \cdots X^j)$ *given* $X^0_D \equiv d$. Of course, from a practical point of view, the *actual computation* of $\operatorname{cum}(i,j)$ is far easier to obtain by formula $\sum_C w_C \operatorname{cum}_c(i,j)$ since the $\operatorname{cum}_c(i,j)$ are locally computed by each clique C during the learning process. We may now interpret $\frac{\partial \ell}{\partial T}$ in the synchronous case.

10.23 PROPOSITION.— *Consider a general synchronous Boltzmann machine, with data units D. Fix a desired mapping F, a loss function L and let $\lambda(x)$ be the associated score function. Let (X^n) be the synchronous Markov chain with clamped inputs $X^n_D \equiv X^0_D$ generated by the environment with a fixed a priori distribution of inputs. Then the expected score $\ell = \lim_{n\to\infty} E[\lambda(X^n)]$ verifies*

$$\text{(10.24)} \qquad \frac{\partial \ell}{\partial T} = \frac{1}{T} \lim_{n,k\to\infty} E\left\{\operatorname{cor}_{X^0_D}\left[\lambda(X^{n+k}), \operatorname{cum}(n, n+k)\right]\right\}$$

where $\operatorname{cum}(n, n+k)$ *is the cumulative transition energy along the sequence* $(X^n X^{n+1} \ldots X^{n+k})$ *defined by* (10.20) (10.22). *As before,* cor_d *denotes conditional correlation given* $X^0_D \equiv d$.

Proof.— Formula (10.24) is a direct consequence of (10.4) and (9.8).

10.25. Interpretation

Note that in view of (10.21) (10.24) (10.1), we see that *the temperature update* $\Delta T = -\eta \frac{\partial \ell}{\partial T}$ *is proportional to the average correlation, at stochastic equilibrium, between the current score* $\lambda(X^n)$ *and the centered log likelihood of the infinite past* $(X^n X^{n-1} \ldots)$.

10.26. Practical temperature adjustment

We point out that the updates ΔT of the temperature computed here deal only with one aspect of the performance, namely the expected score. However another aspect of the performance is crucial in applications, *namely the speed of stabilization of the Boltzmann machine.* Indeed, *for very low temperatures, stochastic equilibrium is reached only after a very long time*, and this second criterion should be taken into consideration when the temperature is adjusted. We shall come back to this problem in a forthcoming paper.

REFERENCES

[A] R. AZENCOTT - -*Markov fields and low-level vision tasks*, Proc. Int. Cong. App. Math. ICIAM, Paris (1987).

- *Gibbs fields, simulated annealing, and low-level vision tasks*, Proc. Cong. Pattern Recognition, AFCET-INRIA, Antibes (1987).

- *Synchronous Boltzmann machines : learning rules*, Proc. Congress "Neural networks", Les Arcs (1989), Springer-Verlag, NATO series (1990), vol. 68, Editors : Fogelman-Herault.

- *Parameter estimation for synchronous Markov fields* (to appear).

[Bo] P. BOURLARD - *Multilayer perceptions and learning*, Proc. Congress "Neural networks", Les Arcs (1989), to appear in Springer-Verlag, NATO series (1990).

[G.G.] D. and S. GEMAN - *Gibbs fields, simulated annealing, and Bayesian reconstruction of images*, IEEE, PAMI (1984).

[H.S.A.] G. HINTON, T. SEJNOWSKI, D.H. ACKLEY - *Boltzmann machines : constraint satisfaction networks that learn*, Technical Report, Carnegie Mellon University (1984).

BAYESIAN 3-D PATH SEARCH AND ITS APPLICATIONS TO FOCUSING SEISMIC DATA

R. Azencott
École Normale Supérieure (Paris) et Université Paris-Sud,
B. Chalmond
Université Paris-Sud,
Ph. Julien
TOTAL-CFP, Geophysical Research.

Abstract

The $3D$-images studied here are essential to the analysis of cubes of seismic focalisation. In the detection of geological horizons, the improvement of migration techniques requires the construction of $3D$ "focal" paths. We start with *blurred* versions of (unknown) $3D$-images consisting ideally of concentrated intensity spots which tend to lie on smooth isolated $3D$-paths. The blur point-spread function is *spatially dependent*, roughly Gaussian in shape, and directly estimated on the blurred image. On the space of admissible paths, we describe the plausibility of a path by an energy function, using thus a $3D$-*Markov random field* model. The adjustment of this Markov field model to the image data relies on an original *interactive robust parameter localization* approach.

Reconstruction of the original paths is based on a maximum (*a posteriori*) likelihood approach, implemented by a *new variant* of Besag's ICM algorithm. Applications to actual $3D$-seismic data are presented.

Acknowledgments

The authors gratefully acknowledge the financial support provided to this study by TOTAL-CFP (Recherche Géophysique), who has also supplied the seismic data and the geological know-how to handle them.

1 Introduction

1.1 Seismic data and horizon detection

The problem we have studied arises in the context of geophysics, and concentrates on the analysis of $3D$-cubes of focalisation (*cf.* Faye and Jeannot [1], Yilmaz and Chambers [2])

One of the main concerns in seismic data analysis is the detection of "geological horizons" separating "homogeneous" layers of rock, sediments, etc. (*cf.* Cheng and Lu [3]). One starts with a $2D$-seismic image (*cf.* Fig. 1), of a geological section, which is composed of hundreds seismic traces. A seismic trace, a sinuoid-like waveform, is a record in reflected wavelets of impedance contrasts between strata. If a cycle can be correlated laterally across many seismic traces on a seismic image, it is called a "seismic horizon". On the seismic plot shown in Fig.1, positive half cycles are darkened to highlight the seismic horizons. A difficult problem is the determination of the correspondance between geological horizons and the seismic horizons detected (Fig.2a) on the $2D$-image. Indeed to compute this correspondance, one needs to know the speeds at which the seismic signals propagate within the various geological strata. This knowledge is required to transform the vertical "time" axis of the image into a "depth" axis and thus to construct the mapping between geological horizons and seismic horizons. The so-called "migration" technique creates a cube of focalisation data, using the $2D$-image and an estimated speed model for subterraneous wave propagation (*cf.* Clearbout [4]). These focalisations data allow the approximate positionning of the main geological horizons, which in turn provide an improved speed model. This is the principle of an iterative algorithm for horizons detection on $2D$-seismic images. Section 8 gives more details on this technique. Here, we are only concerned by the treatment of such a cube without considering the $2D$-image.

1.2 Focalisation cubes

Start with a $2D$-seismic image I^0 (Fig. 1), whose points are denoted (x, t) where x denotes depth and t the (discrete) horizontal distance between the origin and the position of the seismic sensors. The focalisation data associated to I^0 is a computed $3D$-image I with generic voxels (x, y, t), where y denotes the "depth error".

The focalisation intensity $I_t(x, y)$ at voxel (x, y, t) represents essentially an energy reflected by an horizon located at the point (X, t), where X can be expressed as a fonction of (x, t), once the speed model is assumed to be known. For each fixed t, the t-section of this $3D$-cube I is called the focalisation image I_t associated to the seismic trace t. Each $2D$-image I_t is perpendicular to the image I^0 through the t-column, and consists essentially of a few more or less diffuse spots, possibly overlapping (*cf.* Fig. 2.b). If the speed model and the migration technique used to compute the cube I were

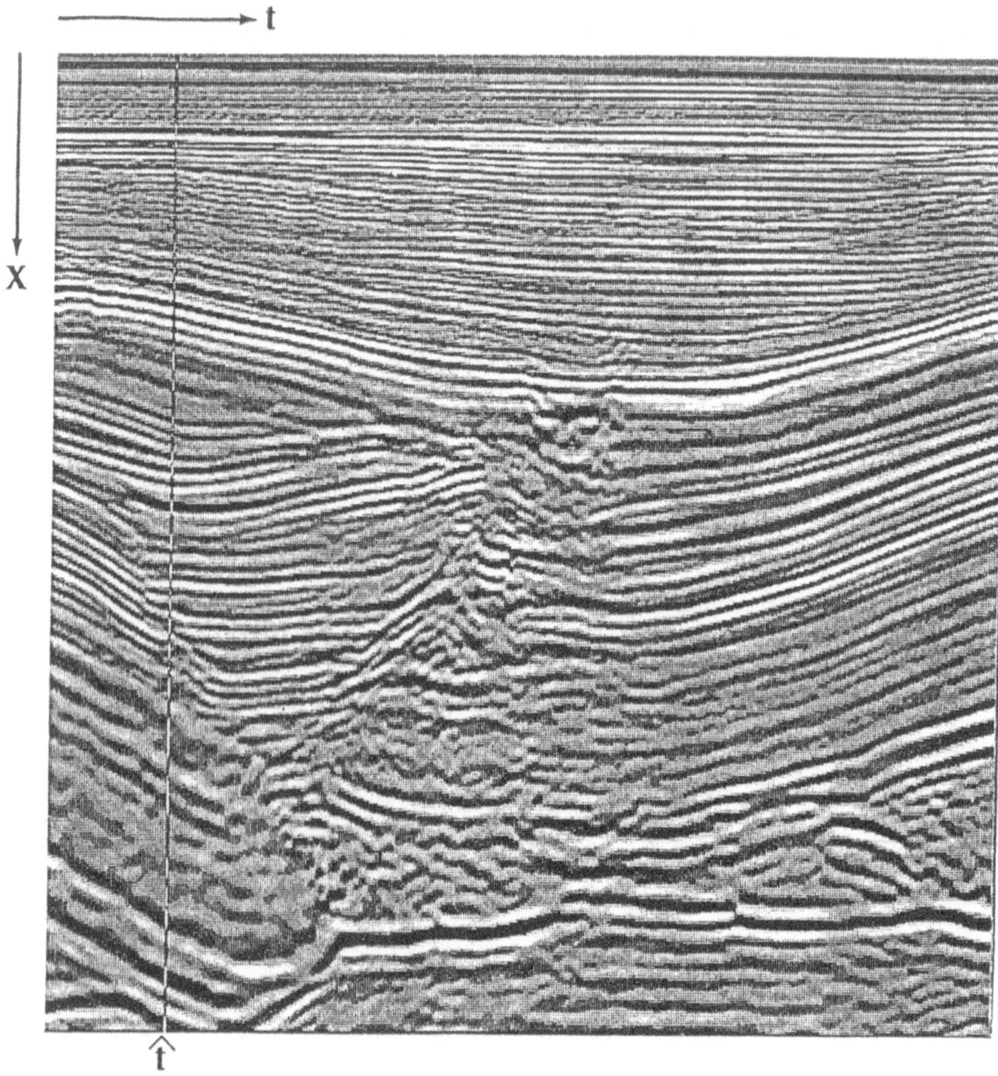

Figure 1
A 2D-seismic image I^0.

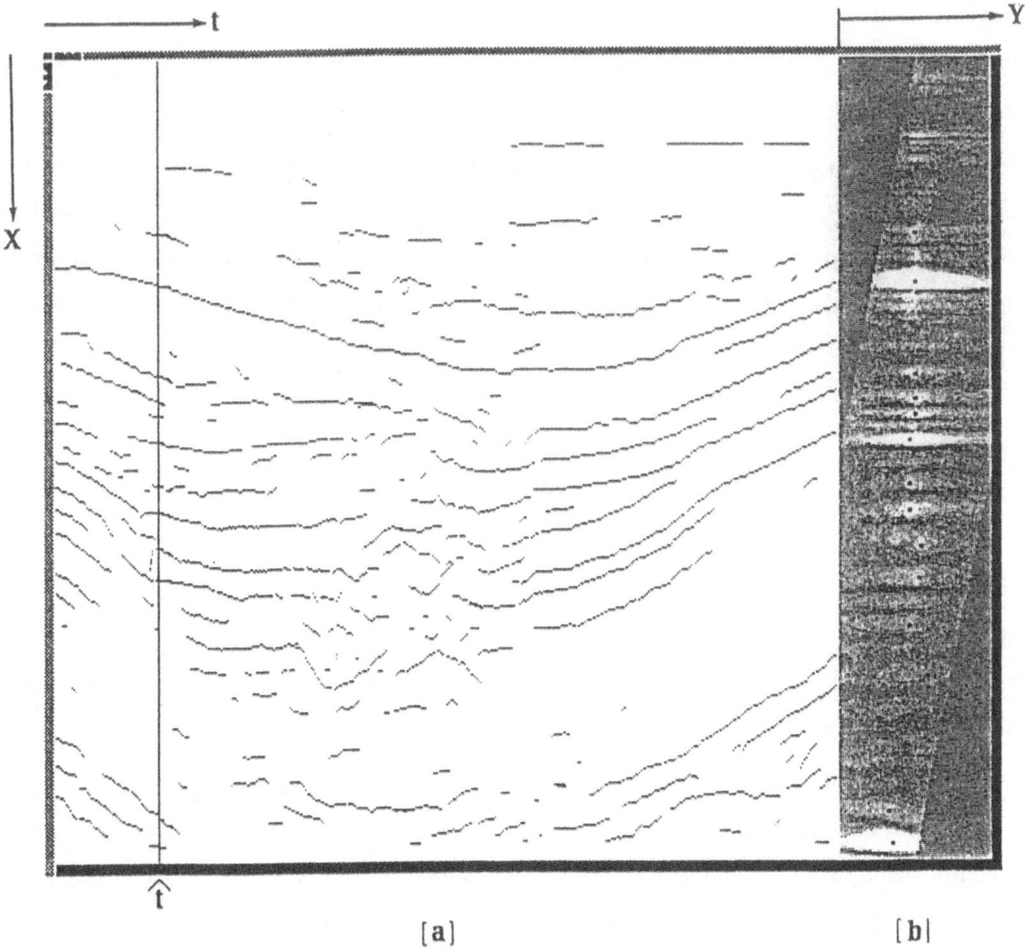

Figure 2

(a) *Exemple of seismic horizons detected on* I^0. (b) *This focalisation image is perpendicular to* I^0 *through the* $\hat{t}$ *column. (Black points in the blob centers emphasized the detected blob centers.)*

perfect, each strongly contrasted seismic horizon $(x_1 t_1) \cdots (x_p t_p)$ on the basic seismic image I^0 would generate a sequence of *theoretical focal points (t.f.p.)* $P_1 \cdots P_p$ of high intensity in the focalisation cube I, $P_j = (x_j, 0, t_j) \in I_{t_j}$. Here, we shall define a t.f.p. as an isolated luminous point on a black background. The horizon itself would simply be the broken line linking these focal points in I^0. But in practical and concrete data, the energy associated to a theoretical focal point P_j is diffused over an ill-defined *focal spot* located within I_{t_j}, and *roughly* centered at $P_j = (x_j, y_j, t_j)$ with $y_j \neq 0$. Using the speed model, the error coordinate y_j, and the erroneous depth x_j, the migration technique gives a simple correction that will (later) yield a corrected depth value $\hat{x}_j$ hopefully closer to the unknown $\overline{x}_j$.

1.3 The 3D-image analysis task

Our data will be the 3D-focalisation cube I with generic voxel (x, y, t) and intensity value $I_t(x, y)$. We want to select a finite family of high intensity sequence $\{Q_1 \cdots Q_p\}$ of points $Q_j = (x_j, y_j, t_j)$ in I, with $t_{j+1} \geq 1 + t_j$ for $j = 1 \cdots p$. Each such sequence will be called a *focal path*, and each Q_j will be a *focal point*.

On our actual data, the intensity at each (unknown) t.f.p. P_j has been *blurred* over a focal spot located with I_{t_j} and roughly centered at P_j. So the task at hand includes a combination of *deblurring* and *selecting "continuous" paths* within a 3D-cube of data. Note that the unknown *blur characteristics* (shape, spread,...) *vary quite a bit from point to point.* Moreover, the overlap of several spots is quite frequent.

The dimensions of a typical focalisation cube being of the order of (800, 500, 1000), while the number of spots fluctuates around 20000, it was natural to seek an automatization of the *search for focal paths,* as proposed in Julien et alt. [5]. In this paper, we present a new algorithm to accomplish this task, using the conceptual framework of Markov Field Models. We have attempted to isolate the image analysis concepts from the one degree remote geophysics context.

1.4 A methodology for path search in blurred 3D-images

Start with an initial (unknown) 3D-image J consisting of the union of a finite family of fairly high intensity focal paths, on a dark background. We observed a *degraded image* I obtained by *blurring* each point of J. The blur function at (x, y, t) will be assumed to be *Gaussian, position dependent*, and concentrated over the $2D$-dimensional t-section J_t of J. This flatness assumption on the blur function was strongly suggested by the geophysics context. Thus I can be viewed a sequence $I_t, t = 1, 2, \cdots$ of $2D$-images *separately* blurred. Our method extends easily to $3D$-blurs. In the wide literature about deblurring (*cf.* Hummel and alt. [6], Chalmond [7], among many others), two problems are considered : identification-estimation of the blur kernel, and efficient deblurring with known blur characteristics. Here the "blur kernels", or "point spread functions" PSF can be estimated by considering (separately) each spot as the blurred image of a point source. This will be done by fitting a $2D$-Gaussian distribution to the observed spot intensity distribution. Goodness of fit will be estimated (and optimized) as a Kullback distance. The difficulty here is that the spots are only *spatially limited observations* of the assumed underlying Gaussians PSF. Moreover, the *t.f.p. intensity and position are also unknown.* We point out that this *blur analysis will only be a pretreatment*, since the goal here is not restauration by *adequate identification* of the t.f.p. and of its "physical" charactistics.

The goal is to select adequately a finite family of focal paths. Each focal path will be a broken line $Q_1 \cdots Q_p$ where the $Q_j = (x_j, y_j, t_j)$ are focal points and $t_{j+1} \geq 1 + t_j$. Using a few qualitative notions about the "look" of good focal paths (suggested by natural geophysics assumptions), we shall define a (conditional) probability distribution P on the space of all admissible path family, quantifying, *given the blurred data* I, the likelihood $P(\omega)$ that an arbitrary path family ω will be a focal path family. Actually P *will be a* $3D$*-Markov field.* The methodology falls thus within the Bayes/Markov approach to $2D$-image reconstruction launched by D. and S. Geman [8], which has since generated a number of research papers (Azencott [9], Besag [10], Bouthemy [11], Chalmond [12,13], Dubois and Konrad [14], Geman and alt. [15], Marroquin [16]).

To *build our* $3D$*-Markov model*, we propose below (§4, 6) *a practical approach* based on three interactive steps : model selection, parameter estimation, validation. Our fairly flexible methodology provides a good adequation between the model and *a priori* qualitative information.

Finally (§5) one has to select actual focal paths by determining good local maxima of the *a posteriori* likelihood P. This search optimization by local likelihood is accomplished by a new algorithm having similarities with Besag's ICM. Note however that the standard version of ICM Besag [10] turns out to be seriously inefficient in our case.

In section 7, we present actual seismic data and comment the results obtained by applying the method described here. Our set of algorithms for the analysis of $3D$-focalisation data has been implemented on a workstation SUN 4-370 ; this methodology is now currently being integrated (at TOTAL-CFP) as one step of a complete chain of seismic images analysis.

2 Mathematical model of the image analysis task

We are given a *degraded 3D-image* denoted by I, indexed by the sampling grid $G = \{(x, y, t), x = 1 \cdots n_1\,,\, y = 1 \cdots n_2\,,\, t = 1 \cdots n_3\}$. We shall study I as the union of the sequence of $2D$-images I_t indexed by $t \in T = \{1, \cdots, n_3\}$, where each I_t is indexed by the sampling grid $S = \{(x, y), x = 1 \cdots n_1\,,\, y = 1 \cdots n_2\}$, $G = S \times T$. We denote respectively by $s = (x, y)$ and $g = (s, t)$ the generic pixel site in S and G.

2.1 The degradation model

The unknown original $3D$-image J is submitted to *blurring with local characteristic dependent on the voxel site*, and to *additive noise with variance depending on local intensity.* Any point of J with nonzero intensity will be called *theoretical focal point (t.f.p.).* So, J_t can be written as :

$$\begin{aligned} J_t(s) &= M_t(s) \quad \text{if } s \text{ is a t.f.p.} \\ J_t(s) &= 0 \qquad\quad \text{otherwise.} \end{aligned}$$

Moreover, the t.f.p. are isolated points. We denote by u the generic t.f.p.

For each $t \in T$, $u \in S$, let $h_t^u : \mathbf{Z}^2 \to \mathbf{R}$ be the *discretized version of a 2D-Gaussian density function with mean μ_t^u and covariance matrix v_t^u.* The kernel h_t^u will be the space-dependant blurring point spread function within the image J_t, such that : $h_t^u = 0$ if u is not a t.f.p. Let $W_t, t = 1 \cdots r$ be a sequence of (independent) $2D$-white noise images with equal variance and indexed by $s \in S$. We state now the *formal degradation model* (slice by slice along the t-coordinate)

$$I_t(s) = \sum_{u \in S} J_t(u)\, h_t^u(s - u) + \sum_{u \in S} J_t(u)\, W_t(s).$$

For instance, if the unknown image J consists of one *single* luminous point P with intensity $M > 0$ (one single t.f.p.) located at (u, t) with $u \in S$, then the degradation model generates an observed image I concentrated on I_t, with intensities lying on a perturbed $2D$-Gaussian shaped surface *centered* at $u + m = \mu$, with total mass M, and covariance matrix v. The additive perturbation is the $2D$-white noise MW_t, with mean zero and variance proportional to M^2. This elementary degraded image will be our model for a *focal spot* of global intensity M, carried by the $2D$-image I_t.

For each observed focal spot, we denote by $\theta = (M, \mu, v)$ the unknown parameters which describe the Gaussian model of its shape : M is the total intensity mass of the focal spot, $\mu \in \mathbf{R}^2$ is the location of its center within the t-slice I_t containing the focal spot, and v is the 2×2 covariance matrix of the Gaussian model of its intensity distribution. Hence we write

$$\mu = u + m, \qquad v = \begin{pmatrix} \alpha & \rho \\ \rho & \beta \end{pmatrix},$$

where $u \in S$ is the location in I_t of the unknown t.f.p. generating the spot. Note that m will typically be of the order of a few pixels in our focalisation data.

2.2 The path space

A *path* γ within the original image J will be the continuous broken line linking by straight line segments a *sequence of t.f.p.* $(s_1 t_1), \cdots (s_p, t_p)$, where the s_j lie in S, and $t_j \in T$.

We impose the following constraints : • $t_{j+1} \geq 1 + t_j$ for $j = 1 \cdots (p-1)$ • the only focal points lying on γ are the (s_j, t_j).

The $3D$-vectors $(s_{j+1} - s_j \, , \, t_{j+1} - t_j)$ are called the *pointers* defining the path γ.

2.3 Pointer fields

Actually J is assumed to be the *union of a finite family of paths.* Any such family can be coded by a *pointer field* ω indexed by the $3D$-sampling grid G. We write $\omega_g = (b_g, c_g)$ with $b_g \in \mathbf{Z}^2$ and $c_g \in \mathbf{Z}^+$ for each $g \in G$. We call c_g the *width* of the pointer. Moreover, we impose for all g in G,

(1.1) $\qquad 0 \leq c_g \leq C \qquad$ where C is a fixed bound

(1.2) $\qquad c_g = 0 \qquad$ if and only if $\qquad \omega_g = 0$

(1.3) $$J_g = 0 \quad \text{implies} \quad \omega_g = 0.$$

We call Ω the set of all pointer fields verifying (1). The associated family of *paths* is then constructed recursively (along the t-coordinate) using the following basic rules :

— if g is a focal point lying on a path γ, and $\omega_g \neq 0$, then $(g + \omega_g)$ is also a focal point in γ, namely the successor of g on γ.

— if $J_g \neq 0$ and if there is no g' in G, such that $\{g' + \omega_{g'} = g$ and $\omega_{g'} \neq 0\}$, then g is the initial point of a path.

In fact we restrict further our families of paths (and hence our pointer fields) by forbidding extreme closeness of paths. More precisely, fixing a rectangular window R in $\mathbf{Z}^2$, of size (n, m), centered at zero, we impose for any two focal points (s, t) and (s', t) lying in the same t-slice the restriction $s' \notin s + R$. We denote Ω_R the set of pointer fiels $\omega \in \Omega$ which verify this restriction.

3 Blur estimation and pre-smoothing

For each t-slice I_t of our observed $3D$-image, we set up an iterative estimation of the blur parameters, and use these estimates for a pre-smoothing of I_t. We have used the following empirical guess : when "noiseless" blurring spots overlap, the *local maxima* of the resulting distribution of intensities give a good first approximation for the location of the centers of these spots.

3.1 Blur estimation

We begin by selecting all the *local maxima* of the observed intensity distribution $s \to I_t(s)$ on slice I_t. Here, the notion of *local maximum* is the usual one, relative to a spatial window of fixed size centered at the local maximum. This size is fixed by the user of the algorithm. All such local maxima will be called "*rough spot centers*". Since neighbouring spots overlap partially, we decided to observe each potential spot within a sequence of increasing windows $D_1, D_2, \cdots$ centered at the spot center. For each such rough spot center u in S, and each such window D, we perform a local estimate of the parameters modelising the spot's shape, and test the quality of fit, according to the following procedure.

Call $f(s) = M\,h(s)$ the unknown distribution of intensities for the spot centered at u, where h is the discretized version of a Gaussian density with mean matrix μ and covariance matrix v. Let

$$M_D = \sum_{s \in D} M\,h(s), \qquad m_D = \sum_{s \in D} I_t(s).$$

We now adjust the unknown parameters $\theta = (M, \mu, v)$ to minimize the Kullback distance d between the two probability distributions on D defined by $\frac{1}{M_D} f$ and $\frac{1}{M_D} I_t$. Let us recall that if f_1, f_2 are two probability distributions on D, the Kullback distance

$$d(f_1\, f_2) = \max\,[K(f_1, f_2)\,,\, K(f_2, f_1)]$$

where $K(f_1, f_2) = \sum_{s \in D} f_1(s) \log\left(f_2(s)/f_1(s)\right)$. We refer to Basseville [17], Kullback [18] for standard statistical consideration justifying its use in parameter estimation problems.

Call then $\widehat{\theta}$ this estimate

$$\widehat{\theta} = \text{Arg}\,\min_{\theta}\, d\left(\frac{f}{M_D}\,,\,\frac{I_t}{m_D}\right)$$

and $\widehat{M_D}$ the corresponding estimate of M_D. We *accept* this estimate of θ on D, iff

$$\text{(2)} \qquad d\left(\frac{\widehat{f}}{\widehat{M_D}}\,,\,\frac{I_t}{m_D}\right) < d_0 \qquad \text{and} \qquad \widehat{M_D} > M_0$$

where d_0 and M_0 are empirical thresholds. When (2) is verified, we stop increasing the size of our estimation window D, and accept the current estimate $\widehat{\theta}$ as our final one (for the particular spot centered at u). To construct d_0, we compute the actual Kullback distance between two Gaussian distributions for which the covariance matrices are submitted to given distorsions, using a classical formula of Kullback [18].

Now if these criterions are not satisfied, we increase the window D and compute a new Kullback estimate etc. Obvious bounds on the size of D are imposed by the location of the neighbouring local minima, as well as by an *a priori* inspection of the data by the user. If the window D reaches this maximal size without a successful verification of criteria (2), we decide that no "real" spot is centered at the currently studied local maximum u.

3.2 Pre-smoothing

Call SPOT_t the subset of the sampling grid S constituted by the local minima $u \in S$ which survived the previously described *estimation and pruning process* on the image I_t. Of course, here t is *fixed* and SPOT_t is relative to this particular t-slice I_t. We now decide to truncate each Gaussian density h, with mean μ, covariance matrix $v = \begin{pmatrix} \alpha & \rho \\ \rho & \beta \end{pmatrix}$ by restricting h to a window centered at μ, and of size $(2\sqrt{\alpha}, 2\sqrt{\beta})$. Denote by $\bar{h}$ the truncation of $\widehat{h}$. To each *surviving* local maximum u, we have thus attached a truncated modelized spot, having the intensity distribution $\widehat{M^u}\,\bar{h}^u$, with an obvious indexation by u of the parameters M, h. We now replace our observed t-slice I_t by the following *smoothed version*

$$\widehat{I}_t = \max_{u \in \text{SPOT}_t} \widehat{M^u}\,\bar{h}^u .$$

We have essentially pruned out the "weak" spots and enhanced the visualisation of the "strong" spots by forcing gaps between neighbouring strong spots. Recall that this whole operation has to be *repeated* separately for each t-slice, and is of course t-dependent.

For each $g = (s,t) \in G$, such that $\widehat{I}_t(s)$ is nonzero, we have singled out a *unique* spot active at g ; we shall denote by $\theta_t(s)$ its estimated parameters, and we shall say that g *lies within* the spot centered at $\mu_t(s)$, with parameters $\theta_t(s) = (M_t(s), \mu_t(s), v_t(s))$, that we denote more briefly by $\theta_g = (M_g, \mu_g, v_g)$.
The smoothed $3D$-image intensities $\widehat{I}_t(s)$ where $s \in S$, $t \in T$ will be recoded more simply by $F_g = \widehat{I}_t(s)$ where $g = (s,t) \in G$. We call FOC_t the subset of the sampling S constituted by the estimated exact centers of the surviving spots in I_t. Each such center will be called a potential focal point of the $3D$-image I.

3.3 An example of pre-smoothing

Using actual focalisation data provided by a well studied benchmark $3D$-cube, we present the results of such a pre-smoothing in illustrations 3a, b, c, d. Sketches 3a and 3c represent the intensity distribution I_t for a given focalisation plane (or t-slice), using standard perspective and intensity level curves respectively. Sketches 3c, 3d represent the smoothed t-slice $\widehat{I}_t$, using the same convention.

For some spots, the pre-smoothing modelizes only the central peak of the spot. This is quite enough in this context since we only want to locate the focal points. Hence

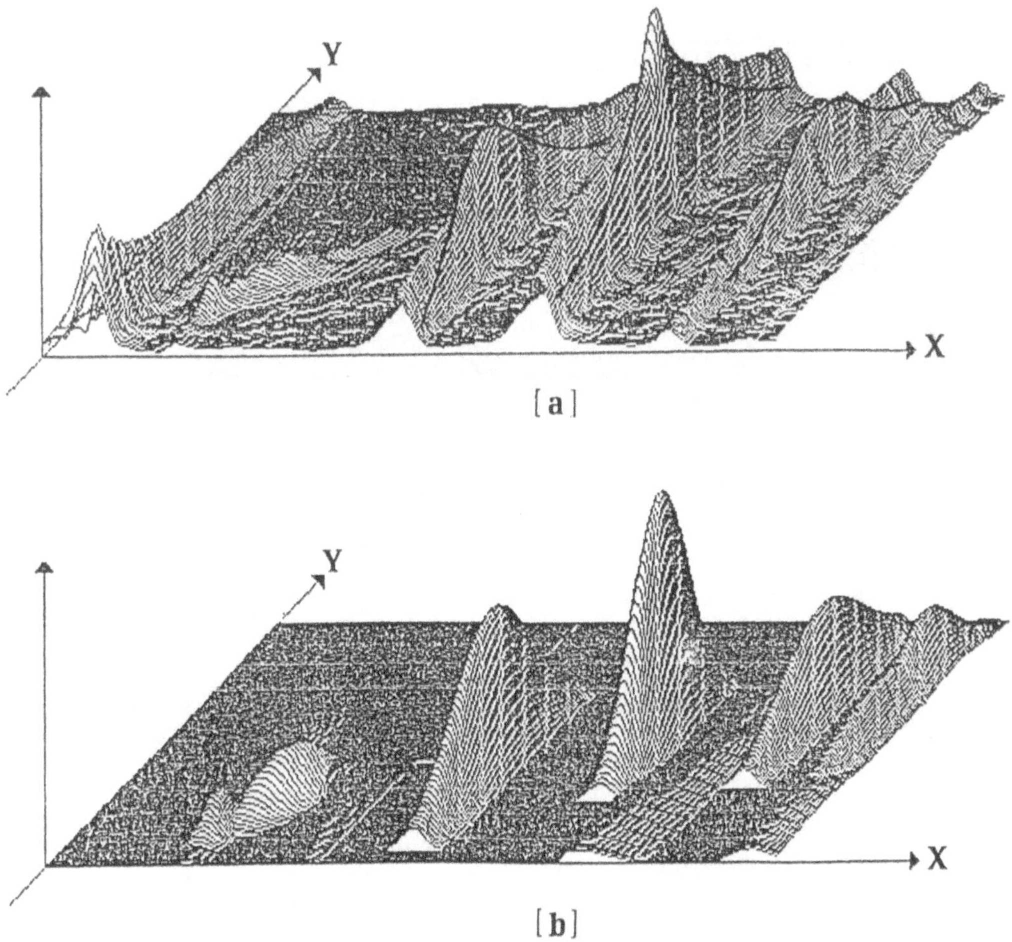

Figure 3 a,b
(a) *Perspective view of an image* I_t. (b) *Perspective view of the smoothed image* $\widehat{I}_t$.

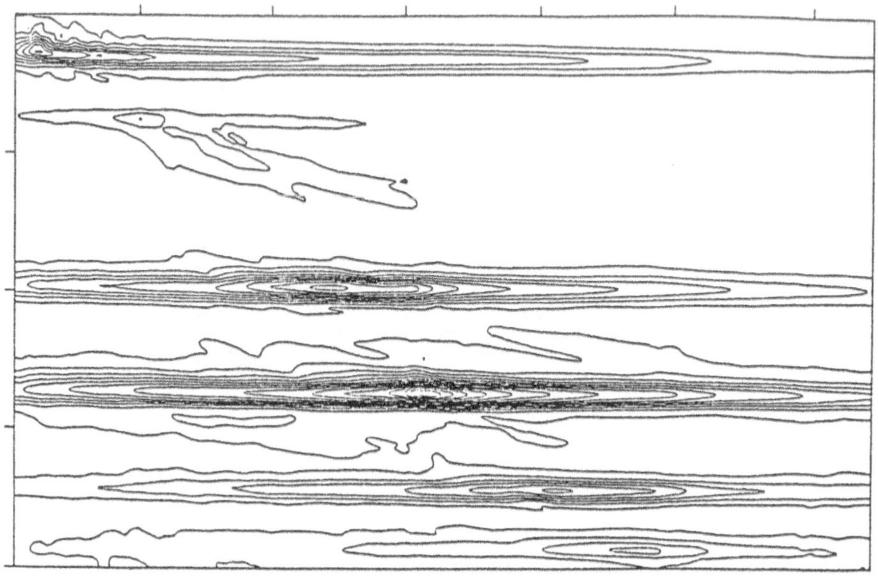

[c]

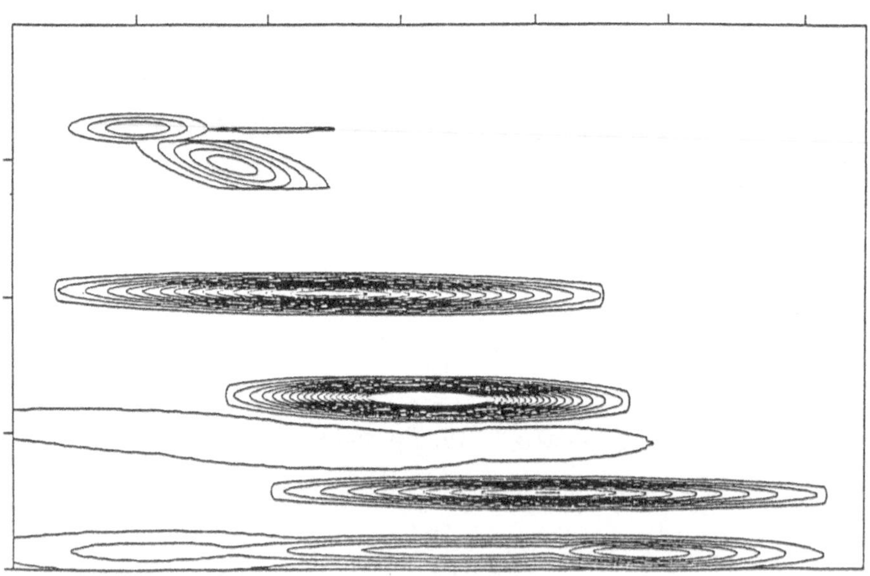

[d]

Figure 3 c,d

(c) *Intensity level curves of the same image I_t.* (d) *Intensity level curves of the smoothed image $\widehat{I}_t$.*

there would have been little to gain if we had tried to build a more precise smoothing modelizing carefully each spot by a *sum* of Gaussian intensity distributions, since our rougher smoothing preserves almost all significant focal points anyway.

4 The Markov fiel model

We now start with the collection $\text{FOC} = \{\text{FOC}_t, t \in T\}$ of focal point candidates obtained by pre-smoothing, the associated estimated spot parameter field $\theta = (\theta_g, g \in G)$ and the *smoothed* 3*D*-image $F = (F_g, g \in G)$. To each possible pointer field $\omega \in \Omega_R$, we are going to attach a plausibility grade $[-U(\omega)]$ which will be high when ω is a quite plausible pointer field *given* F, θ, and low in the converse situation. The function U which we call the (conditional) energy of ω given F, θ, will be a linear combination of four partial energies U_i

$$U = \sum_{i=1}^{4} \beta_i U_i$$

where the $\beta_i > 0$ are unknown parameters.

Given a pointer field ω, we define for all g in G the indicator function δ of its support by

$$\delta_g = 1_{\{\omega_g \neq 0\}}.$$

The first energy U_1 will force pointer fields to build focal paths of high global intensity :

(3) $$U_1(\omega) = -\sum_{g \in G} \delta_g \, \tau_1(F_g)$$

where $\tau_1 : \mathbf{R}^+ \to \mathbf{R}$ is an increasing scale transformation to be optimally selected later.

The second energy U_2 will force homogeneity of the *shapes* of focal spots along focal paths, by minimizing the average Kullback distorsion d between successive Gaussian shapes on paths

(4) $$U_2(\omega) = \sum_{g \in G} \delta_g \, \tau_2 \left[d(\theta_g \, , \, \theta_{g+}) \right]$$

where $g+ = g + \omega_g$ is the successor of g, τ_2 is an increasing scale transformation to be selected later, and d is the Kullback distance between the Gaussian models θ_g , θ_{g+} for the spots centered at μ_g and μ_{g+}.

The third energy U_3 tends to regularize the change of "speed" along focal paths :

$$(5) \qquad U_3(\omega) = \sum_{g \in G} \delta_g \, \delta_{g+} \, \tau_3 \, [|\ell_g(\omega) - \ell_{g+}(\omega)|]$$

where τ_3 is an increasing scale transformation to be selected later, and where for any pointer field ω, the "*speed*" at g is defined by $\ell_g(\omega) = b_g / c_g$ whenever $\omega_g = (b_g, c_g)$ and $c_g \neq 0$ with $b_g \in \mathbf{Z}^2$, $c_g \in \mathbf{Z}^+$. The norm $|\ |$ on $\mathbf{Z}^2$ is the L_1-norm.

The fourth energy U_4 forces the focal paths to be rather long and sparse :

$$(6) \qquad U_4(\omega) = \sum_{g \in G} \delta_g (1 - \delta_{g+}) - \tau_4 \sum_{g \in G} \delta_g \, \delta_{g+}$$

where the first sum is simply the total number of focal paths created by the pointer field ω. The second sum in U_4 is the total length of all these focal paths and $\tau_4 \in]0, 1[$.

We point out that U_3 and U_4 do not depend on the smoothed data F nor on the estimated shape parameters θ. On the other side, $\beta_1 U_1 + \beta_2 U_2$ represents the constraints imposed on the focal paths by the output (F, θ) of our data smoothing.
The parameters to be estimated below (in §6) to finalize the specification of U are the weights $\beta = (\beta_1 \cdots \beta_4)$ and the scaling factors $\tau = (\tau_1 \cdots \tau_4)$.

We adopt a now well grounded probabilistic setup by imposing on the pointer fields ω in Ω_R the Markov field distribution associated to the (conditional) energy U given F, θ, by the Gibbs formula :

$$P(\omega) = \frac{1}{Z} \exp[-U(\omega)] \qquad \text{with} \qquad Z = \sum_{\omega \in \Omega_R} \exp[-U(\omega)].$$

For all $g \in G$, let N_g be the parallelipipedic $3D$-window in G, centered at g, of size $(n, m, 2C+1)$ and deprived of its center g. Recall that in §2.4 we introduced the window R of size (n, m) and the bound C on pointers widths. As is easily seen N_g is the set of sites in G which are in interaction with g, either through the actual expression of U or due to the implicit constraints of §2.4, imposed on widths for pointers (1) and closeness for paths given by the restriction to Ω_R. Actually it is easily proved that, given F, θ, we have

$$P(\omega_g \,|\, \omega_{N_g}) = P(\omega_g \,|\, \omega_{G-g})$$

where $\omega_A = \{\omega_{g'}, \, g' \in A\}$ and hence P is a Markov field with respect to the neighborhood system $(N_g, \, g \in G)$ (*cf.* Besag [19]).

5 Estimation of focal paths

We assumed for the moment that after pre-smoothing of our focalisation data, the energy parameters β_i, τ_i, $i = 1 \cdots 4$ have been adjusted as described in section §6 below. The question tackled in the current section is the search of a good feasible estimate $\widehat{\omega}$ for the unknown actual pointer field.

The standard maximum likelihood estimator maximizing $P(\omega)$ in $\omega \in \Omega_R$ is unrealistic, given the huge size of the data. We have introduced instead a local optimization algorithm which resembles the ICM estimator introduced by Besag [10]. For comparison purposes, let us recall the standard ICM algorithm : one starts with an initial estimate ω^1, and then ω^1 is iteratively modified by refreshing one site at a time in a given arbitrary periodic order. Call g_{n+1} the site modified at the time $n+1$, and ω^n the existing pointer field at time n. Then the standard ICM selects ω^{n+1} as follows :

(7.1) $$\omega_g^{n+1} = \omega_g^n \qquad \text{for} \qquad g \neq g_{n+1}$$

(7.2) $$\omega_g^{n+1} = \text{Arg}\max_{\omega_g} P(\omega_g \,|\, \omega_{N_g}^n) \qquad \text{for} \qquad g = g_{n+1}.$$

This algorithm converges after a few complete sweeps of the $3D$-grid G, and the limit configuration ω^∞ is a local maximum of $P(\omega)$. How good this estimate ω^∞ can be, strongly depends on the context.

In our case, the *initial configuration* is defined as follows : for every g in FOC = {potential focal points}, we define $\omega_g^1 = \widetilde{g} - g$ where :

$$\widetilde{g} = \text{Arg}\min_{g' \in N_g} \{\|g' - g\|\,,\, g' \,\text{posterior to}\, g\}.$$

By convention g' is posterior to g if the t-coordinates t', t of g', g verify $t' > t$. For $g \notin$ FOC, we let $\omega_g^1 = 0$. However the standard ICM is not efficient enough here : indeed due to the constraint preventing extreme closeness of paths, the ICM algorithm (7) will never modify the support of ω^1, and hence will not prune the initial ω^1, although it may reorganize it a bit.

We thus prefer to rely on a local optimisation algorithm which we call BLOCK-ICM, and which we now describe. Fix a current pointer field ω. For every $g \in G$, such that g lies on a path γ defined by ω, and such that g is not the starting point of γ, we

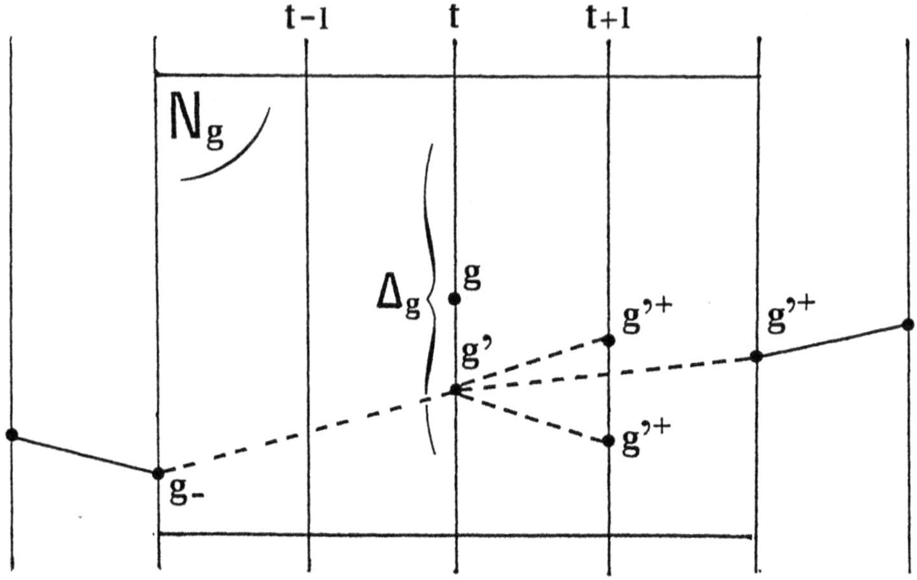

Figure 4
Three instances of local paths configurations (in dot lines) branding out of a tentative particular choice for g' in Δ_g (in projection on the (x,t) plane with $C = 2$).

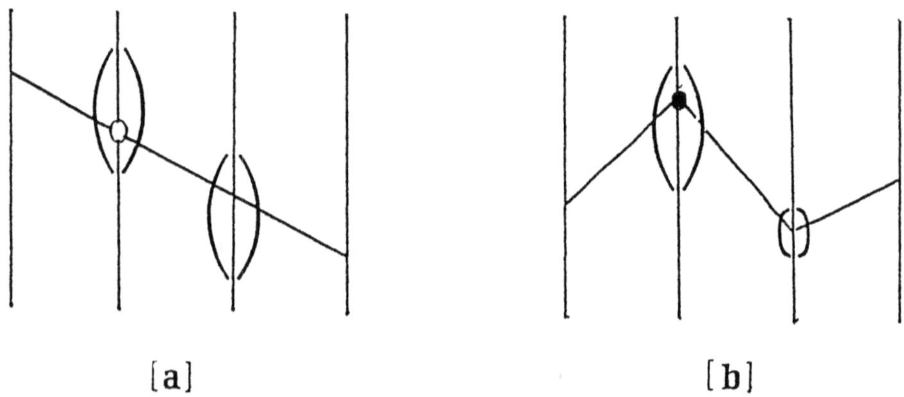

Figure 5
The first two extreme configurations of Table 2, in projection on the plane (x,t). (A white pixel indicates a high intensity and a black one a low intensity.)

let g_- be the predecessor of g, so that $g = g_- + \omega_{g_-}$. Fix $n' < n$, $m' < m$ and for $g = (s,t) \in G = S \times T$, call Δ_g the $2D$-window (within $S \times t$) centered at g, and with size (n', m'). If g lies on a path defined by ω, define :

(8.1) $$B_g = \{g_-\} \cup \Delta_g \qquad \text{if } g_- \text{ exists}$$

(8.2) $$B_g = \Delta_g \qquad \text{if } g \text{ is a starting point.}$$

Call NB_g the union of all $N_{g'}$ for $g' \in B_g$. In our BLOCK-ICM algorithm, we fix an arbitrary infinite periodic sequence g_n visiting all sites $g \in G$. At time n, the current configuration ω^n will be modified only on the sites which belong to B_{g_n}. Let then $\omega^n = \omega$ and $g_n = g$ to simplify the notations and select for the modified block configuration ω'_{B_g} replacing ω_{B_g} as the block configuration realizing :

$$\max_{\omega'_{B_g}} P\left[\omega'_{B_g} \mid \omega_{NB_g - B_g}\right].$$

Then define ω^{n+1} by

$$\omega^{n+1}_{B_g} = \omega'_{B_g} \qquad \omega^{n+1}_{G-B_g} = \omega^n_{G-B_g}.$$

Due to the constraint on mutual distances between paths and on the size of the window Δ, then whenever g_- exists, the possible paths created by a single block of pointers ω'_{B_g} can intersect Δ_g at only one (arbitrary) point $g' \in \Delta_g$, which restricts immediately the list of acceptable ω'_{B_g}. Sketch 4 above represents, in $2D$-projection, three instances of local path configurations branding out of a tentative particular choice for g' in Δ_g.

6 Actual Markov field modelization

6.1 Methodology

The first step is the selection of an energy U, which was described in section §4. The second step is the *adjustment* of the parameters of this particular energy, taking account on *a priori* specifications of performance for the estimator $\widehat{\omega}$ (of the unknown pointer field ω) produced by our BLOCK-ICM algorithm.

This has been achieved by the following interactive scheme :

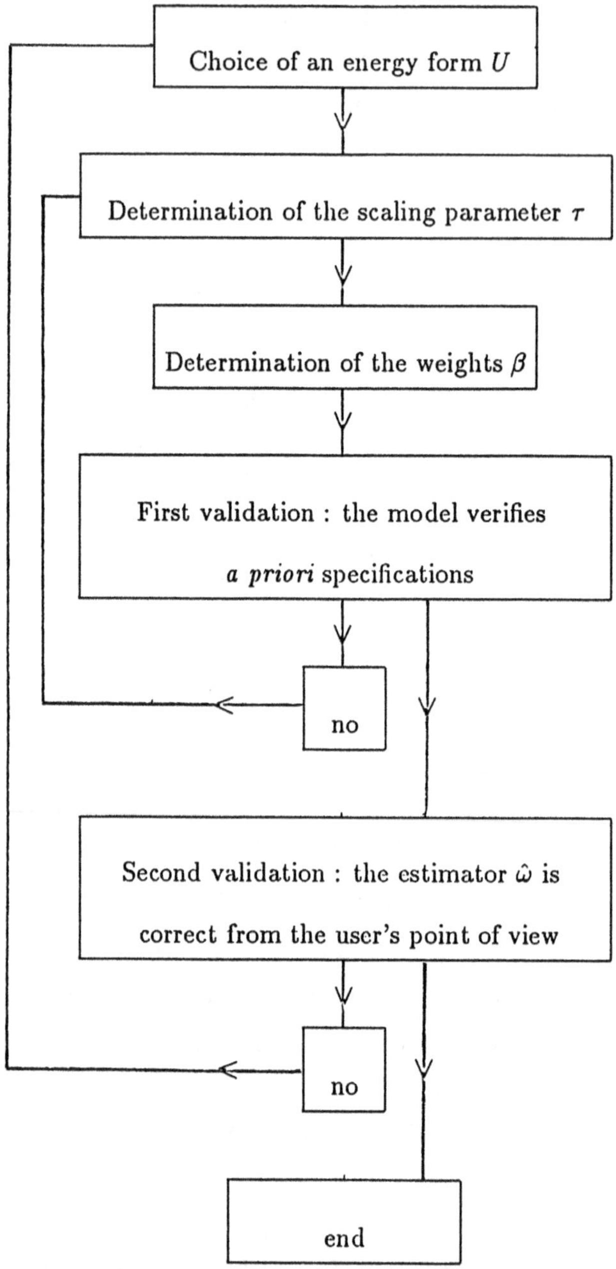

Table 1

Interactive scheme for Markov field modelization

We point out that at the end of this interactive process, the parameters β and τ are selected once and for all, and use to handle incoming new focalisation data. The only parameters left to the choice of the actual user of the algorithm are the integers (n, m, C) which determine the dimensions of N_g, and may be selected according to the "look" of the data.

Initially, we tried linear scale changes for τ_2 τ_3 τ_4 and a logarithm scale change for τ_1. We had to give up this combination in the validation study and our final (and efficient) choice has been to use *only linear scale changes* for $\tau_1 \cdots \tau_4$.

6.2 Determination of the weights β

Assume the scale parameters τ to be already estimated. We shall constrain $\beta = (\beta_1 \cdots \beta_4)$ by imposing *a priori* bounds on conditional probabilities $P(\omega_g \,|\, \omega_{N_g})$, using "*extreme local configurations*" for which such probabilities should clearly be quite close to 0 or 1. This yields for β a system of linear equations which are easily solved. This approach is in the spirit of the "qualitative boxes" approach sketched by Azencott [9] ; we also refer on this point to Possolo [20]and Derin [21].

Up to a normalizing constant, the local specifications are given by :

$$P(\omega_g \,|\, \omega_{N_g}) \propto \exp - \nabla_g(\omega)$$

$$\nabla_g(\omega) = U(\omega) - U(\widetilde{\omega})$$

$$\widetilde{\omega}_{G-g} = \omega_{G-g} \text{ and } \widetilde{\omega}_g = 0.$$

For a given pointer field ω, let $\xi_g(\omega) = 1$ if g is a focal point of ω, having a predecessor g_-, and let $\xi_g(\omega) = 0$ otherwise. Rewrite the energies U_1, U_2, U_3, U_4 as $U_i = \sum_{g \in G} U_{i,g}$ with obvious local components $U_{i,g}$ specified by formulas (3-6) in section §4. Then an immediate computation yields :

$$\nabla_g = \sum_{i=1}^{4} \beta_i Z_{i,g} \qquad \text{with} \tag{9}$$

$$\begin{aligned} Z_{1,g} &= U_{1,g} \\ Z_{2,g} &= U_{2,g} \\ Z_{3,g} &= \xi_g U_{3,g_-} + U_{3,g} \\ Z_{4,g} &= \xi_g U_{4,g_-} + U_{4,g}. \end{aligned}$$

Locally, $Z_1 \cdots Z_4$ quantify respectively the path intensity, the non homogeneity of the spots, the speed irregularity, and the presence/absence of path break. We may thus rewrite (9) as :

$$\sum_{i=1}^{4} \beta_i Z_{i,g}(\omega) = -\log \frac{P(\omega_g|\omega_{N_g})}{P(\omega_g = 0|\omega_{N_g})}. \tag{10}$$

Let us now select a finite set ELC of "extreme local configurations". Each such configuration selects an arbitrary set of sites $(g \cup N_g)$, and specifies arbitrarily *on this set* the values taken by the fields F, θ, ω (which represent resp. the smoothed image, the shape parameters, the pointer field).

Note that for each extreme local configuration $e \in$ ECL, the $Z_{i,g}(\omega)$ are easily computed and yield numbers Z_i^e, $i = 1 \cdots 4$. Moreover, the conditional probabilities appearing on the right hand side of (10) will be *a priori set at values* :

$$p^e = P(\omega_g|\omega_{N_g}) \qquad p_0^e = P(\omega_g = 0|\omega_{N_g})$$

which will be very close to zero or one. Of course, this requires an *adequate choice* of the extreme configuration e. We now have a linear system :

$$\sum_{i=1}^{4} \beta_i Z_i^e = -\log \frac{p^e}{p_0^e} \tag{11}$$

of cardinal equal to Card (ECL) > 4, to have more equations than unknown. The determination of β is then obtained by standard least squares resolution of the system (11).

6.3 Examples of extreme local configurations

We have actually used 12 such configurations. The following table lists the values of the Z_i^e and the p^e. Actual values of the Z_i^e have been replaced by *qualitative evaluations H = High or L = Low* to enhance readability. On Fig.5, we present the first two extreme configurations.

e *configuration*	Z_1^e *intensity*	$1/Z_2^e$ *homogeneity*	$1/Z_3^e$ *regularity*	Z_4^e *path break*	p^e
1	H	H	H	No	0.99
2	L	L	L	No	0.01
3	L	H	H	No	0.1
4	L	H	L	No	0.01
5	H	H	L	No	0.9
6	L	L	H	No	0.1
7	H	L	H	No	0.1
8	H	L	L	No	0.01
9	L	H	H	Yes	0.1
10	H	H	H	Yes	0.99
11	L	L	H	Yes	0.01
12	H	L	H	Yes	0.1

Table 2

Extreme configurations (H = High, L = Low)

6.4 Determination of the scale parameters

Start with an "arbitrary" guess for the scale parameters τ. Apply the method just described in 6.2 to estimate β by $\widehat{\beta}$. The overdetermined linear system (11) above yields for each extreme local configuration e in ELC, a residual error term ε_e given by

$$\varepsilon_e = -\log\left(\frac{p^e}{p_0^e}\right) - \sum_{i=1}^{4} \widehat{\beta}_i\, Z_i^e .$$

We want these residuals ε_e, $e \in$ ELC to be *small*. *If not* we attempt to modify the scale parameters τ. This is done by inspecting the *large* residuals ε_e and more precisely of their corresponding components $\widehat{\beta}_i\, Z_i^e$. This suggests immediate qualitative modifications of the Z_i^e, which are achieved by changing the τ_i.

This methodology, which can easily be formalized, is analogous to interactive modelization in multiple linear regression (Draper-Smith [22]). In our context, we complete this examination of large residuals by one last validation step. Whenever all residuals ε_e are small, we actually check how the $\widehat{\beta}$ estimate of β performs on another *control set* of local configurations, using a similar point of view. Thus we are validating our linear system by an adjoined extra set of linear equations of the same type, with less tightly specified right hand sides.

7 Applications to actual $3D$-focalisation data

The data are real seismic data produced by the algorithmic chain provided in the MIGPACK software package [1]. The actual $3D$-image has dimensions (800, 160, 400). We observe roughly 20 spots per t-slice I_t. Our selected dimensions for the N_g window are $n = 17$, $m = 8$, $C = 8$, and the BLOCK-ICM algorithm uses Δ_g windows of size $n' = 9$, $m' = 5$.

On sketch 6a, we represent the projections on the (x, t) plane of the focal paths detected by the algorithm developed by Julien et alt. [5]. Clearly, the paths appear to be noisy, essentially due to irregular local slopes. Sketch 6b on the other hand represents the projections on the (x, t) plane of the focal paths detected by the Markov field approach developed in this paper. We notice an evident smoothing of the paths, and the erasing of most of the noise. Let us emphasize that the initial configuration ω^1, used by the BLOCK-ICM algorithm, is very closed to the configuration of Fig. 6a. Computations using the BLOCK-ICM estimation have required 6 complete sweeps of the $3D$-image, which amounted to about one hour of CPU time on a CONVEX. The total number of detected focal paths is of the order of 200. Fig. 2a depicts the focal paths obtained with another parameters n, m, C.

The second advantage of this refined automatic picking is that it provides an interpretable image directly comparable to the seismic section shown on Fig. 1. One can notice that the chief horizons, in terms of focused energy, have been determined correctly. A lot of qualitative and pseudo-quantitative information has been calculated as well. This complementary knowledge refers to the shape and the amplitude of the spots, the length of the paths and better describes the seismic data and further the geological subsurface. This kind of information was of course already contained in the raw dataset but it was not explicitely computed.

A very important characteristic of such an automatic picking is that it is continuous and, as a consequence, it leads to the dip of focusing paths. This new knowledge is of great interest. It allowed us to derive a new equation for computing the velocity correction in the case of dipping layers. Until now, the velocity updating was only carried out in the simple horizontal case. The new velocity field varies continuously, laterally and in depth. In this way, non-linear variations of the velocity field may be described within a layer.

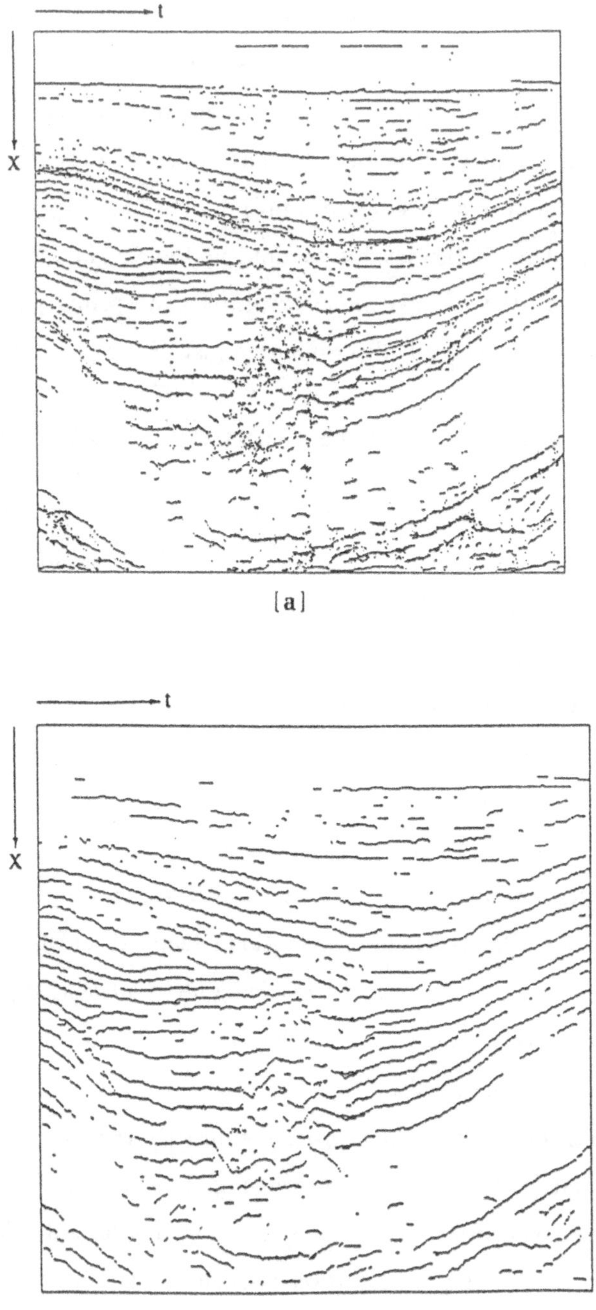

Figure 6

Projection on the (x, t) plane of the focal paths. **(a)** *detected by the standard algorithm of [5].* **(b)** *detected by our Markov field approach.*

The last topic which may be pointed out is the problem of multiples (Wiggins [23] ; Julien and Raoult [24]). Multiple reflections are composed of energy which has been upward reflected more that once. Multiples constitute a coherent noise which often presents a lower apparent velocity than the one of primary events. Presently, no general method is able to suppress all kinds of multiples. If focusing paths are projected and displayed on the plane of the seismic section, at their right (corrected) locations with a color depending on the velocity model, multiples and primary events which present a differential velocity may be easily discriminated just by looking at the colors of the paths.

8 More details on the seismic problem

The aim of seismic processing is to help the geologist to gain a better (and easier) understanding of the subsurface from seismic data study. Seismic sources located at the surface of the earth generate a wavefield which is reflected by the interfaces between layers of different kinds. The reflected wavefield is recorded at the surface, as a function of time, by a network of receivers. These seismic data provide an image of the subsurface reflectivity (see Fig. 1).

In the seismic processing approach, a very important stage is probably $2D$ migration [4], a process which reduces the distance between a $2D$ seismic section and the corresponding geological section. The problems which are solved or reduced by migration are those connected with wavefield propagation effects. If for instance, a seismic (stack) section is migrated, time diffraction hyperbolas (due to diffracting points such as extremities of technical faults) disappear and geological horizons are then located at their right positions on the processed result.

Prestack depth migration, which processes directly the raw dataset recorded on the field, is the ultimate in structural $2D$ processing. It is a good way to produce a correct image of a complex subsurface in the presence of strong lateral velocity variations. The shot-geophone presstack depth migration (Denelle et al. [25,26]) involves two stages. First, the pressure field is downward extrapolated. Given the recorded wavefield at the surface and the acoustic wave equation, the pressure is computed for all depths by using a finite differences algorithm (Clearbout and Doherty [27] ; Schultz and Sherwood [28]). Each elementary downward continuation consists in applying two operators : the phase-shift operator which corrects the pressure field for the vertical wave propagation and the focusing operator which focuses the diffracted energy to the scattering points. In the second stage (imaging step), the migrated result is obtained by mapping the wavefield at a propagation time equal to zero. Really, when the propagation is null, the source and the receiver are located at the same place. The energy that is recorded by the receiver is exactly the energy emitted by the source times the local reflectivity of the medium. Hence, at a null time, the downward continued recorded wavefield is a direct measurement of the local reflectivity of the medium.

The shot-geophone prestack depth migration proves to be effective and affordable : from time recorded seismic data, it leads to a depth image. That means that it requires prior knowledge of the propagation velocities *i.e.* the shape and interval velocities of each layer.

Now, all authors agree to say that the most important input information for depth migration procedures is the velocity distribution. Generally, this information is only approximately known and errors in the migration output due to errors in the velocity input occur in a lot of practical applications. So the following paradox has to be faced : performing a correct pre-stack depth migration requires that the main features of the resulting seismic section in terms of layer geometry plus the interval velocities within each layer be input as processing parameters. Faye and Jeannot [1] proposed a specific procedure to solve this problem. This procedure refines the velocities using depth focusing analyses. Such an approach had already been introduced by Doherty and Clearbout [29] and by Yilmaz and Chambers [2] for time migration. The influence of velocity errors on the focusing aspects of migration has been theoretically studied by De Vries and Berkhout [30]. More recently, some new developments have been derived by Julien et al. [5] who proposed a depth continuous velocity analysis method based on an automatic picking of all the relative maxima of the whole focusing $3D$ block computed over an entire seismic line. MacKay and Abma [31] proposed to obtain directly a refined prestack depth migration image from the focusing block.

The general principle of the focusing analyses is the following. When the velocity is exact, all the diffracted energy focuses, for a null propagation time (imaging criterion), at the location of the scattering point. If the migration velocity is greater than the exact velocity, the energy focuses at a greater rate than expected, at a positive time. For each shot point, a focusing plane is the result of the extrapolation of seismic data at every depth and every time for the zero offset *i.e.* for a geophone located at the source. Each interface leads to a focusing point. The location of the points on a focusing plane is related to the used velocity model. If a focusing point is not located on the imaging curve (corresponding to a null propagation time), the velocity model is changed to bring this point on this curve. In fact, due to numerical problems, the focusing pick is not exactly a point but is rather a spot with a certain length, width and amplitude. To provide the explorationist with a good tool for imaging structures in such hard conditions, a full prestack depth migration program combined with an interactive focusing analyses based software, has been developed and made operational in an industrial environment in house (Julien et al. [32]).

Until now, due to the manual picking of relative maxima, few focusing planes were processed in a seismic line. Hence, it was difficult to achieve geological coherency from one focusing plane to the next one. Too simple velocity models were obtained compared to what was expected from the prestack depth migration ability to take into account large velocity variations. If we process all the focusing planes (a focusing plane for each shot point), we get a $3D$ focusing block. With such a volume, it becomes possible to carry out a continuous velocity analysis. For a real case, this data volume is very large (about four hundred megabytes) and there are several thousands of spots. The goal of this presented paper is to present a refined automatic picking. This picking leads to focusing paths of relative maxima. These paths represent geological interfaces and the associated depth errors. Generally, these paths must be smooth and continuous. With such an automatic picking, lateral velocity variations will be faster and better described.

References

[1] J.-P. FAYE and J.-P. JEANNOT - *Prestack migration velocities from focusing depth analysis*, 65th Annu. Int. Meet., Soc. Expl. Geophys., Expanded Abstract (1986), 38-440.

[2] O. YILMAZ and R. CHAMBERS - *Migration velocity analysis by wave field extrapolation*, Geophysics, vol. 49 (1984), 1664-1674.

[3] Y.C. CHENG and S.Y. LU - *The binary consistancy checking scheme and its applications to seismic horizon detetion*, IEEE Trans. Pattern Anal. Machine Intell., vol.

PAMI-11 (April 1989), 439-448.

[4] J.F. CLEARBOUT - *Imaging the earth's interior*, Palo Alto, Blackwell Scientific Publications, Inc. (1985).

[5] Ph. JULIEN, Y. VUJASINOVIC and J.-J. RAOULT - *Depth continuous velocity analysis based on prestack migration*, 58th Annu. Int. Meet., Soc. Expl. Geophys., Expanded Abstracts (1988), 437-441.

[6] R. HUMMEL, K. ZUCKER and S. ZUCKER - *Deblurring Gaussian blur*, Comp. Vision, Graphics and Image, Proc., vol. 38 (1987), 66-80.

[7] B. CHALMOND - *PSF estimation for image deblurring*, Comp. Graphics, Vision and Image Processing, July 1991.

[8] S. GEMAN and D. GEMAN - *Stochastic relaxation, Gibbs distributions and the Bayesian restoration of images*, IEEE Trans. Pattern Anal. Machine Intell., vol. PAMI-6 (1984), 721-741.

[9] R. AZENCOTT - *Image analysis and Markov field*, Proc. of Int. Conf. on Ind. Appl. Math. SIAM (1987).

[10] J. BESAG - *On the statistical analysis of dirty pictures*, J.R. Statist. Soc. B-148 (1986).

[11] P. BOUTHEMY and P. LALANDE - *Detection and tracking of missing objects based on a statistical regularization method in space and time*, Computer Vision - ECCV90, Lecture Notes in Computer Sciences, vol. 427, Springer-Verlag (1990), 307-314.

[12] B. CHALMOND - *Image restoration using an estimated Markov model*, Signal Processing, vol. 15 (1988), 115-129.

[13] B. CHALMOND - *An iterative Gibbsian technique for reconstruction of m-ary images*, Pattern recognition, vol. 22 (1989), 747-761.

[14] J. KONRAD and E. DUBOIS - *A comparison of stochastic and deterministic solution methods in Bayesian estimation of 2-D motion*, Computer Vision - ECCV90, Lecture Notes in Computer Sciences, vol. 427, Springer-Verlag (1990), 149-160.

[15] D. GEMAN, S. GEMAN, C. GRAFFIGNE and P. DONG - *Boundary detection by constrained optimization*, IEEE Trans. Pattern Anal. Machine Intell., vol. PAMI-12 (1990), 609-628.

[16] J.-L. MARROQUIN, S. MITTER and T. POGGIO - *Probabilistic solution of ill-posed problem in computer vision*, Journal of the Amer. Stat. Ass., vol. 82 (1987), 76-89.

[17] M. BASSEVILLE - *Distance measures for signal processing and pattern recognition*, Signal Processing, vol. 18 (1989), 349-369.

[18] S. KULLBACK - *Information theory and statistics*, J. Wiley (1959).

[19] J. BESAG - *Spatial interaction and the statistical analysis of lattice systems (with discussion)*, J.R. Statist. Soc., vol. B-36 (1974), 192-236.

[20] A. POSSOLO - *Estimation of binary random Markov field*, University of Washington, Technical report 77, Seattle (1986).

[21] H. DERIN and H. ELLIOT - *Modeling and segmentation of noisy and textured images using Gibbs random fields*, IEEE Trans. Pattern Anal. Machine Intell., vol. PAMI-9 (1987), 39-55.

[22] N.R. DRAPER and H. SMITH - *Applied regression analysis*, John-Wiley (1966).

[23] J.W. WIGGINS - *Attenuation of complex water-bottom multiples by waves-equation based prediction and subtraction*, Geophysics, vol. 53 (1988), 1527.

[24] Ph. JULIEN and J.-J. RAOULT - *Adaptive subtraction of emulated multiples*, 59th Annu. Int. Meet., Soc. Expl. Geophys., Expanded Abstracts, vol. II (1989), 1118.

[25] E. DENELLE, Y. DZARD and J.-J. RAOULT - *Implementation of a 2D-prestack depth migration scheme on a* CRAY-1S, 55th Annu. Int. Meet., Soc. Expl. Geophys., Expanded Abstracts (1985), 318-320.

[26] E. DENELLE, Y. DZARD and J.-J. RAOULT - *2D-prestack depth migration in the* (S-G-W) *domain*, 56th Annu. Int. Meet., Soc. Expl. Geophys., Expanded Abstracts (1986), 327-330.

[27] J.F. CLEARBOUT and S.M. DOHERTY - *Downward continuation of moveout corrected seismograms*, Geophysics, vol. 37 (1972), 741-768.

[28] P.S. SCHULTZ and J.W.C. SHERWOOD - *Depth migration before stack*, Geophysics, vol. 45 (1980), 376-393.

[29 S.M. DOHERTY and J.F. CLEARBOUT - *Velocity analysis based on the wave equation*, Stanford Exploration Project Rep. 1 (1984), 160-178.

[30] D. De VRIES and A.I. BERKHOUT - *Influence of velocity errors on the focusing aspects of migration*, Geophys. Prosp., vol. 32 (1984), 629-648.

[31] S. MAC KAY and K. ABMA - *Refining prestack depth-migration images without remigration*, 59th Annu. Int. Meet., Soc. Expl. Geophys., Expanded Abstracts (1989), 1258-1261.

[32] Ph. JULIEN, M. KLEIN and T. THOMAS - *In quest of the base* 59th Annu. Int. Meet., Soc. Expl. Geophys., Expanded Abstracts, vol. II (1989), 1275-1278.

EDGE DETECTION AND SEGMENTATION OF TEXTURED PLANE IMAGES

R. Azencott, C. Graffigne, C. Labourdette
Université Paris-Sud, Laboratoire de Statistiques Appliquées
91405 Orsay Cedex, France.

Abstract

We use a Markov framework for finding edges and for partitioning scenes into homogeneous regions. The images are airplane images with a fine resolution. They have been chosen according to the presence of textures, some of these textures being macro-textures. We are working in a supervised context, and we assume the existence of samples for each of the textures.

Segmenting these textures, and due to their resolution, the problem of edge detection proves to be very important. The definition of edges between textures may not always be straightforward since these edges are sometimes materialized by fences, or roads, but some other times, they are only implicit. We use a statistical definition of the edges, and then, after positioning them, we extract informations for the segmentation. The use of these informations improves greatly the results.

At the same time, one of our main purpose is to keep the computation time within reasonable bounds. This was done by selecting carefully the model energy.

Acknowledgments : This work was supported by DRET grant 88/215.

1 Introduction.

Texture is a very important feature in remotely sensed images, and regions cannot be distinguished by method based only on shading. Several authors have been working on textured images, sometimes real images and sometimes images made of patches of textures (see for example [4, 6, 11, 8, 9, 13, 12, 7, 16]). Applications include the analysis of medical images, automated navigation, mapping of an image on a map, cartography, and so on.

This problem is originally quite difficult, and becomes even harder when there are macro-textures in the image. Let us first clarify the meaning of micro and macro-textures: We define the resolution of a texture to be the size of the smaller square necessary to characterize the texture. A micro-texture is then a texture with a small resolution, say less or equal than 7 pixels. And a macro-texture is a texture with a large resolution, greater than 7 pixels.

Independently of the method, it is always necessary to compute a sufficient number of features, or to estimate some model parameters (if there are some texture models). This induces large computation time, essentially related to the resolution of the textures. In the algorithm presented here, we have used Markov random fields which are often considered as too greedy in cpu time, howether the model energy has been carefully chosen to reduce the cpu time, and in particular, we do not need to use simulated annealing but only a few iterations of the Gibbs sampler algorithm.

We tested our algorithm on two series of images, each of them are grey level images of some region of France taken from an airplane. The resolution of these images is fine enough to allow the observation of natural textures: vineyard, orchard and other types of large spacing farming. Most of the time, these natural textures lack homogeneity and this is one of the main difficulty.

The aim of this work is to segment each image into homogeneous regions, each of the regions containing a texture belonging to a preselectioned set of textures. We use a Markov framework which allows general and uniformly defined models.

In this article, we present first Markov random fields and the texture feature used, then the boundary model and the label model, and finally the experimental results.

2 Markov random fields, simulated annealing and ICM.

A field $X = (X_s, s \in S)$ is a Markov random field of random variables (MRF) associated to the neighborhood system G if, for any configuration x (a configuration is a realization of the procces X, and G_s will denote the neighborhood of a site s):

1. $P(X = x) > 0$.

2. $P(X_s = x_s | X_r, r \neq s) = P(X_s = x_s | X_r = x_r, r \in G_s)$.

The MRF-Gibbs equivalence provides an explicit formula for the joint probability distribution of X: $P(X = x) = \frac{1}{Z} exp(-U(x))$ where Z is a normalisation constant and U is the energy function (or the cost function).

Simulation of these fields is not straightforward, but requires special algorithms like the Gibbs sampler algorithm (see [17, 2]). Note that highly probable configurations have low energy, so that in order to appreciate the model's efficiency, it is necessary to consider configurations with low energy.

Determining the segmentation requires the optimization of an energy function derived from U. Optimization can be obtained using simulated annealing (see [8]). Although, in practice, this algorithm only ends up into a local minimum of U, one can expect that, if the initial temperature is high enough and the decreasing schedule slow enough, this local minimum is close enough of an optimal state. (This temperature schedule is linked to the complexity of the energy function.) So that, if the energy landscape is only composed of a few deep wells, an ICM (Iterated Conditional Mode algorithm, see [2]) with a correct starting point will probably give as good results as an annealing. (The ICM is simply a Gibbs sampler at temperature 0 and realizes a random descent to a local minimum.)

Of course, in a high dimensional space, it is quite impossible to determine the qualitative features of the energy landscape, but knowing the composition of the energy, it is sometimes possible to understand the behavior of each algorithm.

3 Texture features.

Most of the time, the first and second order statistics are sufficient to distinguish micro-textures, but the problem is harder for macro-textures. We used a set of classical features defined in [15], and the first order statistics: the mean and the logarithm of variance, ended up to be sufficient for the concerned micro-textures. For the macro-textures (particularly for figure 8), none of these features was sufficient and we constructed some new characteristics.

We used interactive algorithms in order to select on the samples the appropriate texture attributes and also the resolution of the textures, according to their ability to distinguish two textures. Each of the attributes is assumed to be gaussian, and the mean and variance of these gaussians are estimated on the samples.

In order to characterize the macro-textures, we chose one patterns corresponding to each macro-texture in figure 8: the texture in the top of the picture and the one in the middle. The test for the existence of a pattern is then translated in terms of classical gaussian tests and will be detailed a little later.

These choices are clearly dependent essentially on the texture samples. For more general use, it would be necessary to acquire automatically the convenient pattern. But for our purpose, since we dispose of samples of the different textures, it is possible to select interactively the interesting patterns (in order to distinguish the macro-textures one from another and from the micro-textures).

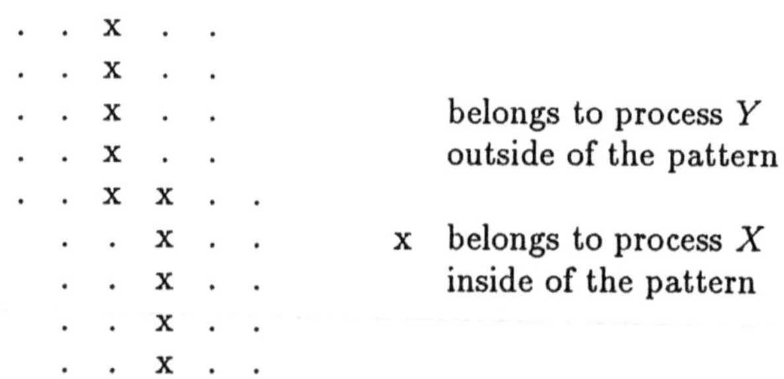

Figure 1: One pattern and the two processes X and Y.

For a given pattern, we define two processes $(X_i, 1 \leq i \leq n)$ and $(Y_i, 1 \leq i \leq p)$, one procces for the inside of the pattern and the other one for the outside (see figure 1). These two processes are supposed to be independent gaussians, and the corresponding hypothesis test is classical: First test the variances of (X_i) and (Y_i), then test the means of (X_i) and (Y_i). if the variances are small enough and the means are different then we have a pattern, otherwise, we do not.

The results obtained with this method are shown in figure 6 and 7. Of course, some isolated patterns are falsely determined and other existing patterns are not found, but the result is generally satisfying.

4 Edge Model.

We are concerned with a problem simpler that a complete detection of texture edges with closure of the boundaries. In fact, since the goal is a segmentation of the image, this segmentation will implicitly supply closed boundaries around the regions. For this first step, we need precise and almost certain informations about edge elements in order to use this information for the segmentation. But the problem is clearly much harder than the usual edge detection problems since we are dealing with textures (see figure 4 for example) which certainly contain themselves classical edge elements that should not appear in the boundary detection.

The edge elements are placed on a dual lattice, between each pair of pixels: see figure 2.

x	|	x	|	x		x	pixel site
—		—		—		—	horizontal edge site
x	|	x	|	x		|	vertical edge site

Figure 2: Dual lattice representing the edge sites.

Two windows are associated to each edge site as shown in figure 3.

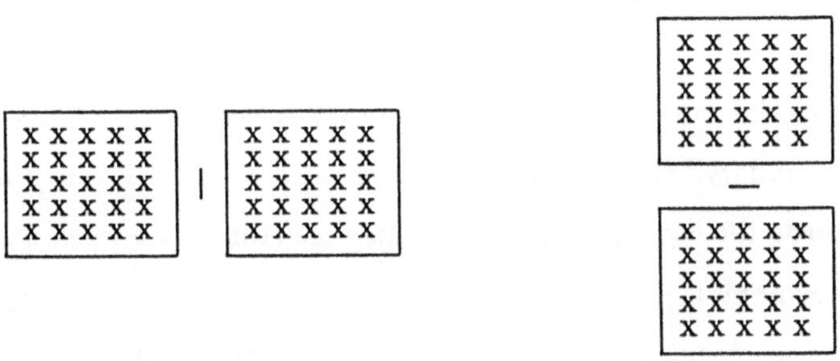

Figure 3: Position of the windows used for the edge detection in both cases: vertical and horizontal edge elments.

The size of these windows may depend on the presence of macro-textures, the algorithm allows the use of two window sizes, one for micro-textures and an other one for macro-textures (for the first experiment, only one resolution of 7 pixels, and for the second experiment two resolutions of 5 and 11 pixels). The choice of the window size is done in the following way: First use the larger windows, if a macro-texture pattern is found in one of them, then compute the attributes using the macro-texture resolution; Otherwise, compute the same attributes but using the micro-texture resolution.

Note that in any case, the two windows associated to an edge element have the same size. Once the attributes are computed, a classical gaussian hypothesis test is done on the law of these two attributes. The statistics Δ corresponding to this test are shown for each experiment (figure 5 et 8) under the name of "distance". After thresholding, we get the edge elements: they are denoted by $(b < i, j >)$, where i and j are two nearest horizontal or vertical neighbors, the clique $< i, j >$ characterizing the corresponding horizontal or vertical edge clique.

The edge elements are fixed and will not be modified in the remaining part of the algorithm.

5 Segmentation Model.

For the segmentation algorithm, the main idea is, at each site, to select the most convenient label according to a given attribute vector. Consequently, we need to define, around each pixel site, a window in which to compute these attributes. Generally this kind of method induces one main difficulty: at the boundary of textures, the windows contain mixed samples of at least two textures, so that the attributes are necessarily non significant.

At this point, the edge information can be very useful, allowing us to truncate the windows containing a boundary. On the other end, it is also possible to use this edge information in order to stop the regularization across an edge (this last part is classically used in restoration, for example).

In order to describe the segmentation algorithm, we will need to define some notations.

Notations.

- $X = (X_i)$ = observed grey levels, i is the pixel site. (Note that a pixel site is also a label site.)

- K is the total number of texture samples (or labels), and the subscript k will denote the k-th texture sample (or label).
- F_i = window of pixel sites, centered around i. The size of F_i will depend on the presence of macro-textures.
- $\Lambda = (\lambda_i)$ = label process which produces the segmentation.
- $\mathcal{V}(i)$ = label neighborhood of i (eight next nearest neighbors).
- $\mathbf{1}$ is the indicator function: $\mathbf{1}_{\lambda_i=\lambda_j} = 1$ if $\lambda_i = \lambda_j$, and 0 otherwise.
- a_i = attribute vector computed on F_i.
- $a_i^{(m)}$ is the m-th attribut. This attribute is computed on the window F_i which may be truncated according to the edge elements: The computation of $a_i^{(m)}$ varies according to two criteria. The first criterion is the size of the window F_i:
  - If there is no macro-texture pattern in the larger window centered around i, then F_i has the size of the resolution of the micro-textures. (the smaller size).
  - Otherwise, F_i has the size of the resolution of the macro-textures.

  The second criterion is the presence of edge elements:
  - If there are enough edge elements in F_i (no matter what the size of this window is), then we select the best line fitting these edge elements according to the mean square error, and we truncate the window at this line. (The attributes are then computed on a smaller sample.)
  - Otherwise, the attributes are computed on the whole window F_i.
- $d(k,a)$ is the "distance" between an attribut vector a and the attribut vector for texture label k: let $\mu_k^{(m)}$ and $\sigma_k^{(m)}$ be the mean and the standard error of the m-th attribute for texture label k, then $d(k,a) = \sup_m \left(\frac{a^{(m)}-\mu_k^{(m)}}{\sigma_k^{(m)}}\right)^2$.

Energy function. We now define the segmentation energy function:

$$U_L(\Lambda) = \sum_{\text{site } i} \sum_{k=1}^{K} \mathbf{1}_{\lambda_i = k} d(k, a_i) - \beta V_L(\Lambda)$$

Where $V_L(\Lambda) = \sum_i \sum_{j \in \mathcal{V}(i)} \mathbf{1}_{\lambda_i=\lambda_j} \mathbf{1}_{b_{<i,j>}=0}$ is the regularization part of the energy function. β is a parameter and will be discussed later. This energy is clearly divided in two parts: the first part represents the basic information extracted from the data, and the second part is only a regularization term. Both terms are minimized in case of perfect data for the perfect segmentation, which explains that we want to minimize this energy given the original data. This is also called the MAP method.

Computation of the parameter β. β characterizes the importance of the regularization terms compared to the informations given by the data. It is then a very important parameter: If β is too large, the label configuration may move too far away of the data, but if it is too small, some defects will remain in the segmentation.

Following R. Azencott ([1]), we computed β using a list of special cases for which we chose the conditional probabilities: For example, assume that a falsely determined label is isolated in the middle of correct labels and assume given values for the attribute vector, then the conditional probability for this label site to keep this false label should be small (say 5% for example). At the same time, we want to accept the possibility that some edge information may be missing, and for example, when there are no edge information, even if the number of neighbors with label λ is 5 (over 8) and the number of neighbors with label μ is 3, we may want to give the priority to the attribute vector and choose μ if necessary. These two simple cases will give a lower and an upper bound for β, and a list of more sophisticated cases will refine these bounds. Any value of β within these bounds will correspond to our characterization and we just select one of them.

Computation time. Since the goal is theoretically to compute a MAP estimator, it should be necessary to use a simulated annealing algorithm in order to optimize the energy function. But the selected energy function is constructed in order to simplify the optimization process: The first term of U_L is directly optimizable because it is sufficient to minimize each term of the sum, these terms being independent one from another. The result of this first minimization will give a very good starting point, and a simple ICM using both terms of U_L will then complete the optimization.

Of course, this algorithm will not converge theoretically to a global minimum of the energy function, but in practice, simulated annealing also converges only to a local minimum. On the other hand, the computation time is much lower in the case of ICM: a few iterations are sufficient (10 at most in our examples). Another important point is that it is not necessary to chose a starting temperature which is always a delicate choice.

The important part of the computation time is the computation of the attributes while the ICM itself only takes a few seconds.

6 Experimental results.

The first experiment is done on an image of vineyards. The main difficulty of this image was not so much to distinguish the textures (in fact the simplest attributes like the mean and the logarithm of the variance are sufficient for that), but the precise selection of the boundaries and even the segmentation of the small roads are not simple at all.

We used only one window size for the example: 7 pixels. The computation time is of the order of 3 minutes for an image 256 by 256 on a Sun4/370.

The second experiment is done on an image of orchards. This image is clearly harder than the precedent one mostly because of the resolution of the macro-textures, and because these textures are pattern-based (repetition of patterns). We constructed patterns for each of the macro-textures, the results are shown in the figure 6 and 7, each of these patterns is contained in a window 9 by 9 pixels. Note that one of the label is used as a trash label, accepting the sites which are difficult to classify.

We used two window sizes for this example: 5 and 11 pixels (on the image reduced to 128 by 128 pixels). The computation time is of the order of 4 minutes for an image 256 by 256 on a Sun4/370.

7 Conclusion.

We use textured plane images to validate a segmentation algorithm using primary information on edge detection between textures. The framework is a Markov random field,

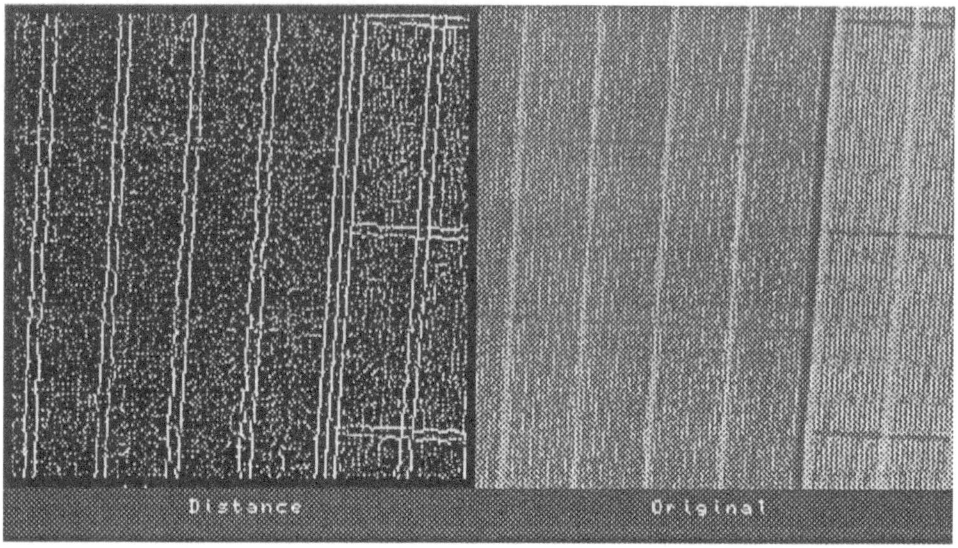

Figure 4: First experiment: Original image, the image called distance represents Δ, the value of the statistic realizing the edge element test.

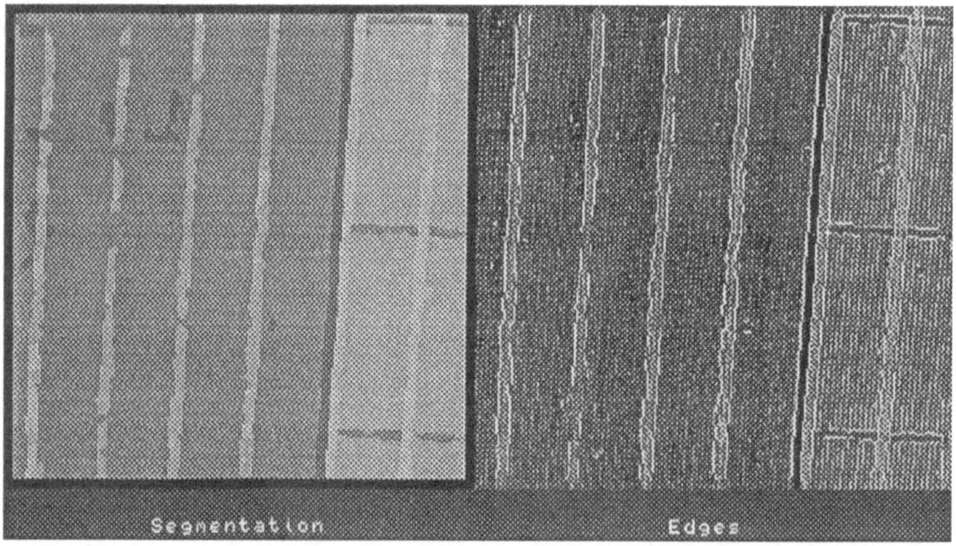

Figure 5: First experiment: Segmentation (4 classes) and superposition of the final edge elements on the original image.

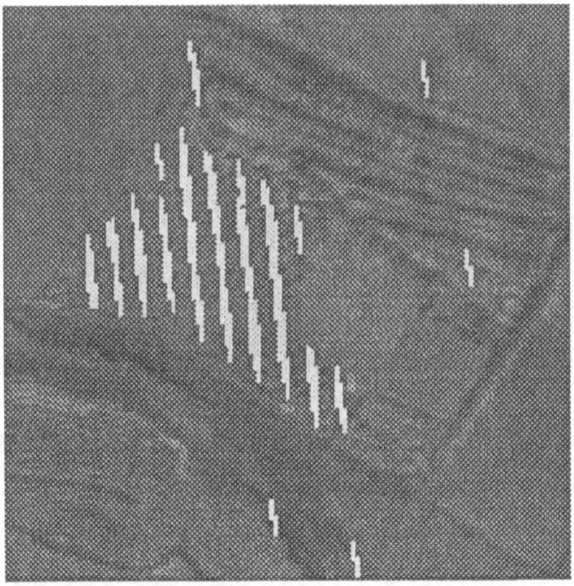

Figure 6: Second experiment: Macro-texture patterns superposed on the original image which is shown in the next figure. Some patterns are isolated and they show the shape of these patterns.

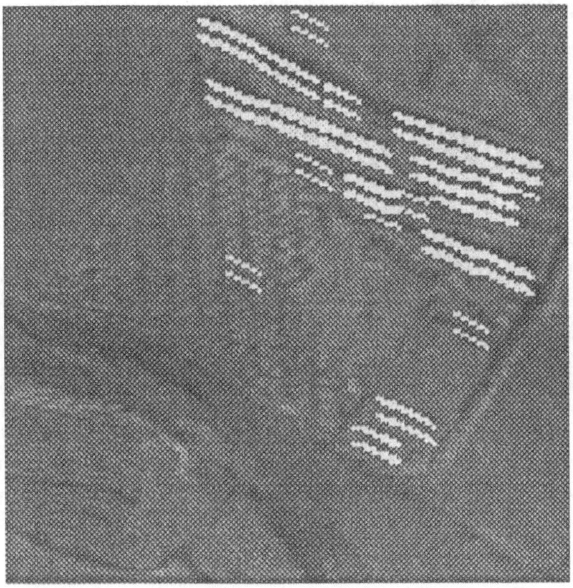

Figure 7: Second experiment: Macro-texture patterns superposed on the original image which is shown in the next figure. Some patterns are isolated and they show the shape of these patterns.

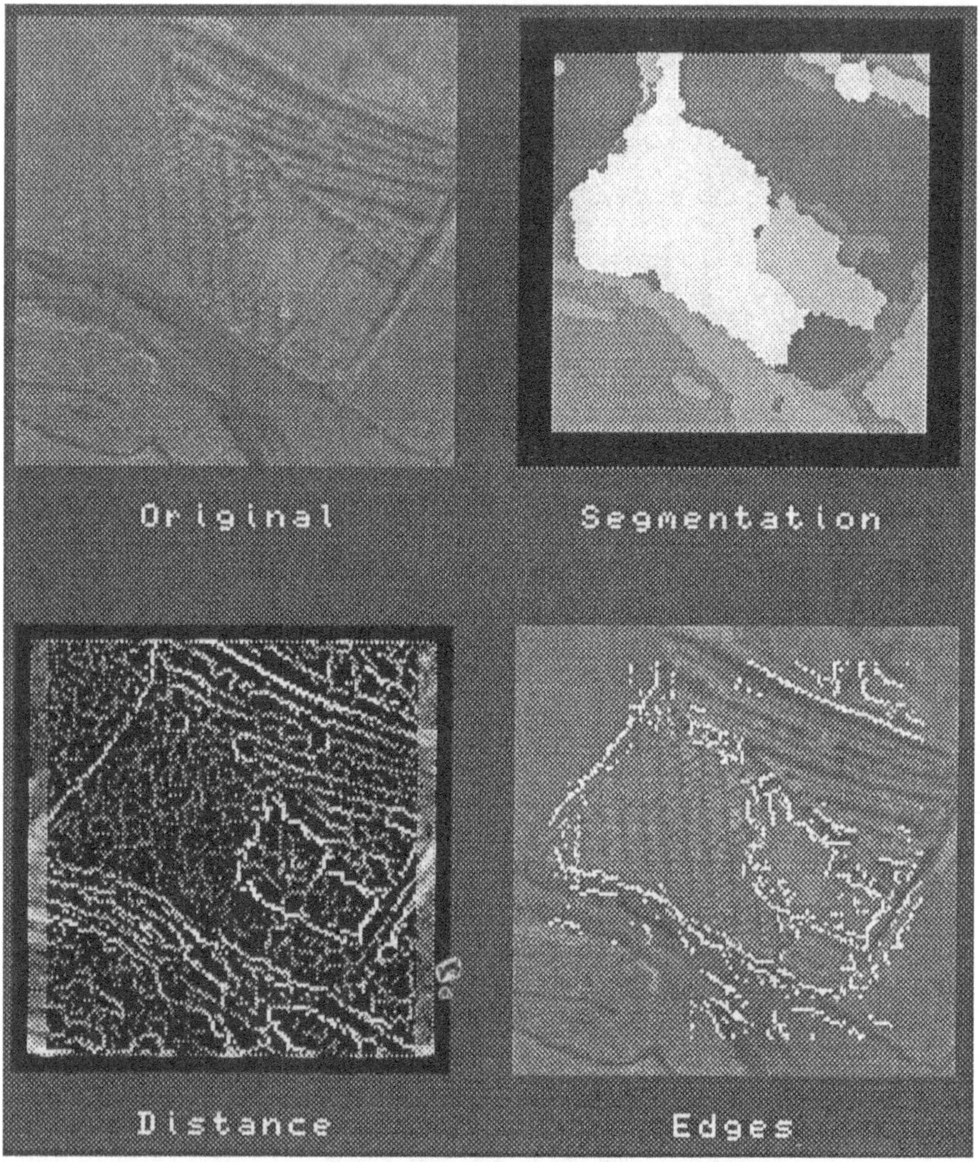

Figure 8: Second experiment: Original image, segmentation (7 classes), the image called distance represents Δ, the value of the statistic realizing the edge element test. The last image is the superposition of the final edge elements on the original image.

but the cpu time is kept reasonable, the optimization being done with only a few iterations of ICM. The results seem quite good considering the difficulty of the second image.

References

[1] R. Azencott, "Markov random fields and image analysis," Proceedings of AFCET, Antices, France, 1987.

[2] J. Besag, "On the statistical analysis of dirty pictures," J. Roy. Statist. Soc., series B, vol. 48, pp. 259–302, 1986.

[3] F. S. Cohen and D. B. Cooper, "Simple parallel hierarchical and relaxation algorithms for segmenting noncausal Markovien random fields," IEEE Trans. Pattern Anal. Machine Intell., vol. PAMI–9, pp. 195–219, 1987.

[4] D. B. Cooper and F. P. Sung, "Multiple-window parallel adaptative boundary finding in computer vision," IEEE Trans. Pattern Anal. Machine Intell., vol. PAMI–5, pp. 299–316, 1983.

[5] R. W. Conners, M. M. Trivedi and C. A. Harlow, "Segmentation of a high-resolution urban scene using texture operators," Computer Vision, Graphics and Image Proc. 25, pp. 273–310, 1984.

[6] H. Derin and H. Elliott, "Modeling and segmentation of noisy and textured images using Gibbs random fields," IEEE Trans. Pattern Anal. Machine Intell., vol. PAMI–9, pp. 39–55, 1987.

[7] A. Gagalowicz and C. Graffigne, "Blind texture segmentation," Proceedings of the 9th International Conference on Pattern Recognition, Rome, Italy, 1988.

[8] S. Geman and D. Geman, "Stochastic Relaxation, Gibbs Distributions, and the Bayesian Restoration of Images," IEEE Trans. on Pattern Anal. Machine Intell., vol. PAMI-6, no 6 November 1984.

[9] S. Geman, D. Geman and C. Graffigne, "Locating texture and object boundaries," Pattern Recognition Theory and Applications, P. A. Devijver and J. Kittler, Eds. Heidelberg, Springer-Verlag, 1987.

[10] S. Geman, D. Geman, C. Graffigne and P. Dong, "Boundary detection by constrained optimization," IEEE Trans. Pattern Anal. Machine Intell., vol. PAMI–12, pp. 609–628, 1990.

[11] S. Geman and C. Graffigne, "Markov random field image models and their applications to computer vision," in Proc. Int. Congr. Mathematicians, 1986, A. M. Gleason, Ed. Amer. Math. Soc., Providence, RI, 1987.

[12] C. Graffigne, "Application des statistiques au traitement d'images," These de troisième cycle, Université Paris-Sud, France, 1986.

[13] C. Graffigne, "Experiments in texture analysis and segmentation," Ph.D. dissertation, Division of Applied Mathematics, Brown Univ., 1987.

[14] U. Grenander, "Tutorial in pattern theory," Division Applied Mathematics, Brown Univ., Lecture Notes, 1984.

[15] R. M. Haralick, K. Shanmugam and I. Denstein, "Textural features for image classification," IEEE Trans. Syst., Man, Cybern., vol. 6, pp. 610–621, 1973.

[16] R. L. Kashyap and K. Eom, "Texture boundary detection based on the long correlation model," IEEE Trans. Pattern Anal. Machine Intell., vol. PAMI–2, pp. 58–67, 1989.

[17] N. Metropolis, A. W. Rosenbluth, M. N. Rosenbluth, A. H. Teller and E. Teller, "Equations of state calculations by fast computing machines," J. Chem . Phys., vol. 21, pp. 1087–1091, 1953.

IFS Algorithms for Wavelet Transforms, Curves and Surfaces, and Image Compression

Marc A. Berger
School of Mathematics
Georgia Institute of Technology
Atlanta, GA 30332 USA

This report concerns research topics discussed while the author was at the Institute Mauro Picone (June 25–28, 1990), under the sponsorship of IAC–CNR. These topics involve applications of affine *iterated function systems* (IFS) [1]. An affine IFS consists of affine transformations $T_i : \mathbb{R}^m \to \mathbb{R}^m$, $i = 1, \ldots, N$; and it generates a discrete–time dynamical system $(\mathbf{X}_n)$ in $\mathbb{R}^m$ according to

$$\mathbf{X}_n = T_{\omega_n}\mathbf{X}_{n-1}$$

where (ω_n) is an appropriately chosen sequence of indices $\omega \in \{1, \ldots, N\}$. This sequence (ω_n) is said to ***drive*** the dynamics. In most IFS applications it is an i.i.d. sequence

$$\mathbb{P}(\omega = i) = p_i > 0, \quad i = 1, \ldots, N$$

where the p_i's are pre–assigned weights. When the transformations T_i are strictly contractive, the IFS process $(\mathbf{X}_n)$ is a recurrent Markov chain, and its orbit is dense in the *attractor* $\mathcal{A}$ for the IFS, with probability one. The attractor is the unique non–empty compact set satisfying

$$\mathcal{A} = \bigcup_{i=1}^{N} T_i\mathcal{A}$$

Sec. 1. Parallel IFS Algorithm for Computation of the Wavelet Transform

Wavelets are functions $\psi : \mathbb{R} \to \mathbb{R}$ with the property that $\{2^{j/2}\psi(2^j x - k) : j, k \in \mathbb{Z}\}$ form an orthonormal basis of $L^2(\mathbb{R})$. Compactly supported wavelets are particularly useful in PDE's and image processing [12], [15]. Daubechies [6] has shown that they can be constructed out of *scaling functions* $f : \mathbb{R} \to \mathbb{R}$, $\operatorname{supp}(f) \subseteq [0, N]$, $\int f = 1$, satisfying *two–scale dilation equations*

$$f(x) = \sum_{k=0}^{N} c_k f(2x - k) \tag{1}$$

Specifically, the coefficients c_k are chosen so as to satisfy (where we take $c_k = 0$ for $k < 0$ or $k > N$)

$$\sum_k c_k c_{k+2\ell} = \begin{cases} 2, & \ell = 0 \\ 0, & \ell \neq 0 \end{cases}$$

$$\sum_k (-1)^k k^\ell c_k = 0, \quad \ell = 0, 1, \ldots, M$$

Then the *quadratic mirror filter* (QMF)

$$\psi(x) = \sum_k (-1)^k c_{1-k} f(2x - k) \tag{2}$$

is a compactly supported wavelet. Moreover by choosing M (and hence N, too) large enough, ψ will have any desired degree of smoothness.

The ***wavelet transform*** of a function $\phi \in L^2(\mathbb{R})$ is the set of coefficients

$$W_{j,k}(\phi) = 2^{j/2} \int \phi(x)\psi(2^j x - k)dx$$

We have developed an IFS algorithm for generating the solution f of (1), $\int f = 1$, which involves only two affine transformations T_0, T_1. Precisely, define $T_0, T_1 : \mathbb{R}^{N-1} \to \mathbb{R}^{N-1}$ by

$$(T_\omega \mathbf{x})_i = \sum_{j=1}^{N-1} c_{2i-j+\omega-1} x_j + c_{2i-N+\omega-1} \left(1 - \sum_{j=1}^{N-1} x_j\right), \tag{3}$$

$\omega = 0, 1$. Let $X_n \in \mathbb{R}$, $\mathbf{Y}_n \in \mathbb{R}^{N-1}$ be the IFS process

$$\begin{cases} X_n = \dfrac{X_{n-1} + \omega_n}{2} \\ \mathbf{Y}_n = T_{\omega_n} \mathbf{Y}_{n-1} \end{cases} \tag{4}$$

where (ω_n) is the bit sequence driving the IFS. Choose $X_0 = 0$ and let $\mathbf{Y}_0$ be the fixed point of T_0. Then for any n

$$\left[\mathbf{Y}_n^t \,\middle|\, 1 - \sum_{i=1}^{N-1} (\mathbf{Y}_n)_i\right] = [f(X_n), f(X_n + 1), \ldots, f(X_n + N - 1)]$$

The IFS (4), along with the QMF relation (2), leads to a fast *parallel* algorithm for generating compactly supported wavelets ψ. The parallel nature of this algorithm can be seen as follows. Suppose one has two processors. As shown in Sec. 3 below, the optimal sequence $\omega_1, \ldots, \omega_L$ for driving the IFS has $L = 2^\ell + \ell - 1$ bits, where ℓ is the desired level of resolution; and it has the special property that every possible ℓ–bit pattern appears somewhere in it. Split this sequence into two subsequences

$$\omega_1, \ldots, \omega_{2^{\ell-1}+\ell-1} \qquad \omega_{2^{\ell-1}+1}, \ldots, \omega_{2^\ell+\ell-1}$$

Let one processor run the IFS driven by the subsequence on the left, and in parallel (asynchronous) let the other processor run the IFS with the right subsequence. Observe that any ℓ–bit pattern appears either in the left or right subsequence, so together the two processors generate all values $f(x+k)$, $x = \dfrac{m}{2^\ell}$ $(m = 0, \ldots, 2^\ell - 1)$. (Note the overlap of $\omega_{2^{\ell-1}+1}, \ldots, \omega_{2^{\ell-1}+\ell-1}$.) In general if there are K processors, K a power of two, then the total work load is $2^\ell + K(\ell - 1)$, the second term being due to the window spill–overs.

The IFS algorithm for computing ψ is shown in Figures 1 and 2 for Daubechies' wavelets D_4 and D_6 [6]. The loop for $n = 1, L$ can be distributed (asynchronously) among many processors. In these Figures the function ψ is simply plotted, but in wavelet transform applications one can put the values

$$\psi\left(\frac{X_n - N + 1}{2}\right), \psi\left(\frac{X_n - N + 2}{2}\right), \ldots, \psi\left(\frac{X_n + N}{2}\right)$$

generated at stage n into quadrature formulas for $\int \phi(x)\psi(2^j x - k)dx$.

Derivation of (4) was carried out in [2], [4]. In order for the IFS algorithm to work we require that T_0, T_1 both be strictly contractive relative to some operator norm. The analysis in [4] is based on results about the *two-scale operator* $T : L_1(\mathbb{R}) \to L_1(\mathbb{R})$ given by

$$Tf(x) = \sum_{k=0}^{N} c_k f(2x - k)$$

What follows is a quick synopsis of parts of this paper.

Let $\mathbf{V} : L_1(\mathbb{R}) \to L_1([0,1); \mathbb{R}^N)$ be the operator

$$\mathbf{V}f(x) = [f(x), f(x+1), \ldots, f(x+N-1)]$$

If f satisfies (1) with $\text{supp}(f) \subset [0, N]$ then

$$\mathbf{V}f(x) = P_{\omega_1}^t \mathbf{V}f(\tau x) \tag{5}$$

where ω_1 is the first bit in the dyadic expansion of x, $\tau x = 2x (\text{mod } 1)$ and P_0, P_1 are the $N \times N$ row stochastic matrices

$$P_0 = (c_{2j-i-1})_{i,j=1}^{N} \quad P_1 = (c_{2j-i})_{i,j=1}^{N}$$

Iterating (5) we come up with a solution to (1) via

$$\mathbf{V}f(x) = (1, 0, \ldots, 0)P_*(x)$$

where $P_*(x) = \lim_{n\to\infty} P_{\omega_n} \cdots P_{\omega_1}$. Here $x = \sum_{n=1}^{\infty} \frac{\omega_n}{2^n}$ has the dyadic code $.\omega_1\omega_2\cdots$.

Let $S : L_1(\mathbb{R}) \to L_1([0,1))$ be the *summation operator*

$$Sf(x) = \sum_{k=-\infty}^{\infty} f(x+k)$$

If $T^n f_0 \to f$ in $L_1(\mathbb{R})$ then $Sf_0 \equiv$ const a.e. If T_0, T_1 defined above in (3) are strictly contractive (relative to some operator norm) and if f_0 is bounded, measurable, $Sf_0 \equiv$ const then $T^n f_0$ converges uniformly to a continuous function f. Conversely if $T^n f_0$ converges uniformly to a continuous function f, where $f_0 = \chi_{[0,1)}$, then T_0, T_1 are strict contractions (relative to some operator norm). A lot of this theory hinges on $\{P_0, P_1\}$ satisfying the *left convergent products* (LCP) property; i.e., $\lim_{n\to\infty} P_{\omega_n} \cdots P_{\omega_1}$ exists for any sequence $\omega_1, \omega_2, \ldots$ of zeros and ones. This property is studied in [3], [9].

```
Initialize r = x = 0; y = 1.366

For n = 1, L
    b1 = -0.183x; b2 = -0.5x - 0.183y; b3 = 0.866x - 0.317y - 0.183
    b4 = 0.5x + 1.183y - 0.5; b5 = 0.683(1 - x - y)
    plot (r - 1, b1), (r - 0.5, b2 - b1), (r, b3 - b2), (r + 0.5, b4 - b3),
    (r + 1, b5 - b4), (r + 1.5, -b5)
    choose ω = ω_n ∈ {0, 1} from the bit string
    if ω = 0
    then r ← 0.5r
         y ← 0.366x + 0.5y + 0.683
         x ← 0.683x
    else r ← 0.5r + 0.25
         z ← x
         x ← 1.183x + 0.683y
         y ← -1.366z - 0.866y + 1.183
endfor
```

Figure 1: Dynamical algorithm to generate Daubechies' wavelet D_4

The coefficients c_k are given by

$$c_0 = \frac{1+\sqrt{3}}{4}, \quad c_1 = \frac{3+\sqrt{3}}{4}, \quad c_2 = \frac{3-\sqrt{3}}{4}, \quad c_3 = \frac{1-\sqrt{3}}{4}$$

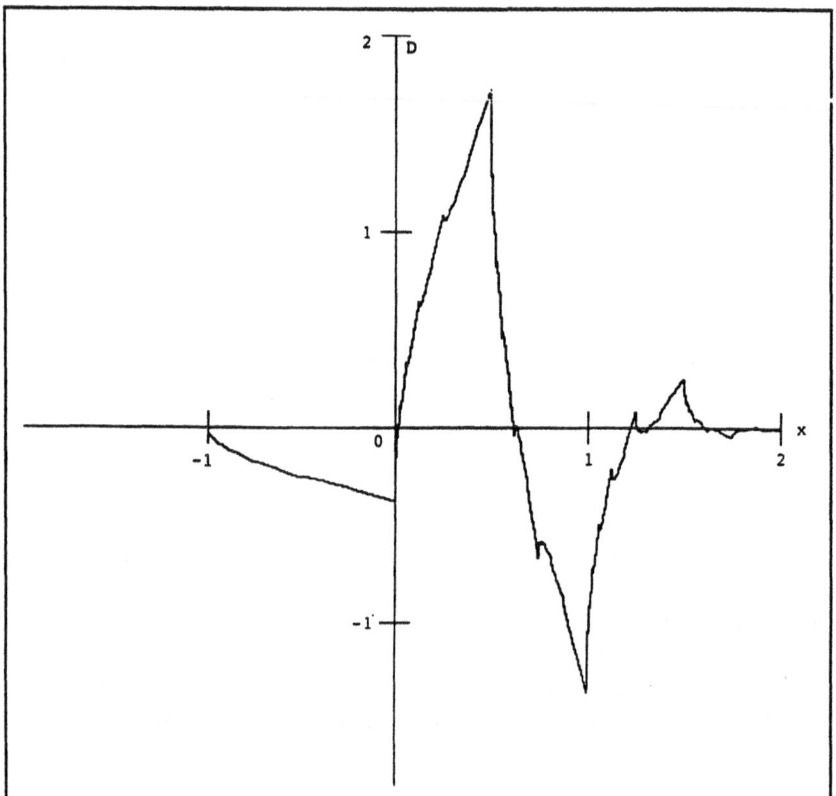

```
Initialize r = z_0 = z_6 = x_1 = 0; x_2 = 1.286; x_3 = -0.386; x_4 = 0.095;
c_0 = 0.470; c_1 = 1.141; c_2 = 0.650; c_3 = -0.191; c_4 = -0.121; c_5 = 0.050

For n = 1, L
    x_5 = 1 - x_1 - x_2 - x_3 - x_4;   c_0 <- -c_0; c_2 <- -c_2; c_4 <- -c_4
    For l = -4, 5
        m_1 = max(0, 1 - l); m_2 = min(5, 5 - l); y = 0
        For i = m_1, m_2
            y <- y + c_i x_{i+l}
        endfor
        plot (r + l/2, y)
    endfor
    choose ω = ω_n ∈ {0,1} from the bit string
    r <- 0.5r + 0.25ω; c_0 <- -c_0; c_2 <- -c_2; c_4 <- -c_4
    For l = 1, 5
        z_l <- x_l; x_l <- 0
    endfor
    For l = 1, 4
        m_1 = max(0, 2l - 6); m_2 = min(5, 2l - 1); j = 2l + ω - 1
        For i = m_1, m_2
            x_l <- x_l + c_i z_{j-i}
        endfor
    endfor
endfor
```

Figure 2: Dynamical algorithm to generate Daubechies' wavelet D_6

The coefficients c_k are given by

$$c_0 = 0.470,\ c_1 = 1.141,\ c_2 = 0.650,\ c_3 = -0.191,$$

$$c_4 = -0.121,\ c_5 = 0.050$$

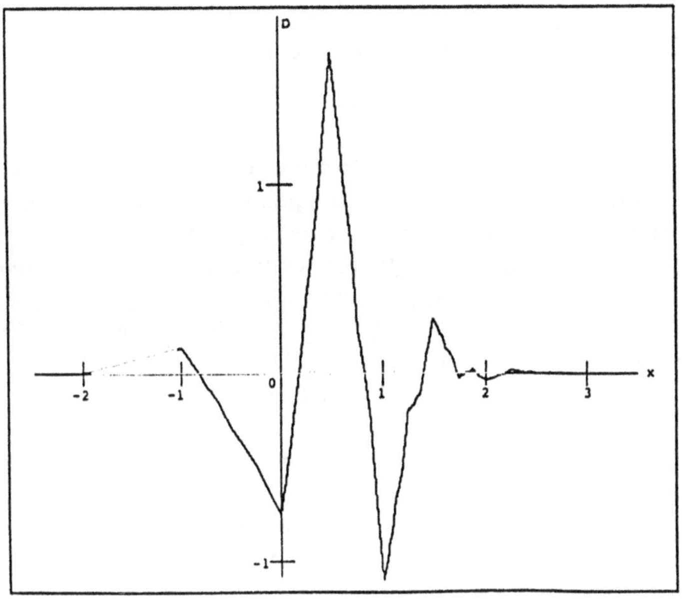

Sec. 2. Curve and Surface Generation with Application to Signal and Image Compression

Sub-division methods for curve and surface generation are governed by a set of control points $\mathbf{v}_1, \ldots, \mathbf{v}_N \in \mathbb{R}^m$ and two $N \times N$ row stochastic matrices P_0, P_1 [2], [5], [14]. One can identify the control polytope C whose vertices are the $\mathbf{v}_i$'s with the $N \times m$ matrix

$$C = \begin{pmatrix} \mathbf{v}_1^t \\ \cdots \\ \mathbf{v}_N^t \end{pmatrix}$$

whose rows are the $\mathbf{v}_i$'s; and one can then identify an action of P_ω on C by $C \leftarrow P_\omega C$ ($\omega = 0, 1$). The sub-division method proceeds in a binary tree fashion by replacing polytopes C at a given depth with the two smaller polytopes $P_0 C$ and $P_1 C$. At depth d we generate all the 2^d polytopes $P_{\omega_n} \cdots P_{\omega_1} C$. Typically in these schemes if d is large enough then these 2^d miniscule polytopes all link-up together like droplets to form the desired curve or surface

$$\mathcal{A} = \cup \left(\lim_{n \to \infty} P_{\omega_n} \cdots P_{\omega_1} C : \omega_1, \omega_2, \ldots \in \{0, 1\} \right)$$

By reducing

$$P_\omega \sim \begin{pmatrix} 1 & \mathbf{b}_\omega^t \\ 0 & A_\omega \end{pmatrix} \tag{6}$$

using a similarity relation, we see that products of the P_ω's evolve as products of the affine transformations $T_\omega : \mathbf{x} \mapsto A_\omega^t \mathbf{x} + \mathbf{b}_\omega$. Indeed

$$P_{\omega_n} \cdots P_{\omega_1} \sim \begin{pmatrix} 1 & \mathbf{b}^t(\omega_1, \ldots, \omega_n) \\ 0 & A_{\omega_n} \cdots A_{\omega_1} \end{pmatrix}$$

where

$$\mathbf{b}(\omega_1, \ldots, \omega_n) = T_{\omega_1} \cdots T_{\omega_n} \mathbf{0}$$

This leads to an IFS algorithm for generating the curve or surface $\mathcal{A}$, as follows. Define processes

$$\mathbf{Y}_n = T_{\omega_n} \mathbf{Y}_{n-1}, \qquad \mathbf{Z}_n^t = \left[1 \mid \mathbf{Y}_n^t \right] B^{-1},$$

where B is the change-of-basis matrix used to effect the similarity transformation (6). Choose $\mathbf{Y}_0$ to be the fixed point of T_0. Then the points $\sum_{i=1}^N (\mathbf{Z}_n)_i \mathbf{v}_i$ are dense in $\mathcal{A}$.

The sub-division methods described above are typically used to generate smooth curves and surfaces. For generating rough curves there is a *fractal interpolation* technique available [1]. Given data points $(x_i, \mathbf{y}_i) \in [0,1] \times \mathbb{R}^{m-1}$, $0 \le i \le N$, this technique constructs an affine IFS $T_i : \mathbb{R}^m \to \mathbb{R}^m$, $1 \le i \le N$, whose attractor collapses to the graph of a continuous function $f : [0,1] \to \mathbb{R}^{m-1}$ passing through them. There are various free parameters in the T_i's (vertical scale factors), which can be used to control the fractal dimension of the attractor [1].

Fractal interpolation can be used as a means of signal and image compression. Suppose one is dealing with a time series (x_i, y_i), $0 \le i \le N$. Decimate this series to obtain a shorter series (x_j', y_j'), $0 \le j \le N'$. For instance one could sample (x_i, y_i) at every fourth or eighth time series point. Construct a fractal interpolant for the decimated series, choosing the vertical scale factors v_j to least squares fit the original full series (x_i, y_i) [13]. The data points (x'_j, y'_j) and the v_j's comprise a compressed file for (x_i, y_i). The decoding algorithm involves running the IFS process, using the appropriate affine transformations constructed from the v_j's, and it is very fast. It produces the fractal interpolation curve described above as an approximate reconstruction of the (x_i, y_i) time series.

These ideas generalize to image compression and surface interpolation, as follows. Let (x_i, y_i, z_i), $0 \le i \le N$, be image or surface data laid down over the 2–D triangular grid $\Delta = \{(x,y) : x, y \ge 0; x + y \le 1\}$. That is, (x_i, y_i) are discrete data points in Δ, and z_i is the color intensity or surface height at (x_i, y_i). In Fig. 3 the (x_i, y_i) are the 133 grid points arranged in the triangular grid. The array is decimated by sampling every fourth point in the $x-$ and $y-$directions. The decimated array has 15 points, and these points form 16 triangles, shown by the dark edges in Fig. 3.

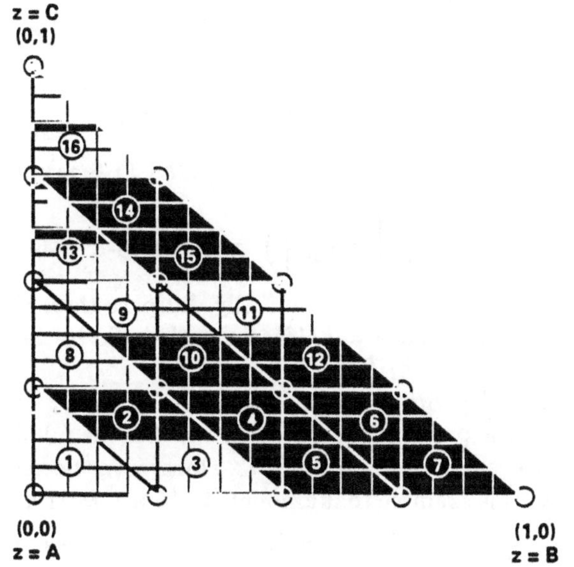

Figure 3. Triangular elements and nodes for image/surface compression. The circled nodes are used to construct the fractal surface interpolant.

Consider one such triangle Δ_j, say

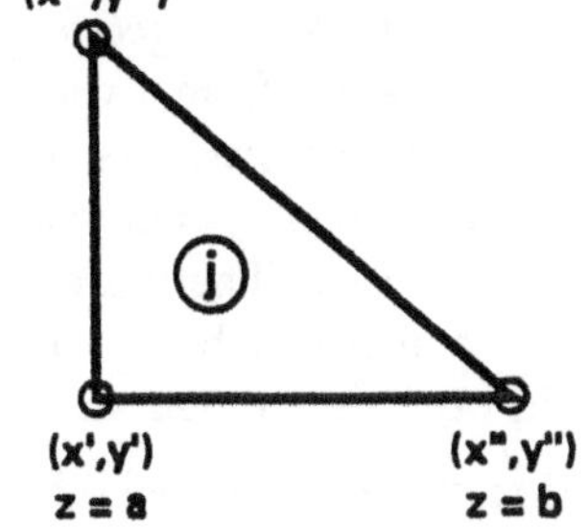

From it construct $T = T_j : \mathbb{R}^3 \to \mathbb{R}^3$ of the form

$$T : \begin{bmatrix} x \\ y \\ z \end{bmatrix} \mapsto \begin{bmatrix} * & * & 0 \\ * & * & 0 \\ * & * & v \end{bmatrix} \begin{bmatrix} x \\ y \\ z \end{bmatrix} + \begin{bmatrix} * \\ * \\ * \end{bmatrix}$$

with the property that

$$\begin{bmatrix} 0 \\ 0 \\ A \end{bmatrix} \mapsto \begin{bmatrix} x' \\ y' \\ a \end{bmatrix}, \quad \begin{bmatrix} 1 \\ 0 \\ B \end{bmatrix} \mapsto \begin{bmatrix} x'' \\ y'' \\ b \end{bmatrix}, \quad \begin{bmatrix} 0 \\ 1 \\ C \end{bmatrix} \mapsto \begin{bmatrix} x''' \\ y''' \\ c \end{bmatrix}$$

Given $v = v_j$ this then determines T as

$$T : \begin{bmatrix} x \\ y \\ z \end{bmatrix} \mapsto \begin{bmatrix} x'' - x' & x''' - x' & 0 \\ y'' - y' & y''' - y' & 0 \\ b - a - v(B - A) & c - a - v(C - A) & v \end{bmatrix} \begin{bmatrix} x \\ y \\ z \end{bmatrix} + \begin{bmatrix} x' \\ y' \\ a - vA \end{bmatrix} \tag{7}$$

In order to determine v, use least squares to best match the image of Δ under T to Δ_j. Precisely, for each data point (x_i, y_i, z_i) define the heights

$$H_i = z_i - [(1 - x_i)(1 - y_i)A + x_iB + (1 - x_i)y_iC]$$

$$h_i = z\left(\begin{bmatrix} x'' - x' & x''' - x' \\ y'' - y' & y''' - y' \end{bmatrix} \begin{bmatrix} x_i \\ y_i \end{bmatrix} + \begin{bmatrix} x' \\ y' \end{bmatrix}\right) - [(1 - x_i)(1 - y_i)a + x_ib + (1 - x_i)y_ic]$$

If the fractal surface generated by the T_j's really did pass through all the data points (x_i, y_i, z_i), then there would hold $v = h_i/H_i$ for each i. So in the least squares sense, find v by minimizing $\sum_i (H_iv - h_i)^2$; i.e., $v = \sum_i h_iH_i / \sum_i H_i^2$.

By running the 3–D IFS with the transformations $T_1, \ldots T_{16}$ we obtain a fractal surface which interpolates the points (x'_j, y'_j, z'_j) and best fits the full data series (x_i, y_i, z_i). The compressed file needs to store the values z'_j, along with the vertical scale factors v_j computed as above. In this example that amounts to 15+16 = 31 data points — yielding a compression ratio of $\frac{133}{31} = 4.3$. The decoding algorithm knows to set up the geometry for the triangular elements and nodes, and so this information need not be part of the compressed file. If the dimensions scale up to a high resolution image on a triangular grid of 1024×1024, and if one samples every 32nd point, say, in the x– and y–directions, the compression ratio would accordingly scale up to be $\frac{525,825}{561 + 1024} = 331.8$.

To obtain a less lossy reproduction of the original image, there are several options. One option is to make the geometry more flexible, in that the triangular elements and nodes need not be uniform. Another option is to use ***hidden variable interpolation*** [1]. In this scheme one constructs an appropriate IFS to interpolate in more than three dimensions, and then obtains the desired fractal interpolation surface by projecting back down onto 3–D. There are additional free parameters for each transformation in this scheme, which can be determined by least squares fitting. In 4–D, for instance, there would work out to be four free parameters per transformation, and then less lossy reproduction would be traded off against the smaller compression ratio of $\frac{525,825}{561 + 4096} =$

112.9. Alternately one could stay with the 3–D IFS, but increase the sampling from every 32nd point up to every 16th point — dropping the compression down to $\dfrac{525,825}{2145+4096} =$ 84.3.

Sec. 3. Shift–Register Bit Sequences and Optimal IFS Generation

In the wavelet and scaling function application from Sec. 1 above, the IFS generates the values $f(X_n + m)$, where $X_n \in [0,1)$ is the dyadic number $X_n = .\omega_n \cdots \omega_1 = \sum_{k=1}^{n} \frac{\omega_k}{2^{n+1-k}}$. Suppose the desired resolution is $2^{-\ell}$, for some level ℓ of resolution, and suppose the IFS process is run for L steps. Then in order to get all values $X_n = \dfrac{k}{2^\ell}$ $(k = 0, 1, \ldots, 2^\ell - 1)$, the driving sequence $\omega_1, \ldots, \omega_L$ must have the property that every possible successive ℓ–tuple appears somewhere in the sequence. In other words,

(P) *if a window of length ℓ slides across the sequence, then every possible ℓ–bit pattern should appear in it.*

Randomly generated bits will indeed determine sequences with this property (P) if L is large enough. However, this is not an efficient way to go. One can instead *engineer* the driving bit sequence $\omega_1, \ldots, \omega_L$ to be optimal with respect to producing every ℓ–tuple with the shortest possible sequence length L. Indeed, since there are 2^ℓ different windows of size ℓ, it is clear that L must be at least $2^\ell + \ell - 1$. (The spill–over of $\ell - 1$ is to accommodate the last $\ell - 1$ windows.) Furthermore sequences of length $L = 2^\ell + \ell - 1$ can be constructed so as to have this property (P). They are known as *shift-register sequences* [10]. For example with $\ell = 4, L = 19$ we can use

0 0 0 1 1 0 1 1 1 1 0 0 1 0 1 0 0 0 0

As a window of length 4 slides across the sequence, one encounters the various 4–tuples

0001, 0011, 0110, 1101, . . . , 0000

and indeed all 16 possibilities occur here.

Constructing such shift–register sequences can be achieved by finding an Eulerian circuit through a directed graph. The $2^{\ell-1}$ vertices of the graph are all of the $(\ell-1)$–bit sequences $\omega_1 \cdots \omega_{\ell-1}$, and a directed edge goes from $\omega_1 \cdots \omega_{\ell-1}$ to $\eta_1 \cdots \eta_{\ell-1}$ if and only if

$$\omega_2 \cdots \omega_{\ell-1} = \eta_1 \cdots \eta_{\ell-2}$$

Each vertex thus has in–degree two and out–degree two, making the graph Eulerian. For the example above, the directed graph is displayed in Fig. 4, together with the Eulerian circuit which generates the bit string.

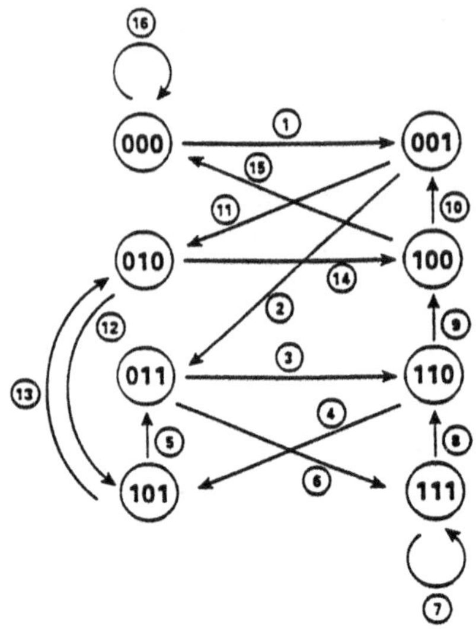

Figure 4. The directed graph for generating shift–register sequences with $\ell = 4$. Every vertex has in–degree and out–degree two, and so this graph is Eulerian. Shown is a particular Eulerian circuit, used to generate the sequence given in the text.

In summary, then, the most efficient way to solve (1) using the IFS approach is to first load in a shift–register bit sequence $\omega_1, \ldots, \omega_L$ of $L = 2^\ell + \ell - 1$ zeros and ones satisfying (P), to be used for driving the dynamics (4), where ℓ is the desired level of resolution. Moreover as explained above in Sec. 1, this can be run in parallel. The bit string $\omega_1, \ldots \omega_L$ is cumbersome to generate when ℓ is large (8–12 in typical applications), but it is only necessary to do this once (and for all), and then it can be packed into a permanent array — to be used for all ensuing applications. This dramatically out–performs the random number method, both in timing per iteration (random number generators take time) and in total number of iterations L.

The same ideas apply to general IFS T_i, $1 \leq i \leq N$, where N is not large. The bits in the driving sequence $\omega_1, \ldots, \omega_L$ take values in $\{1, \ldots, N\}$. For IFS with probabilities p_i, the point generated at stage n gets assigned weight $\prod_{k=1}^{\ell} p_{\omega_{n+1-k}}$. Cf. the scheme in [11].

For the compression scheme described in Sec. 2 the situation is different. There one has $N = 1024$ transformations, and needs to construct a driving sequence $\eta_1, \ldots, \eta_L$ of numbers $\eta \in \{0, 1, \ldots, 1023\}$ such that every successive 2–tuple appears somewhere in the sequence. (It can be seen in Fig. 3 that $\ell = 2$ suffices to fill in every pixel.) In this case the directed graph of 1024 vertices is complete, and so it has an Eulerian circuit— which leads to a sequence of $1024^2 + 1$ 10–bit integers η. There is, however, a shorter sequence which can be used to generate the full dynamics. One can use instead a sequence $\omega_1, \ldots, \omega_L$ of $L = 2^{20} + 19$ binary bits, which has the property that every window of length 20 appears somewhere in it. In particular every *pair* of length 10 windows also appears in it. The dynamics can be generated by the *double-pass*

$$\eta_1 = \omega_1 \cdots \omega_{10}, \quad \eta_2 = \omega_{11} \cdots \omega_{20},$$
$$\eta_3 = \omega_2 \cdots \omega_{11}, \quad \eta_4 = \omega_{12} \cdots \omega_{21},$$
$$\eta_5 = \omega_3 \cdots \omega_{12}, \quad \eta_6 = \omega_{13} \cdots \omega_{22}, \text{ etc.}$$

In this way only $L = 1024^2 + 19$ single–bit integers need to be stored — but this involves $2L$ iterations of the IFS, on account of the double–back sweep. This raises the important question of finding the shortest length bit string which can be used to drive the full IFS dynamics in roughly 1024^2 iterations.

4. OPEN PROBLEMS

An outstanding problem is to find continuous fractal interpolation functions for surfaces. Unlike the case for curve interpolants $f : [0,1] \to \mathbb{R}^{m-1}$, surface interpolation functions $f : \Delta \to \mathbb{R}^{m-2}$ constructed via transformations (7) need not be continuous. Match–up at the triangle element boundaries is not automatic. Discontinuous surfaces have to be allowed for in image compression, so as to model changes in texture, boundaries, etc. Nonetheless it is also important to be able to construct continuous and smooth surfaces.

Another outstanding problem is to produce optimal driver bit sequences for generating IFS attractors and invariant measures. For N small this involves finding Eulerian circuits like the one in Fig. 4, but for much larger scale graphs. For N large this involves resolving combinatorial issues relating to the IFS driving strategy, like the one posed at the end of Sec. 3.

Another problem is to continue the analysis of the IFS (4) and solutions of (1), described at the end of Sec. 1 above. How do conditions on the c_k's for ergodicity/boundedness of the IFS relate to conditions for existence of solutions for (1) and convergence of approximants $T^n f_0$? If the IFS is ergodic, do the approximants converge? If the approximants converge, is the IFS ergodic? are $\{P_0, P_1\}$ LCP? If (1) has a nontrivial continuous compactly supported solution $f \neq 0$, is the IFS ergodic? are $\{P_0, P_1\}$ LCP? are they even product–bounded? A partial analysis of these issues appears in [4], [5], but it is far from complete. Also, how smooth is the normalized solution of (1)? What is the Hölder modulus of continuity of its highest derivative? This is partially analyzed in [7], [8], but there is more work to be done.

References

1. Barnsley, M. F., *Fractals Everywhere*, Academic Press, New York, 1988.

2. Berger, M. A., Random affine iterated function systems: smooth curve generation, *SIAM Review*, in press.

3. Berger, M. A. and Wang, Y., Bounded semi–groups of matrices, to appear.

4. Berger, M. A. and Wang, Y., Multi–scale dilation equations and iterated function systems, to appear.

5. Cavaretta, A. S., Dahmen, W. and Micchelli, C. A., Stationary subdivision, *Memoirs Amer. Math. Soc.*, in press.

6. Daubechies, I., Orthonormal bases of compactly supported wavelets, *Comm. Pure Appl. Math.* 41 (1988), 909–996.

7. Daubechies, I. and Lagarias, J. C., Two-scale difference equations I: existence and global regularity of solutions, *SIAM J. Math. Anal.*, in press.

8. Daubechies, I. and Lagarias, J. C., Two-scale difference equations II: local regularity, infinite products of matrices and fractals, *SIAM J. Math. Anal.*, in press.

9. Daubechies, I. and Lagarias, J. C., Sets of matrices all infinite products of which converge, to appear.

10. Golumb, S., *Shift Register Sequences*, Holden–Day, San Francisco, 1967.

11. Hepting, D., Prusinkiewicz, P. and Saupe, D., Rendering methods for iterated function systems, in *FRACTAL 90: Proceedings of the First IFIP Conference on Fractals*, Peitgen, H.-O., Henriques, J. M. and Peneda, L., eds., Elsevier, Amsterdam, 1990.

12. Lawton, W., Wavelet discretization methods for surface estimation and reconstruction, *SPIE Vol. 1251: Curves and Surfaces in Computer Vision and Graphics*, 1990, 242–253.

13. Mazel, D. and Hayes, M., Iterated function systems applied to discrete sequences, *IEEE Trans. ASSP*, in press.

14. Micchelli, C. A. and Prautzsch, H., Uniform refinement of curves, *Lin. Alg. Appl.* 114/115 (1989), 841–870.

15. Zettler, W. R., Huffman, J. and Linden, D. P., Application of compactly supported wavelets to image compression, *SPIE Vol. 1244: Image Processing Algorithms and Techniques*, 1990, 150–160.

Image restoration by stochastic dichotomic reconstruction of contour lines

Olivier Catoni
DIAM - Intelligence Artificielle et Mathématiques
Laboratoire de Mathématiques de l'Ecole Normale Supérieure
U.A 762 du C.N.R.S
45, rue d'Ulm 75230 Paris Cedex 05.

Introduction

The problem addressed in this paper is to recover a grey level image f from some noisy observation g. The drawback of naïve restoration techniques, such as linear filtering, is that the edges get blurred. This loss of spatial localization makes further processing more difficult and even sometimes unsuccessful.

In the framework of stochastic Bayesian estimation many attempts have been made to take into account explicitly the presence of edges (Marroquin [11], Chalmond [3], Geman, Graffigne and Dong [6], Dinten [4], Azencott [1]). This is done by introducing explicit edge sites and edge values and a coupled edges - grey levels process. It allows to introduce refined geometrical constraints on the shape of expected edges and the nature of their intersections.

The idea developed here is coarser in the sens that it introduces less a priori knowledge on the image to be reconstructed. Rather than introducing an explicit edge model, we work with a Bayesian stochastic model of the contour lines of the image. The representation of an image by its contour lines is well known in cartography. Contour lines are simpler linear objects than edges because they are closed and without intersections. They are defined without ambiguity as the boundaries of the level sets and are easy to detect. Smoothing the contour lines will translate into imposing regularity constraints on the edges, though edges do not follow necessarily contour lines. On the other hand it will not necessarily smooth the gradient: regions with straight edges will remain untouched, even if the edges correspond to sharp discontinuities. Hence there is a possibility to erode only "peaks" leaving "cliffs" unaltered.

In order to implement this idea, we have to circumvent the algorithmic complexity due to the large number of the contour lines (at least as much as grey levels). This can be done by introducing a coupled model on the family of all contour lines instead of treating them separately, taking into account the inclusion relation. Contour lines are accessed through level sets. We start with a global model of independent level sets and condition it in different ways to get a global model where the inclusion relation on level sets is satisfied.

The model of one contour line takes into account two criteria. We do not want to disturb too much the spatial localization of the contour lines. This is controlled by measuring the area between the reconstructed contour line and the original one. We want the contour lines to be smooth. This is controlled by the euclidean length of contours per area unit.

This competition between two terms one of which has the dimension of an area and the other is one dimensional is classical in the variational approach to image restoration. In our stochastic framework, it will translate into a noisy Ising model on the level set corresponding to the contour line under study. Up to a constant term the hamiltonian of an Ising model is proportional to the length of the boundary. By choosing appropriately the connectivity of the model and the weights on the different links, we can make this length be a satisfactory approximation of the euclidean length. The noise model reflects the distance between the observation and the reconstructed picture. A distance proportional to the area corresponds to an independent identically distributed noise. Noisy Ising models are discussed into the interesting paper of Marroquin, Mitter and Poggio [10]. Discrete approximation of euclidean length is also a classical topics in computer science.

Some real and synthetic images are shown and discussed at the end of the paper.

I am pleased to acknowledge useful discussions with Robert Azencott during the course of this work. I want also to thank Pierre Vincens from the Laboratoire de Biochimie de l'Ecole Normale Supérieure who helped me to numerize the pictures and incited me to experiment on biological images.

1 Boundary smoothing of binary images

As explained in the introduction, we choose to model the observation $Y : S \longrightarrow \{-1, 1\}$, where $S \subset \mathbb{Z}^2$ is the set of sites of the image, by a noisy Ising model.

The law of the original binary image X knowing the observation Y is supposed to be

$$P(X \mid Y) = \frac{1}{Z} \exp\left(\frac{1}{T}(H(X) + N(X, Y))\right) \tag{1}$$

where

$$Z = \sum_{\tilde{X}} \exp\left(\frac{1}{T}(H(\tilde{X}) + N(\tilde{X}, Y))\right), \tag{2}$$

and where

$$\begin{aligned} H(X) &= \sum_{(i<j)\in S^2} \xi_{i,j} X_i X_j \\ & \quad \text{with } \xi_{i,j} = \begin{cases} 1 & |i-j| = 1; \\ 1/\sqrt{2} & |i-j| = \sqrt{2}; \\ 0 & \text{otherwise} \end{cases} \\ N(X, Y) &= \beta \sum_{i\in S} X_i Y_i. \end{aligned} \tag{3}$$

Up to a constant $-H(X)$ is a discrete approximation of the euclidean length of the boundary of X. The constant $\xi_{i,j} = 1/\sqrt{2}$ when $|i-j| = \sqrt{2}$ is chosen so that the computation of the length is exact for straight boundaries with orientations $0, \pi/4, \pi/2, 3\pi/4$ with respect to the x axis.

The coefficient of the noise β can be taken so that circular blobs of diameter inferior to d pixels should disappear at low temperature.

$$\beta = \frac{4(1+\sqrt{2})}{d}. \tag{4}$$

(If Y is a circular blob say of 1s on a background of -1s with diameter d, it is easy to see that the maximum value of $H(X) + N(X,Y)$ is reached either for $X = Y$ or for $X_i = -1$ everywhere, using the isoperimetric inequality.)

The temperature $1/T$ determines what neighbourhood of the minima of the a posteriori energy $H(X)+N(X,Y)$ will be explored during the Monte-Carlo simulation of the law of X.

Our estimate of X is based on the cost function suggested by Marroquin, Mitter and Poggio [10]:

$$C(X,\hat{X}) = -\sum_{i\in S} X_i\hat{X}_i. \tag{5}$$

We estimate X knowing Y by minimizing

$$E(C(X,\hat{X}) \mid Y). \tag{6}$$

This comes to taking at each site $i \in S$ $\hat{X}_i$ to be the maximum likelihood estimator of the marginal law of X_i at site i knowing Y. This marginal law is estimated with the help of a Monte-Carlo simulation.

2 Global models for grey levels images

We will build two global models. In both cases the method will be the same. Let us say that our grey levels images have 2^n grey levels. We start with $2^n - 1$ independent level sets L_k, $k = 1,\ldots,2^n - 1$. These level sets do not represent an image, because the inclusion relation may not be satisfied. Then we condition this independent model so that the inclusion relation should be satisfied.

2.1 First global model: dichotomic conditioning

Let g be the observation of the grey levels image f, $f : S \longrightarrow [0, 2^n[$. Let us say that $L_s = \{i \in S \mid f_i \geq s\}$, $s = 1,\ldots,2^n - 1$. We can group the level sets L_s in a dichotomic way putting

$$\mathcal{L}_k = \{L_s,\ s \in 2^k I\!N\}. \tag{7}$$

The family of all level sets is $\mathcal{L}_1$. $\mathcal{L}_1$ is an element of $\Omega = \mathcal{P}(S)^{2^n-1}$. We start with a product law $Q = \prod_{s=1}^{2^n-1} q_s$ on Ω. Then we consider the finite Markov chain $\mathcal{L}_{n-1}, \mathcal{L}_{n-2}, \ldots, \mathcal{L}_1$ (it is a Markov chain for any law on Ω). We can write

$$Q(\mathcal{L}_1) = Q(\mathcal{L}_{n-1})Q(\mathcal{L}_{n-2} \mid \mathcal{L}_{n-1})\ldots Q(\mathcal{L}_1 \mid \mathcal{L}_2). \tag{8}$$

For the law Q the inclusion relation $L_1 \geq L_2 \geq \ldots \geq L_{n-1}$ is not satisfied with probability one. Hence it is not a good model for level sets. Nevertheless we can condition step by step the above Markov chain to ensure that this relation is satisfied: we replace the probability law Q on Ω by the law $\tilde{Q}$ defined by

$$\begin{aligned}\tilde{Q}(\mathcal{L}_1) &= Q(\mathcal{L}_{n-1})Q(\mathcal{L}_{n-2} \mid \mathcal{L}_{n-1} \text{ and } L_{2^{n-2}} \geq L_{2^{n-1}} \geq L_{2^{n-2}+2^{n-1}})\cdots \\ &\quad Q(\mathcal{L}_k \mid \mathcal{L}_{k+1} \text{ and } L_{2^{k+1}s} \geq L_{2^{k+1}s+2^k} \geq L_{2^{k+1}(s+1)},\ s = 0,\ldots,2^{n-k-1})\cdots.\end{aligned} \tag{9}$$

Applying this scheme to the a posteriori model of the previous section, we get

$$P(f \mid g) = P(\lfloor f/2^{n-1}\rfloor \mid \lfloor g/2^{n-1}\rfloor)P(\lfloor f/2^{n-2}\rfloor \mid \lfloor f/2^{n-1}\rfloor, \lfloor g/2^{n-2}\rfloor)\cdots \tag{10}$$

with

$$\begin{aligned} &P(\lfloor f/2^k\rfloor \mid \lfloor f/2^{k+1}\rfloor, \lfloor g/2^k\rfloor) = \\ &\frac{1}{Z_k(\lfloor f/2^{k+1}\rfloor, \lfloor g/2^k\rfloor)} \exp\Bigg(\sum_{s\in(2^k+2^{k+1}\mathbb{N})\cap[1,2^n[} 1/T\sum_{i<j}\xi_{i,j}\chi(f_i \geq s)\chi(f_j \geq s) \\ &+ \beta/T\sum_i \chi(f_i \geq s)\chi(g_i \geq s)\Bigg), \end{aligned} \tag{11}$$

where χ is the characteristic function with values in $\{-1,1\}$ and $\lfloor x\rfloor$ is the lower integer part of x. The partition function is

$$\begin{aligned} Z_k(\lfloor f/2^{k+1}\rfloor, \lfloor g/2^k\rfloor) &= \sum_{h\in I(f,k)} \exp\Bigg(\sum_{s\in(2^k+2^{k+1}\mathbb{N})\cap[1,2^n[} 1/T\sum_{i<j}\xi_{i,j}\chi(h_i \geq s)\chi(h_j \geq s) \\ &+ \beta/T\sum_i \chi(h_i \geq s)\chi(g_i \geq s)\Bigg), \end{aligned} \tag{12}$$

where

$$I(f,k) = \{h \in ([0,2^n[\cap 2^k\mathbb{N})^S \;/\; \lfloor f/2^{k+1}\rfloor = \lfloor h/2^{k+1}\rfloor\}. \tag{13}$$

As constant terms may be cancelled from numerator and denominator, these formulas are simpler than they may appear. This reflects in the expression of the conditional law of one site $i \in S$ knowing the others, where, as a matter of fact, the summation on the levels s reduces to one term (namely $s = 2^{k+1}\lfloor f_i/2^{k+1}\rfloor + 2^k$).

$$\begin{aligned} &P(\lfloor f_i/2^k\rfloor \mid \lfloor f_i/2^{k+1}\rfloor, \lfloor f_j/2^k\rfloor, j \neq i,\ g) = \\ &\frac{1}{Z_i}\exp\Bigg(\frac{1}{T}\sum_j \xi_{i,j}\chi(f_i \geq 2^{k+1}\lfloor f_i/2^{k+1}\rfloor + 2^k)\chi(f_j \geq 2^{k+1}\lfloor f_i/2^{k+1}\rfloor + 2^k) \\ &+ \frac{\beta}{T}\chi(f_i \geq 2^{k+1}\lfloor f_i/2^{k+1}\rfloor + 2^k)\chi(g_i \geq 2^{k+1}\lfloor f_i/2^{k+1}\rfloor + 2^k)\Bigg), \end{aligned} \tag{14}$$

with

$$Z_i = \sum_{h\in\{-1,+1\}} \exp\left(\frac{1}{T}\sum_j \xi_{i,j}h\chi(f_j \geq 2^{k+1}\lfloor f_i/2^{k+1}\rfloor + 2^k) + \frac{\beta}{T}h\chi(g_i \geq 2^{k+1}\lfloor f_i/2^{k+1}\rfloor + 2^k)\right) \tag{15}$$

Hence it is possible to perform a simulation of $\lfloor f/2^k\rfloor$ knowing $\lfloor f/2^{k+1}\rfloor$ and g dealing with all the level sets $s \in 2^k + 2^{k+1}\mathbb{N}$ in the same sweep. If we had simulated independently each level set we would have had instead one sweep for each level, or 2^{n-1-k} sweeps. Hence the speed-up increases exponentially with $k = n-1, n-2, \ldots, 0$, and we can perform the reconstruction of all the level sets in n steps instead of $2^n - 1$ steps: we reconstruct successively the n bits of the binary representation of f, from left to right.

Let us notice that we would have obtained a slightly different a posteriori global model if we had applied our conditioning scheme separately to the a priori law and to the law of the noise, instead of applying it directly to the a posteriori law. Our choice has been motivated by the opinion that our dichotomic conditioning is a trick which should be delayed as long as possible and also by the fact that the chosen method leads to a slighlty simpler algorithm.

In this first global method, the estimation itself is performed step by step in a dichotomic way:

The successive bits of $\hat{f}$ are estimated from left to right by minimizing successively

$$E(C(\lfloor \hat{f}/2^{n-1} \rfloor, \lfloor f/2^{n-1} \rfloor) \mid \lfloor g/2^{n-1} \rfloor),$$

$$\ldots$$

$$E(C(\hat{f} - 2\lfloor f/2 \rfloor, f - 2\lfloor f/2 \rfloor) \mid \lfloor \hat{f}/2 \rfloor, g). \tag{16}$$

The marginal laws are estimated by Monte-Carlo simulation at constant temperature (Gibbs sampling).

2.2 Second global model: global conditioning

Instead of conditioning step by step by the inclusion relation larger and larger sets of level sets, we can do it all at a time. We get

$$P(f \mid g) = \frac{1}{Z(g)} \exp\left(\sum_{s \in [1, 2^n[} \frac{1}{T} \sum_{i<j} \xi_{i,j} \chi(f_i \geq s) \chi(f_j \geq s) + \frac{\beta}{T} \sum_i \chi(f_i \geq s) \chi(g_i \geq s) \right), \tag{17}$$

where

$$Z(g) = \sum_{h \in [0, 2^n[^S} \exp\left(\sum_{s \in [1, 2^n[} \frac{1}{T} \sum_{i<j} \xi_{i,j} \chi(h_i \geq s) \chi(h_j \geq s) + \frac{\beta}{T} \sum_i \chi(h_i \geq s) \chi(g_i \geq s) \right). \tag{18}$$

We have the following identity

$$P(f \mid g) = \frac{1}{R(g)} \exp\left(-\frac{1}{T} \sum_{i<j} \xi_{i,j} 2|f_i - f_j|) - \frac{\beta}{T} \sum_i 2|f_i - g_i|) \right), \tag{19}$$

with

$$R(g) = \sum_{h \in [0, 2^n[^S} \exp\left(-\frac{1}{T} \sum_{i<j} \xi_{i,j} 2|h_i - f_j|) - \frac{\beta}{T} \sum_i 2|h_i - g_i|) \right). \tag{20}$$

Here again it must be said that a different result would have been obtained by conditioning separately the a priori model and the law of the noise, the reason being that the partition function for the global law of the noise conditioned by the inclusion relation depends on f (it is not the case for the independent model).

We have now to choose a cost function. Let it be the sum of the cost functions for the level sets:

$$C(\hat{f}, f) = - \sum_{s=1}^{2^n - 1} \sum_{i \in S} \chi(f_i \geq s) \chi(\hat{f}_i \geq s) = \sum_{i \in S} -2^n + 1 + 2|f_i - \hat{f}_i|, \tag{21}$$

which is equivalent to the cost function

$$C_1(f, \hat{f}) = \sum_{i \in S} |f_i - \hat{f}_i|. \tag{22}$$

We will keep this last definition which is more simple.

In order to minimize $E(C(f, \hat{f}) \mid g)$, it is easy to see that we have to take at each site i $\hat{f}_i$ to be the median of the marginal distribution of f_i knowing g.

The law of f knowing g is estimated by the empirical ditribution of a Gibbs sampler. As the computation of the median of the marginal laws requires to store histograms, we have approximated it by the mean

$$\hat{f}_i = E(f_i \mid g), \tag{23}$$

which would correspond to a quadratic cost function

$$C_2(f, \hat{f}) = \sum_{i \in S} (f_i - \hat{f}_i)^2. \tag{24}$$

3 Examples and comments

Though the second global model may seem to be simpler and more elegant from the mathematical point of view it leads to a more complicated and slower computer program. The dichotomic conditional model requires less memory allocation, because only one bit is flipped at each step. Hence for 256 grey levels we need 8 bits/pixel for the observation g, one bit/pixel for the simulation and one counter/pixel (say 8 bits) to store the empirical law. We can at the end of the kth step replace $\lfloor g/2^{n-k} \rfloor$ by $\lfloor \hat{f}/2^{n-k} \rfloor$, or more precisely replace g by

$$g \vee (2^{n-k} \lfloor f/2^{n-k} \rfloor) \wedge (2^{n-k}(\lfloor f/2^{n-k} \rfloor + 1) - 1). \tag{25}$$

(This change will not affect the logical value of the tests "$g_i \geq 2^{r+1} \lfloor f_i/2^{r+1} \rfloor + 2^r$", $r < n - k$, hence will not affect further calculations). Thus the restoration can be done "in situ" and does not require extra memory to store the result.

The global conditional model requires 8 bits/pixel for the observation g, 8 bits/pixel for the simulation of f and say 16 bits/pixels for the empirical mean (which is also the result), that is 32 bits/pixel instead of 17.

Moreover the tabulation of the exponential is easy in the dichotomic model: no more than 50 values are needed. In the second model only a partial tabulation is possible: we have to tabulate separately $\exp(n_1/T)$, $\exp(n_2/(\sqrt{2}T))$ and $\exp(n_3\beta/T)$. Hence we have to perform 2 multiplications per pixel and per sweep more than for the dichotomic model.

The fact that the exponential can be tabulated is an argument in favour of the use of a cost function which allows to do the simulation at constant temperature instead of running a simulated annealing algorithm to compute the maximum likelihood estimate.

Not surprisingly, the dichotomic algorithm and the global one give comparable but not identical results. The parameters $1/T$ and β have to be given different values in order to obtain approximatly the same effect. Even though, some slight differences remain in the computed solutions: the dichotomic algorithm performs a more drastic regularization than the global one. This could be a weakness when the noise is small, and a favourable feature when it is strong. Of course it is not obvious to tell whether these dissimilarities are due to the differences in the model or to the differences in the choice of the moves and the exploration of the states space. However it is interesting to point out that using the dichotomic moves of the first algorithm with the energy of the second one does not work satisfactorily, as is illustrated in the last example (this is not surprising because the appropriate methodology would have been to compute the marginal laws of the first bits of the binary representation of the image, which is unworkable).

We give examples where we use 11 sweeps per bit in the dichotomic algorithm (that is 88 sweeps in all) and 105 in the global one. No significant improvement was observed in increasing the number of sweeps.

In order to demonstrate the robustness of the choice of parameters, we have used the same set of values for all pictures:

- $1/T = 1$ and $\beta = 2$ for the dichotomic model.
- $1/T = 2$ and $\beta = 1$ for the global algorithm.

We show two kinds of examples:

- Computed geometrical pictures with gaussian additive noise or with independent corruption of some random pixels (the grey level of any pixel is changed to a uniform random value with some given probability corresponding to the ratio of altered pixels).
- Pictures acquired from a video camera (a bicycle, a queen and some biological tissue viewed under the microscope), with the same kind of noise as in the first case.

These are 256 grey levels pictures of various sizes ranging from 64×64 to 256×256.

The results are not so good when the noise is gaussian as when it is of the second kind. This could be expected because independent corruption of pixels matches the noise we have hypothesized on contour lines, whereas a gaussian noise does not.

The global algorithm could be modified to take into account a gaussian noise, changing the energy to

$$-\frac{1}{T}\sum_{i<j}\xi_{i,j}|f_i - f_j| - \beta\sum_i (f_i - g_i)^2. \tag{26}$$

This has not been tried, since our main interest in this paper was in the dichotomic approach.

As for the performances, the dichotomic algorithm is much faster than the global one: for a 128×128 picture it takes about 69 seconds of CPU time instead of 5 minutes and 54 seconds (when implemented in C on a Sun-4 workstation). It is thus up to five times faster !

References

[1] Azencott R., (1987) Markov Fields and Image Analysis, *Congrès AFCET 1987, Antibes, France (organized by INRIA and AFCET).*

[2] Catoni O. and Gaudron I., (1989) Détection de contours par seuillage adaptatif et restauration stochastique d'images binaires, *Proceedings of Pixim 1989, sept 25-29 1989, Gagalowicz A. (ed.), Hermes, 1989, p. 341.*

[3] Chalmond B., (1988) Image Restoration Using an Estimated Markov Model, *Signal Processing, vol 15, sept 1988.*

[4] Dinten J.-M. and Tatar M., (1989) Application des champs de Markov à la restauration d'images, *Proceedings of Pixim 89, sept 25-29 1989, Gagalowicz A. (ed.), Hermes 1989, p. 325.*

1 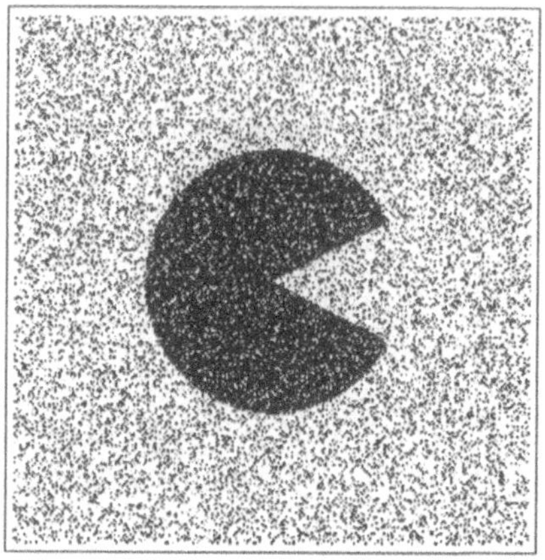 2 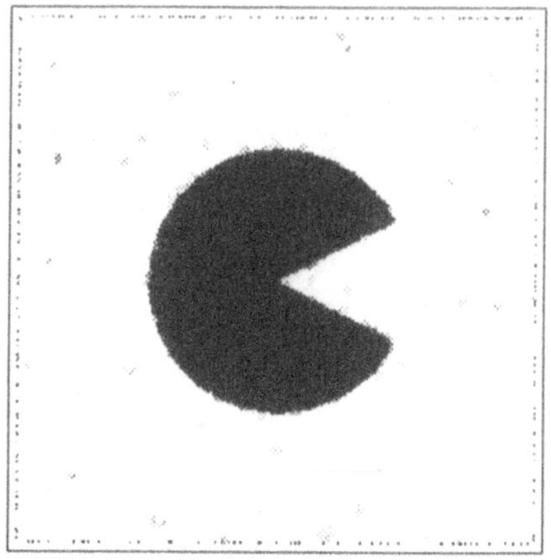

1. Geometrical shape with 30/100 of the pixels corrupted, size 128×128. 2. Result of dichotomic restoration.

3 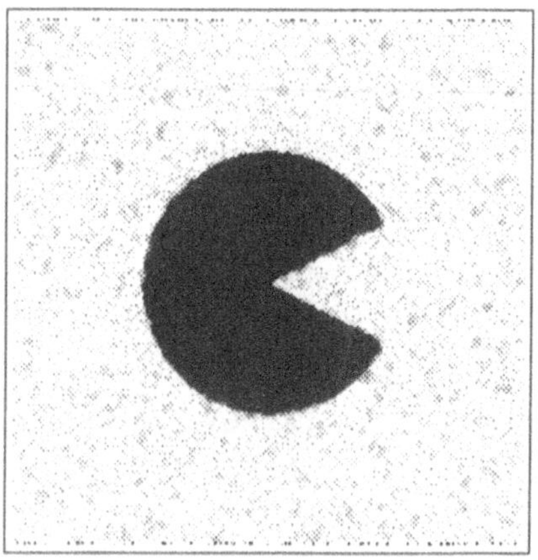 4 

3. Result of global restoration. 4. Another geometrical picture: $f(x, y) = \lfloor x^2 + y^2 \rfloor$, original image of size 128×128.

5

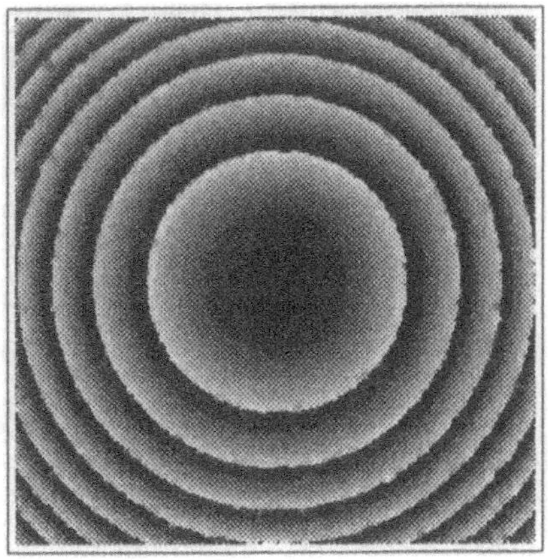

6

5. 30/100 of the pixels have been corrupted. **6.** Result of dichotomic restoration.

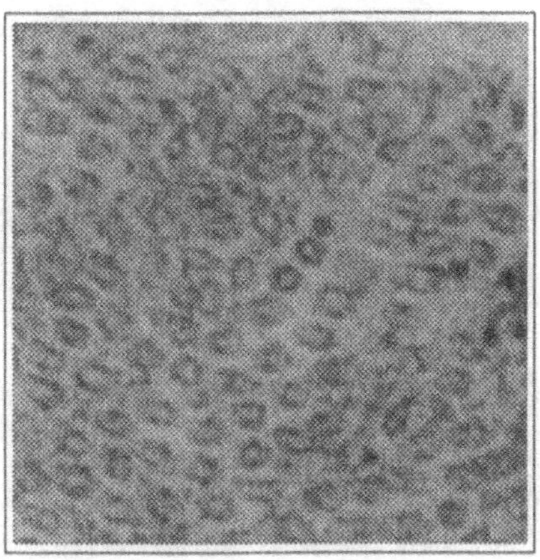

7

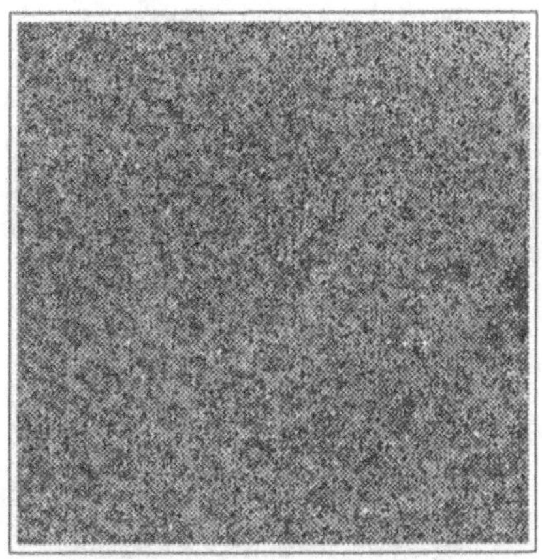

8

7. View of a cellular tissue kindly supplied by Pierre Vincens from the Laboratoire de Biochimie de l'Ecole Normale Supérieure, original image of size 128 × 128. **8.** 30/100 of the pixels have been corrupted.

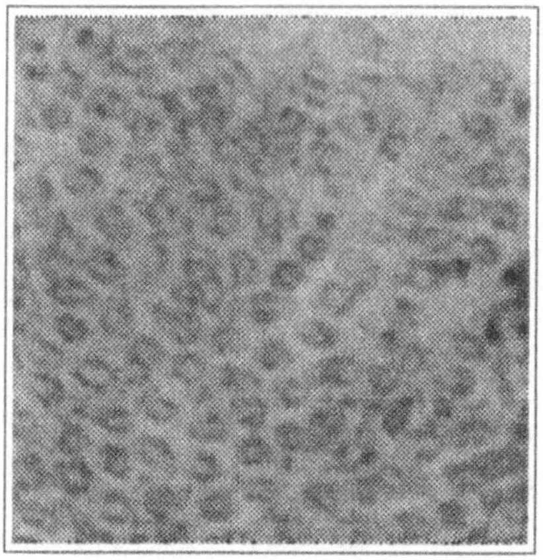

9

10

9. Result of global restoration. **10.** A bicycle in the rue Soufflot, original image of size 256×256.

11

12

11. 30/100 of the pixels have been corrupted. **12.** Global restoration.

13

14

13. Additive gaussian noise of standard deviation 30. **14.** Dichotomic restoration.

15

16

15. A queen in the jardin du Luxembourg, original picture of size 200 × 500. **16.** 30/100 of the pixels have been corrupted.

17

18

17. Dichotomic restoration. **18.** Additive gaussian noise of standard deviation 60.

19

20
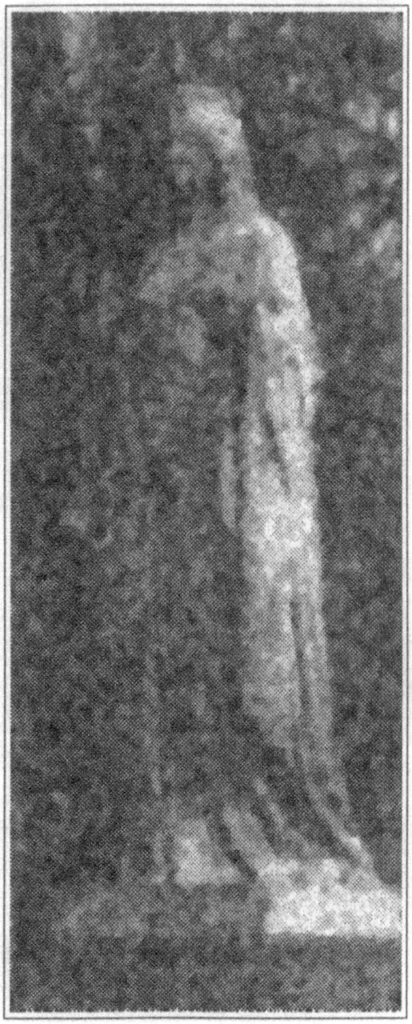

19. Dichotomic restoration. **20.** Global restoration.

21

22

21. Gaussian noise of standard deviation 30. **22.** Dichotomic restoration.

23

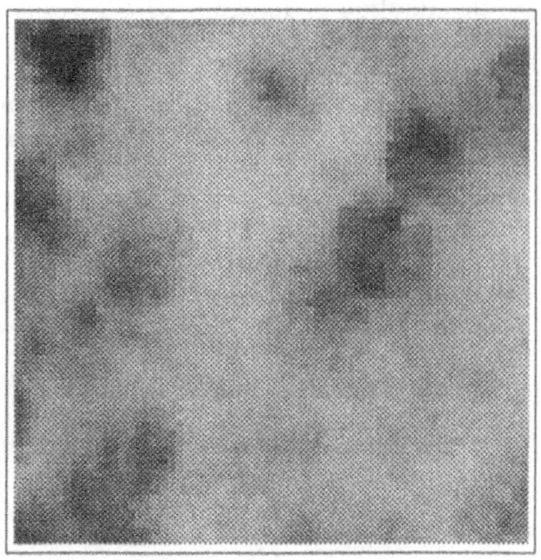

24

23. Global restoration. **24.** Some small detail of a cellular tissue, original image of size 64 × 64.

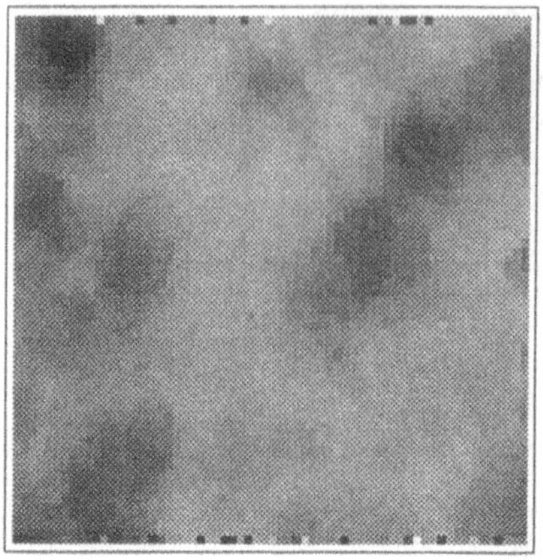
25

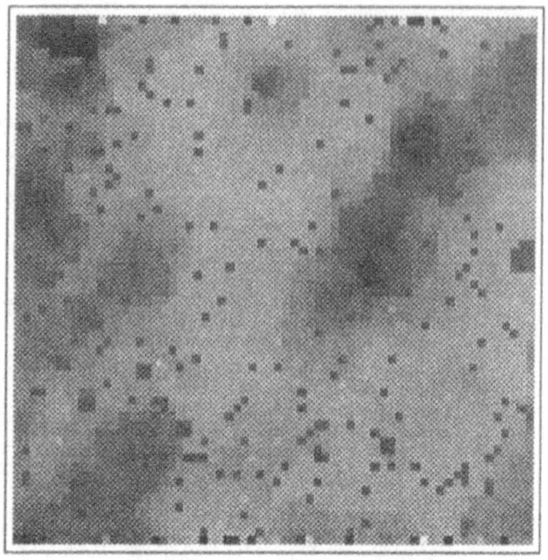
26

25. Global restoration of the 30/100 corrupted image. **26.** For comparison, global restoration with the moves of the dichotomic algorithm (it is not very satisfactory, iteration of the algorithm would have brought an improvement).

[5] Geman D. and Geman S., (1985) Stochastic Relaxation, Gibbs Distributions and the Bayesian Restoration of Images, *IEEE Trans. on Patt. An. and Mach. Int., vol 6, nb 6, June 1985, pp. 721-741.*

[6] Geman D., Geman S., Graffigne C. and Dong P., (1990) Boundary Detection by Constrained Optimization, *IEEE Trans. on Patt. An. and Mach. Int., vol 12, nb 7, July.*

[7] Geman D., Geman S. and Graffigne C., (1987) Locating Textures and Object Boundaries, *Patt. Rec. Theor. and Appl., Devijver and Kittler J. Eds., Heidelberg, Springer Verlag.*

[8] Geman S. and Graffigne C., (1987) Markov Random Field Image Models and their Applications to Computer Vision, *Proc. Intl. Congress of Mathematicians, preprint, A. M. Gleason (ed.).*

[9] Hammersley and Handscomb, (1964) Monte-Carlo Methods, *Methuen and Co., London.*

[10] Marroquin J., Mitter S. and Poggio T. (1987) Probabilistic Solution of Ill Posed Problems in Computational Vision, *Journal of the American Statistical Association, vol. 82, nb 397, Theory and Methods, pp. 76-89, March 1987.*

[11] Marroquin J. L., (1989) A Markovian Random Field of Piecewise Straight Lines, *Biological Cybernetics.*

A COMPARISON OF SIMULATED ANNEALING OF GIBBS SAMPLER AND METROPOLIS ALGORITHMS

Chiang, Tzuu–Shuh* ** and Chow, Yunshyong**
Institute of Mathematics, Academia Sinica
Taipei, Taiwan

Abstract. We prove that Gibbs sampler and Metropolis algorithm are asymptotically equivalent in annealing for lattices. They are not equivalent in general if there is no lattice structure of the state space.

§ 1. Introduction

Let L be a finite lattice and for each site $i \in L$, there is a finite level set ℓ_i with size $|\ell_i| > 1$. Let $\Omega = \prod_{i \in L} \ell_i$ be the configuration space and let $U:\Omega \longrightarrow [0,\infty)$ be a potential function. The following is usually referred to as the Gibbs distribution of U at temperature T.

$$(1.1) \qquad P_T(x) = \frac{1}{Z_T} \exp(-\frac{U(x)}{T}), \qquad x \in \Omega$$

where Z_T is a normalizing constant. One way to generate the Gibbs distribution (1.1) is to run a Markov chain on Ω whose invariant distribution is given by (1.1). Two frequently used such Markov chains are Gibbs sampler (GS) and Random Metropolis (RM) (see [5],[6]). We shall give their definitions below.

For a site $i \in L$ and $x \in \Omega$, let $N_i(x) = \{y \in \Omega: y(k) = x(k) \text{ for all } k \neq i, k \in L\}$ and $N(x) = \bigcup_{i \in L} N_i(x)$. N(x) is called the neighborhood system of x. At a fixed site $i \in L$ the following transition is the a local updating scheme:

$$(1.2) \qquad G_i^T(x,y) = \begin{cases} \exp(-\frac{U(y)}{T}) \,/\, \sum_{z \in N_i(x)} \exp(-\frac{U(z)}{T}) & \text{if } y \in N_i(x) \\ 0 & \text{otherwise.} \end{cases}$$

AMS 1980 Subject classification. Primary 15A51. Secondary 15A18.
Key words and phrases: Gibbs sampler, Metropolis algorithm, simulated annealing.

*Work done while the author was at the Center for Stochastic Processes. He is grateful for their hospitality.

**Partially supported by National Science Council, Taiwan and the Air Force Office of Scientific Research Contract No. F49620 85C 0144.

GS is obtained from (1.2) with randomized initial value. Thus, at temperature T, GS is a Markov chain on Ω with the following transition,

(1.3) $$G_T(x,y) = \frac{1}{|L|} \sum_{i \in L} G_i^T(x,y).$$

RM is similar to GS except for the local updating schemes:

(1.4) $$M_i^T(x,y) = \begin{cases} \frac{1}{|\ell_i|-1} \exp\left(-\left(\frac{U(y)-U(x)}{T}\right)^+\right) & \text{if } y \in N_i(x),\ y \neq x, \\ 0 & \text{if } y \notin N_i(x). \end{cases}$$

The transition of RM is:

(1.5) $$M_T(x,y) = \frac{1}{|L|} \sum_{i \in L} M_i^T(x,y).$$

The simulated annealing of GS or RM (SGS or SRM) is to run the Markov chain (1.3) or (1.5) with temperature T slowly decreasing to 0 as time going to ∞. To be precise, the simulated annealing of GS and RM are inhomogeneous Markov chains whose transition functions at time t are taken to be:

(1.6) For $x,y \in \Omega$, $P_t^G(x,y) = G_{T(t)}(x,y)$ and

(1.7) For $x,y \in \Omega$, $P_t^M(x,y) = M_{T(t)}(x,y)$

respectively. Here, $T(t) = c/\log t$ where c is some positive number to be specified later. SGS and SRM have been used to search for the ground states of the potential U ([6], [9], [10]). It is therefore important to know if they converge to the same states and natural to compare SGS with SRM to see which one has a faster convergence rate. In Section 2, we shall state a general and well–known theorem regarding the convergence rate of inhomogeneous Markov chains with transitions like (1.6) or (1.7) and prove that SGS and SRM reach the ground states (and any state) at the same rate. Another application of our result (Theorem 3.6) can be found in [4]. Some examples are also given to illustrate that SGS and SRM are not equivalent if L is not a lattice.

§2. Cycles and simulated annealing.

For a non–negative function $U: \Omega\times\Omega \to [0,\infty]$, let $Q^T = (Q^T_{x,y})$ be a stochastic matrix with

(2.1) $$Q^T_{x,y} \sim \exp\left(-\frac{U(x,y)}{T}\right) \quad \text{for} \quad x,y \in \Omega,\ x \neq y.$$

Let X_t be the inhomogeneous Markov chain on Ω with transition $Q^{T(t)}_{x,y}$ at time t where $T(t) = c/\log t$ with $c>0$. We shall describe in Theorem 2.1 the asymptotic behavior of X_t as $t \to \infty$. First, we need some definitions. The function U is always assumed to have minimum 0 and irreducible in the sense that for any two states $x,y \in \Omega$, there are $x_0 = x, x_1, \ldots, x_m = y$ such that $U(x_v, x_{v+1}) < \infty$ for each v. We say that $x \geq y$ (relative to U) if, in the above sequence

$$U(x_v,x_{v+1}) = \min_{z\in\Omega,\ z\neq x_v} U(x_v,z) \quad \text{for each } v.$$

A state x is called minimal if $y \geq x$ whenever $x \geq y$. Two different states $x,y \in \Omega$ are said to be equivalent ($x \sim y$) if (i) x is minimal and (ii) $x \geq y$ and $y \geq x$.

We always assume $x \sim x$. An equivalent class of Ω under the equivalence relation " $\sim$ " is called a cycle of (Ω,U). A cycle is non–trivial if it contains more than one state.

For a pair (Ω,U), we can successively define (Ω^n,U^n,V^n), n=0,1,2,... as follows: Let $(\Omega^0,U^0)=(\Omega,U)$ and $V^0(x) = \min_{x\neq y\in\Omega} U(x,y)$ for each $x \in \Omega$. Having defined $(\Omega^{n-1},U^{n-1},V^{n-1})$, let $\Omega^n = \{\text{cycles of } (\Omega^{n-1},U^{n-1})\}$. For any $C^n = \{C_i^{n-1}\}_i$ and $\tilde{C}^n = \{\tilde{C}_j^{n-1}\}_j$ in Ω^n, let:

$$d_{n-1}(C^n) = \max_i V^{n-1}(C_i^{n-1}), \qquad R(C^n) = \{C_i^{n-1}: V^{n-1}(C_i^{n-1}) = d_{n-1}(C^n)\}$$

(2.2) $$U(C^n,\tilde{C}^n) = d_{n-1}(C^n) + \min_{i,j} \{U^{n-1}(C_i^{n-1},\tilde{C}_j^{n-1}) - V^{n-1}(C_i^{n-1})\}$$

(2.3) $$V^n(C^n) = \min_{C^n\neq\tilde{C}^n\in\Omega^n} U^n(C^n,\tilde{C}^n).$$

The double arrow "$C^n \Rightarrow \tilde{C}^n$" means $U^n(C^n,\tilde{C}^n) = V^n(C^n)$.

A notion of height is necessary to describe our results. Since U is irreducible, there exists an N such that $\{C^N: C^N \in \Omega^N\}$ forms a single cycle, i.e., $|\Omega^{N+1}| = 1$ but $|\Omega^N| \geq 2$. For each $x \in \Omega$, there exists uniquely $C^n \in \Omega^n$, n=0,1,...,N+1 such that $x = C^0\in C^1\in\ldots\in C^{N+1}$. The height of a state $x \in \Omega$ is given by

(2.4) $$h(x) = \sum_{n=0}^{N} (d_n(C^{n+1}) - V^n(C^n)).$$

Let $B = \{x \in \Omega: h(x) = 0\}$. The following constant is crucial for the annealing rates c/log t in (1.6) and (1.7):

(2.5) $$\Gamma = \max_{x\in\Omega} \min_{y\in B,\ y\neq x} V^k(C^k)$$

where k is the first integer such that $x = C^0 \in C^1 \in \ldots \in C^k \in C^{k+1}$ and $y = \tilde{C}^0 \in \tilde{C}^1 \in \ldots \in \tilde{C}^k \in C^{k+1}$. The following theorem can be found in [1],[2],[3],[7],[8].

<u>Theorem 2.1</u>. Let X_t be the Markov chain with inhomogeneous transitions $Q_{x,y}^{T(t)}$ as in (2.1) where $T(t) \doteq c/\log t$ for some constant $c > \Gamma$. Then there exist positive constants β_x such that $\lim_{t\to\infty} P(X_t=x) / \exp(-\frac{h(x)}{T(t)}) = \beta_x$ for each $x \in \Omega$.

Obviously, one needs to know Γ and the height function h in order to compare the convergence of Markov chains corresponding to different U's. The following easy theorem serves that purpose.

Theorem 2.2. Let U and $\bar{U}$ be two functions from $\Omega\times\Omega \longrightarrow [0,\infty]$. Let (Ω^n,V^n) and $(\bar{\Omega}^n,\bar{V}^n)$ be defined in (2.3), h and $\bar{h}$ in (2.4) and Γ and $\bar{\Gamma}$ in (2.5) for (Ω,U) and $(\Omega,\bar{U})$ respectively. If $(\Omega^n,V^n) = (\bar{\Omega}^n,\bar{V}^n)$ for each n, then $h(x) = \bar{h}(x)$ for each $x \in \Omega$ and $\Gamma = \bar{\Gamma}$.

Proof: The proof is trivial by the observation that the only quantities involved in the definitions of h (2.4) and Γ (2.5) are C^n's and V^n's. Since $\Omega^n = \bar{\Omega}^n$ and $V^n = \bar{V}^n$, the theorem follows trivially.

§3. Comparison of RGS and SRM.

Let U be a potential function from Ω to $[0,\infty)$ as in Section 1 and let SGS and SRM be the inhomogeneous Markov chains $X_G(t)$ and $X_M(t)$ with transition functions $P_t^G(x,y)$ and $P_t^M(x,y)$ ((1.6) and (1.7)) respectively. They obviously are of the form (2.1) with

$$(3.1)\qquad U_G(x,y) = \begin{cases} U(y) - \min\limits_{z\in N_i(x)} U(z) & \text{if } y \in N_i(x) \text{ for some } i \\ \infty & \text{otherwise} \end{cases}$$

and

$$(3.2)\qquad U_M(x,y) = \begin{cases} (U(y) - U(x))^+ & \text{if } y \in N(x),\ y \neq x \\ \infty & \text{otherwise.} \end{cases}$$

Let h_G and Γ_G be defined in (2.4) for U_G. Similarly for h_M and Γ_M. We say that SGS and SRM are asymptotically equivalent if $h_G=h_M$ and $\Gamma_G = \Gamma_M$. Note that for Ising models, i.e., $|\ell_i|=2$ for each $i \in L$, $U_G = U_M$ and SGS and SRM are therefore identical. In general, since $U_G \neq U_M$, it is non-trivial to compare their convergence rates. We shall indicate a proof (the detail will appear in [4]) in this section that $h_G = h_M$ and $\Gamma_G = \Gamma_M$ and therefore conclude that SGS and SRM converge to the ground states of U at the same rate. First, we state some properties regarding the cycle structures of (Ω,U_M).

Lemma 3.1. Let $x,y \in \Omega$. Then $x\Rightarrow y$ in U_M implies that $U(x)+V_M(x) \geq U(y)+V_M(y)$.

Let [x] be the cycle in (Ω,U_M) which contains x. Then the following is obvious.

Lemma 3.2. $U(y) + V_M(y) = U(x) + V_M(x)$ for any $y \in [x]$.

We next define the neighborhood system of cycles in Ω^1. For $[x] \in \Omega^1$, let

(3.3) $$N([x]) = \{[y] \in \Omega^1:\ \text{there exist } x' \in [x] \text{ and } y' \in [y] \cap N(x')\} \text{ and}$$

(3.4) $$U^1([x]) = \min_{z \in [x]} U(z).$$

Let $(\Omega^1,U_M^1),\dots,(\Omega^n,U_M^n),\dots$ be the successive cycles in (2.2) for (Ω,U_M).

Theorem 3.3. For $[x]$ and $[y] \in \Omega^1$, $U_M^1([x],[y])$ has the following representation:

(3.5) $$U_M^1([x],[y]) = \begin{cases} (U^1([y]) - U^1([x]))^+ & \text{if } [y] \in N([x]) \\ \infty & \text{otherwise.} \end{cases}$$

Proof: By the definition of a cycle and (3.2), we only need to prove the case $[y] \in N([x])$ and $[x]$ is not trivial. If $x' \in [x] \cap N(y)$, then $U_M(x',y) = (U(y) - U(x'))^+ > V(x')$ and

$$U_M^1([y],[x]) = \min_{x' \in [x]} U(y,x') = 0 = (U_M^1([x]) - U_M^1([y]))^+.$$

In general, if $[x]$ is nontrivial and $[y] \in N([x])$, then

$$\begin{aligned} U_M^1([x],[y]) &= \max_{z \in [x]} V_M(z) + \min_{x' \in [x]} (U_M(x',y) - V_M(x')) \\ &= \max_{z \in [x]} V_M(z) + \min_{x' \in [x]\cap N(y)} (U(y)-U(x')-V_M(x')) \\ &= \max_{z \in [x]} V_M(z) + U(y) - (U(x) + V_M(x)) \quad \text{by Lemma 3.2} \\ &= U(y) - \min_{z \in [x]} U(z) \quad \text{by Lemma 3.2} \\ &= (U^1[y] - U^1[x])^+. \end{aligned}$$

By an induction argument we conclude that U_M^n can be written as the positive part of a potential function as in (3.2) for any n. We next list without the trivial proof all the properties of $(\overline{\Omega},U_G)$ that are necessary for our concerns. Here, $\overline{\Omega}$ is the same as Ω only to indicate that U_G is considered.

Lemma 3.4. Let $(\overline{\Omega},U_G)$ be the pair defined in (3.1). Let $N(x)$ be the neighborhood system of x. Then the following properties hold:

(3.6) $$N(x) = \bigcup_{i \in L} N_i(x) \quad \text{and} \quad x \in N_i(y) \quad \text{if and only if} \quad y \in N_i(x).$$

(3.7) For each $x \in \Omega$ and $i \in L$, there exists a state $m_i(x) \in N_i(x)$ such that

$$U_G(x,m_i(x)) = (U(m_i(x)) - U(x))^+ = 0 \quad \text{and}$$

$$U_G(m_i(x),x) = [U(x) - U(m_i(x))]^+.$$

(3.8) $\qquad U_G(x,y) \geq U_M(x,y).$

Lemma 3.5. Let $(\overline{\Omega},U_G)$ be a pair satisfying (3.6)–(3.8). If $x \Rightarrow y$ then $U(x) + V_G(x) \geq U(y) + V_G(y)$. The converse is true if $y \in N_i(x)$ and $x = m_i(x)$. Moreover, $V_G(x) = V_M(x)$ for any $x \in \Omega$.

We now come to the main theorem of this section. Let U be a function from Ω to $[0,\infty)$ and let $N([x])$ and U^1 be defined as in (3.3) and (3.4).

Theorem 3.6. Let $(\overline{\Omega},U_G)$ and (Ω,U_M) be defined in (3.1) and (3.2) respectively. Then $\overline{\Omega}^1 = \Omega^1$ and $(\overline{\Omega}^1,U^1{}_G)$ satisfy (3.6)–(3.8). Moreover, $V_G^1([x]) = V_M^1([x])$ for any $[x] \in \Omega^1$.

Proof: Let $[x]$ and $\boxed{x}$ be the cycles in (Ω,U_M) and $(\overline{\Omega},U_G)$ containing x respectively. By Lemma 3.5, it is easy to see that x is minimal in $(\overline{\Omega},U_G)$ if and only if x is minimal in (Ω,U_M).

We next show that $[x] = \boxed{x}$. Since $[x]$ or $\boxed{x}$ is trivial if and only if x is not minimal, by the above we thus only need to consider the case where x is minimal. Suppose $y \in N_i(x)$ and $x \Rightarrow y$ in $(\overline{\Omega},U_G)$. By the same argument above, we have $x \Rightarrow m_i(x) \Rightarrow y \Rightarrow m_i(x) \Rightarrow x$ in (Ω,U_M). This in particular implies $x \geq y$ in (Ω,U_M). Thus $y \in [x]$ and $\boxed{x} \subseteq [x]$. The proof of $[x] \subseteq \boxed{x}$ is the same. Hence $[x] = \boxed{x}$ for every x and $\Omega^1 = \overline{\Omega}^1$. Let $N_i([x]) = \{[y]: y' \in N_i(x') \text{ for some } x' \in [x] \text{ and } y' \in [y]\}$. Obviously $N([x]) = \bigcup_i N_i([x])$ and (3.6) is satisfied. Let $m_i([x])$ be a cycle in $N_i([x])$ such that $U^1(m_i[x]) = \min_{[y]\in N_i([x])} U^1([y])$. We have to show $m_i([x])$ satisfies (3.7) and (3.8). Let $x' \in [x]$ and $y' \in m_i([x])$ such that $y' \in N_i(x')$. If $U(y') \geq U(x')$, then $U_M(y',x') = 0$ and $[x] = m_i([x])$ by the definition of a cycle. Thus $U(x') > U(y')$. This implies that $[x]$ is trivial because x is not minimal in (Ω,U_M). Since $y' \in N_i(x)$ and $y' \Rightarrow m_i(x)$, it is clear that $m_i(x) \in m_i([x])$ and therefore

$$U_G^1([x],m_i([x])) = \min_{y\in m_i([x])} U_G(x,y) \leq U_G(x,m_i(x)) = 0.$$

On the other hand,

$$U_G^1(m_i([x]),[x]) = \max_{y\in m_i([x])} V_G(y) + \min_{y\in m_i([x])} (U_G(y,x)-V_G(y)).$$

If $y \in m_i([x])$ but $y \neq m_i(x)$, then $U_G(y,x) \geq U_M(y,x) = U(x) - U(y)$. Hence

$$U_G(y,x) \geq U(x) - U(m_i(x)) + U(m_i(x)) - U(y)$$
$$= U(x) - U(m_i(x)) + V_G(y) - V_G(m_i(x)) \quad \text{by Lemma 3.5}$$
$$= U_G(m_i(x),x) - V_G(m_i(x)).$$

Therefore

$$U_G^1(m_i([x]),[x]) = \max_{y \in m_i([x])} V_G(y) + U_G(m_i(x),x) - V_G(m_i(x))$$
$$= \max_{y \in m_i([x])} V_G(y) + U_M(m_i(x),x) - V_M(m_i(x))$$
$$= \max_{y \in m_i([x])} V_G(y) + \min_{y \in m_i([x])} (U_M(y,x) - V_M(y)) \quad \text{by Lemma 3.5}$$
$$= U_M^1(m_i([x]),[x]).$$

This proves (3.7). (3.8) is trivial.

Because of Theorem 3.3 and Theorem 3.6, the induction can be applied to yield that $(\Omega^n, V_G^n) = (\Omega^n, V_M^n)$ for any n. Thus by Theorem 3.2, SGS and SRM are asymptotically equivalent.

The following examples show that if the state space does not satisfy (3.6)–(3.8), then SGS and SRM are in general not equivalent.

Example. Let $S = \{1,2,3\}$ and U be a potential with $U(1)=0$, $U(2)=2$, $U(3)=1$. The neighborhood systems are as follows: $N(1)=\{2\}$, $N(2)=\{1,3\}$ and $N(3)=\{2\}$. For the Metropolis algorithm, it is easy to see that $U_M(2,1) = U_M(2,3) = 0$ and (S,U_M) forms a cycle. Thus $V_1^M - V_2^M = 1$ and $h^M(2) = 2$, $h^M(3) = 1$. On the other hand, for GS, $U_G(2,1) = 0$ and $U^G(2,3) = 1$. Thus $S^1=\{1,3\}$ and $U_G^1(3,1)=1$, $U_G^1(1,3) = 3$. Thus $V_1^G - V_2^G = 1 = V_1^M - V_2^M$, and $h^G(2)=2=h^G(3)$. Hence $P(X_t^M=1) \sim 1-t^{-1}$ and $P(X_t^G=1) \sim 1-t^{-2}$. Consequently, GS is faster than Metropolis.

Example. Let $S = \{1,2,3,4,5\}$ and U be a potential function with $U(1) = 0$, $U(2) = U(4) = 2$, $U(3) = 3$ and $U(5) = 1$. The neighborhood system is the nearest neighbors, i.e. $N(1) = \{2\}$, $N(2) = \{1,3\}$, $N(3) = \{2,4\}$, $N(4) = \{3,5\}$ and $N(5) = \{4\}$. For the Metropolis algorithm, one can easily see that $V_1^M - V_2^M = 2$ and $h^M(x) = U(x)$. For the GS, the cycle structures are as follows: $S^1 = \{1,3,4\}$ with $U_G^1(1,3) = 5$, $U_G^1(3,1) = U_G^1(3,4) = 0$ and $U_G^1(4,3) = 3$. Here, again, a cycle is represented by its smallest number. Thus $v_1^G - v_2^G = 3 > v_1^M - v_2^M$. However, in this case, $h^G(2) = g^G(5) = 2$,

$h^G(3) = 5$ and $h^G(4) = 4$. Since

$$P(X_t^G=1) \sim 1-t^{-2/3} \geq 1-t^{1/2} \sim P(X_t^M=1),$$

we conclude that GS is better than Metropolis algorithm even though the difference of the first and second eigenvalues is larger for U_M than U_G.

References

[1] R. Azencott (1989), Simulated annealing, Seminaire Bourbaki, 1987–88, no. 697.

[2] O. Catoni (1988), Grandes déviations et décroissance de la temperature dans les algorithmes de recuit, C.R. Acad. Sci. Paris, t. 307, Serie I, 535–539.

[3] T.–S. Chiang, Y. Chow (1989), A limit theorem for a class of inhomogeneous Markov processes, Ann. Probab. 17, 1483–1502.

[4] T.–S. Chiang, Y. Chow (1990), Asymptotic behavior of eigenvalues and random updating schemes. Center for Stochastic Processes, University of North Carolina Technical Report No. 307.

[5] A. Frigessi, C.R. Hwang, S.J. Sheu, P. Stefano (1990), On the Gibbs sampler, the Metropolis algorithm and other single–site updating dynamics, preprint.

[6] S. Geman, D. Geman (1984), Stochastic relaxation, Gibbs distribution, and the Beysian restoration of images, IEEE Trans. Pattern Analysis and Machine Intelligence 6, 721–741.

[7] B. Hajek (1985), Cooling schedules for optimal annealing, Math. Oper. Res. 13, 311–329.

[8] C.–R. Hwang, S.–J. Sheu (1988), Singular perturbed Markov chains and exact behaviors of simulated annealing process, Technical Report, Institute of Math., Academia Sinica.

[9] P.M. van Laarhoven, E.H.L. Aarts (1987), Simulated Annealing: Theory and Applications, D. Reidel, Dordrecht.

[10] D. Mitra, F. Romeo, A. Sangiovanni–Vincentelli (1986), Convergence and finite time behavior of simulated annealing, Adv. Appl. Prob. 18, 747–771.

Some Limit Theorems on Simulated Annealing

T. S. Chiang, Y. Chow and J. Hsieh
Institute of Mathematics, Academia Sinica
Taipei, Taiwan

Many combinatorial optimization problems can be described as finding the global minimum of a certain function $U(\cdot)$ over a finite state space S, say, $\{1, 2, ..., N\}$. A commonly used approach is the gradient method. It takes "downhill" movements only. This guarantees a fast convergence. But it usually ends up with a local minimum, which might depend on the initial state.

Borrowing ideas from statistical physics, Cerny [6] and Kirkpatrick et.al. [27] proposed a new method: simulated annealing. It can be formulated as a time inhomogeneous Markov process $\{X(t): t \geq 0\}$ with state space S and transition rate $q_{ij}(t)$ from i to j at time t satisfying

$$(1)\quad q_{ij}(t) = \begin{cases} p_{ij}\exp\{-(U(j)-U(i))^+/T(t)\}, & \text{if } i\neq j\in N(i), \text{ the neighborhood of } i, \\ 0 & \text{, if } i \neq j \text{ and } j \notin N(i), \\ 1_{\{\text{discrete time}\}} - \sum_{k\neq i} q_{ik}(t), & \text{if } j = i, \end{cases}$$

where $T(t)$ is the "temperature" function converging to 0 as time $t \to \infty$.

The key feature of simulated annealing is that "uphill" movements are allowed but less and less favored as time goes by. By choosing a proper cooling schedule $T(\cdot)$ one hopes that whatever the initial state is,

$$(2)\qquad \lim_t P(X(t) \in \underline{S}) = 1, \text{ where } \underline{S} = \{i \in S: U(i) = \min U\}.$$

Without loss of generality we assume $\min U = 0$. Thus with a high probability $\underline{S}$ is achieved after running the algorithm a sufficiently long period.

In the study of image restoration problems Geman and Geman [20] applied the annealing algorithm to compute the maximum a posteriori (MAP) estimate of the original image given the degraded image. They proved (2) holds for $T(t) = c/\log(t+1)$ with c sufficiently large. Since then there has been a wide interest in the analysis and applications of simulated annealing.

The validity of (2) can be reasoned heuristically as follows. Suppose there is a reversible probability measure $\mu = \{\mu_i: i \in S\}$ on S. That is the detailed balance relation $\mu_i p_{ij} = \mu_j p_{ji}$ holds for all i, j. It is easy to check that $\mu_T = \{\mu_i \exp(-U(i)/T)/[\Sigma_j \mu_j \exp(-U(j)/T)]: i \in S\}$ is the limiting measure as $t \to \infty$, if the temperature $T(t) \equiv T$. Now let $T \downarrow 0$. We see that $\lim \mu_T$ is concentrated on

$\underline{S}$. By the diagonal argument, (2) should hold if $T(t)$ is well chosen.

Among the many contributors to the proof of (2) Hajek [23] introduced some new concepts and gave a necessary and sufficient condition for (2) in the discrete time case. In order to describe his results we need some terminologies.

State j is said reachable from state i (at level L) if there is a path $\mathscr{P} = [i_0 = i, i_1, i_2, \ldots, i_k = j]$ from i to j with $p_{i_s,i_{s+1}} > 0$ for all $0 \leq s < k$ (and $\max_s U(i_s) \leq L$). The number $e_{\mathscr{P}} = \max_{0 \leq s \leq k} U(i_s) - U(i)$ can be regarded as the "energy" required for i to reach j along path $\mathscr{P}$. For $i \notin \underline{S}$ the depth $d(i) = \min e_{\mathscr{P}}$, where the minimum is taken over all paths from i to some state j with $U(j) < U(i)$. It is the minimum "energy" needed for i to reach a lower state. For example, $d(i) = 0$ if i is not a local nor global minimum. Define

$$d_H = \max_{i \notin \underline{S}} d(i).$$

Physically speaking, d_H is the escape energy allowing any non–global minimum state to reach some state in $\underline{S}$. The process $\{X(t)\}$ is irreducible if any two states can be reached from each other. It is said weakly reversible if j can be reached from i at level L whenever i can be reached from j at level L. In particular, the process is so if $p_{ij} > 0$ implies $p_{ji} > 0$.

Theorem 1. Let $\{X(t)\}$ be given in (1). Assume it is irreducible and weakly reversible. Let $\{T(t)\}$ be decreasing to 0. Then (2) holds iff $\Sigma \exp(-d_H/T(t)) = \infty$. In the special case $T(t) = c/\log(t+1)$, (2) holds iff $c \geq d_H$.

Some questions arise naturally. What is the convergence rate in (2)? What is the limit distribution on $\underline{S}$? The existence of such a distribution is apparent from the heuristic reasoning above. In fact, it was found in Geman and Geman [20] and Gidas [22] for the cases they considered.

By analyzing the Kolmogorov forward equations

$$dP(X(t) = i)/dt = \Sigma_j q_{ji}(t) \cdot P(X(t) = j), \quad 1 \leq i \leq N, \tag{3}$$

both questions were answered in Chiang and Chow [7]. In order to have a limit distribution on $\underline{S}$ independent of the initial state, states in $\underline{S}$ should "communicate" with each another. The escape energy d_H is not sufficient for this purpose. For $i \in \underline{S}$ let $d(i) = \max_{j \neq i, j \in \underline{S}} \{\min e_{\mathscr{P}}\}$, where the minimum is taken over all paths from i to j. It is the minimum energy for i to reach every other state in $\underline{S}$. In case $\underline{S} = \{i\}$, $d(i) = 0$. Define

$$d_V = \max_{i \in S} d(i).$$

This amount of energy is what we require. Note that $d_V \geq d_H$ and $d_V = d_H$ if $|\underline{S}|$

= 1. Constant d_V can be computed through Ventcel's W–graphs [32]. It was first introduced to simulated annealing by Gidas [22].

Besides its convergence to 0, some regularity assumptions on $\lambda(t) = \exp(-1/T(t))$ are needed for using differential equation techniques,

(A.1;E) $$\int_0^\infty (\lambda(t))^E dt \quad [\text{or } \Sigma_{t=0}^\infty (\lambda(t))^E] = \infty,$$

(A.2;E) $$\lambda'(t) \quad [\text{or } \lambda(t+1) - \lambda(t)] = o((\lambda(t))^{E+1}).$$

Theorem 2. Let $\{X(t): t \geq 0\}$ be as in Theorem 1. Assume (A.1; d_V) and (A.2; d_H) hold. Then there exist positive constants β_i such that

(4) $$\lim_t P(X(t) = i)/(\lambda(t))^{U(i)} = \beta_i \quad \text{for } i \in S.$$

In particular, $\{\beta_i: i \in \underline{S}\}$ is the limit distribution on $\underline{S}$ and $\lim_t P(X(t) \notin \underline{S}) /(\lambda(t))^b$ exists, where $b = \min_{i \notin \underline{S}} U(i)$.

The linear differential system (3) is singular. The asymptotic behavior of $P(X(t) = i)$ cannot be obtained, as expected, from analyzing the process with transition rate $(q_{ij}(\infty))$, for which all local and global minima are absorbing states. Through a series of merging and boosting procedures, the estimate of $P(X(t) = i)$ was improved successively from $O(1)$, $O(\lambda(t))$, ..., to the desired $O(\lambda^{U(i)}(t))$. Such a method, called Ventcel's cycle method, becomes clearer when the weak reversible condition is removed and the exponent $(U(j)-U(i))^+$ in (1) is replaced by a general "cost" $U(i,j) \geq 0$. See Chiang and Chow [8]. Note that in this case $\underline{S}$ is not clear in advance. The constants d_H and d_V are redefined and a height function $h(\cdot)$ is introduced to replace the exponent $U(i)$ in (4). Then $\underline{S} = \{i: h(i)=0\}$ by definition.

The proof of (4) is very technical and complicated. But the idea behind is simple. Two examples are provided at the end to help readers understand how it can be explained in terms of balance of mass flows among states in S.

Recently Hwang and Sheu [26] gave a probabilistic proof to (4). It was based on Ventcel's theory of small perturbations [15].

Let $\tau = \inf\{t \geq 0: X(t) \in \underline{S}\}$. By (2) $P(\tau < \infty) = 1$. From the practical viewpoint it is interesting to know $E\tau = \int_0^\infty P(\tau > t)dt$. Let $F_i(t) = P(X(t) = i, \tau > t)$. Since $P(\tau > t) = \Sigma_{i \notin \underline{S}} F_i(t)$, it suffices to study $F_i(t)$. Similar to (3), we have the following differential systems,

(5) $$F_i'(t) = \Sigma_{j \notin \underline{S}} q_{ji}(t) F_i(t), \quad \text{for } i \notin \underline{S}.$$

Regarding $\underline{S}$ as an absorbing state, the system $\{F_i: i \notin \underline{S}\}$ is no longer conservative. That is $\Sigma_{i \notin \underline{S}} F_i(t) \neq 1$. This causes some difficulties unencountered before. Firstly, all the estimates in the proof of (4) are of polynomial orders $O(\lambda^n(t))$. From the experiences in time homogeneous Markov processes we expect an exponential type estimate for F_i. Secondly, the cycle method usually breaks down at certain stage where no new cycle can be formed. States in a cycle have their probabilities well proportioned. They can thus be merged and treated as a single state in the next stage. In the previous case new cycle always exists till all states are merged into one state. There the ratios among $P(X(t) = i)$ are known. Then (4) follows from the mass conservation formula $\Sigma P(X(t) = i) = 1$.

The first difficulty above is a technical one and can be overcome rather easily. The second difficulty is more intricate. Under conditions similar to (A.1;E) and (A.2;E), it is not hard to show

Theorem 3. $\lim_t [\log P(\tau > t)]/[\int_0^t \lambda^M(s)ds]$ exists for some $M \geq 0$.

In order to get ratio limit laws like (4) an additional condition, called Perron–Frobenius property, is assumed. It allows us to locate a dominant state, say i_0, when the cycle method can no longer be applied. Then the ratio laws can be obtained [9].

Theorem 4. There are constants $\beta_i > 0$ and $h(i) \geq 0$ such that

$$\lim_t F_i(t)/[\lambda^{h(i)}(t)F_{i_0}(t)] = \beta_i, \quad \text{for } i \notin \underline{S}. \tag{6}$$

Some remarks are necessary here. The PF property above, and thus (6), depend on how positive $\{p_{ij}\}$ are. Formula (4) for the conservative case depends only on the signs of $\{p_{ij}\}$. If the PF property fails, something strange may happen. See example 3.

Roughly speaking, Theorem 2 means states outside $\underline{S}$ can be ignored and the process can be regarded as if it is time homogeneous on $\underline{S}$. In this way, it is natural to expect some time homogeneous results still hold. This leads us to study [12] the occupation time $T_i(t) = \int_0^t 1_{\{X(s)=i\}}ds$ (or $\Sigma_{s=0}^t 1_{\{X(s)=i\}}$) up to t at state i.

Theorem 5. Besides $(A.1;d_V)$ and $(A.2;d_V)$, assume there are constants $u \leq v < 1/d_V$ such that $t^{-v} \leq \lambda(t) \leq t^{-u}$ for t large. Then w.p.1

$$\lim T_i(t)/t = \begin{cases} 0 & \text{for } i \notin \underline{S}, \\ \beta_i & \text{for } i \in \underline{S}, \end{cases} \tag{7}$$

where β_i is given in (4).

Similar results have been obtained by Gantert [18] and Winkler [33]. They seem motivated by studying the ergodic properties of the process:

$$\lim \Sigma_{s=1}^t f(X_s)/t = \int f d\mu_\infty ,$$

which is equivalent to (7) due to the finiteness of S.

A fundamental issue in simulated annealing is the study of finite time behaviors [19]. From the applications viewpoint it is important to know how to choose an optimal cooling schedule if one intends to run the algorithm for a predetermined period α of time. Take $Y_\alpha = \min_{0 \le s \le \alpha} U(X(s))$ as a criterion. It is unknown how to find $\min \Sigma_{i=1}^N E(Y_\alpha | X(0) = i)/N$, where the minimum is taken over all temperature functions $T(\cdot)$ on $[0,\alpha]$. D. Geman mentioned that it is unknown even if the minimum is taken over constant temperature functions. Catoni [5] has recently obtained some remarkable results conconcerning optimal cooling schedules.

Finally we would like to remark that the updating scheme in (1) is usually called the Metropolis sampler. Another commonly used scheme is the Gibbs sampler. Furthermore, updating sites can be picked up randomly or in a systematic way. Comparison of different schemes is an important topic in statistical physics. Readers are referred to Sokal [30] for a recent survey and [2, 4, 11, 16, 17] for some recent progresses.

The following examples demonstrate Theorems 2 and 4.

Examples 1. $S = \{1, 2, 3\}$. All the finite-valued $U(i,j)$ are shown in Figure 1. A boldface arrow there means a cost $U(i,j) = V(i) \equiv \min_k U(i,k)$. It is clear from (1) that a boldface arrow is more likely used to reach out. States 1, 2 and 3 form a cycle. Because they can reach each other via boldface arrows only. That plain arrows can be ignored will become evident in the following argument.

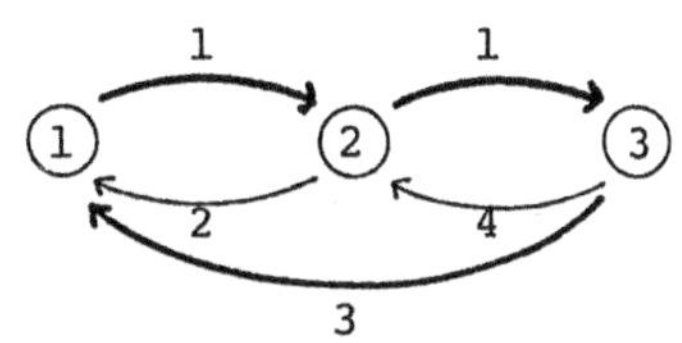

Fig.1

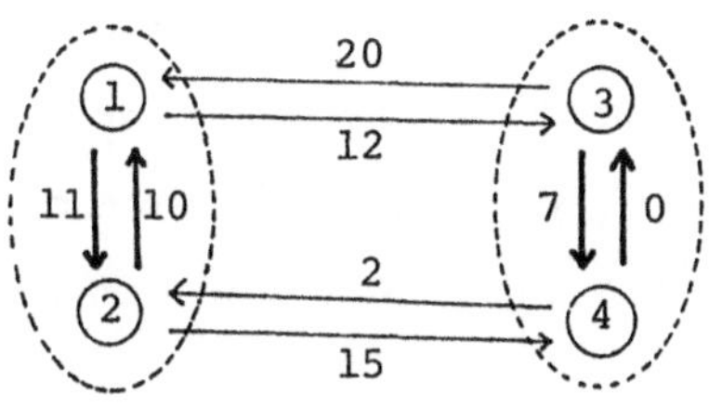

Fig.2

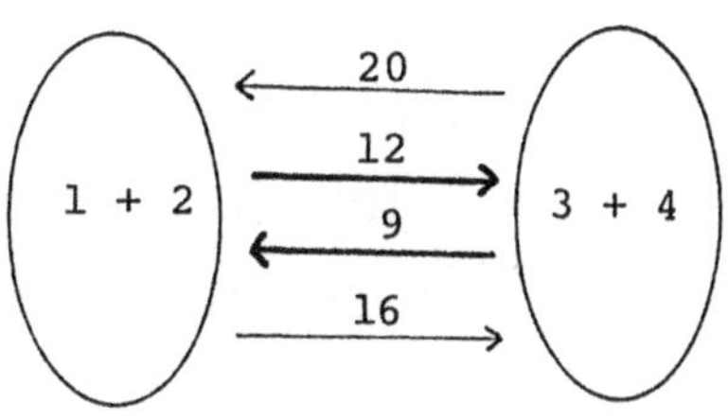

Fig.3

Since all $q_{ij}(t)$ are powers of $\lambda(t)$ we naturally guess $P(X(t) = i) \approx \beta_i \lambda^{a_i}(t)$. Though the system is time inhomogeneous, we hope some kind of equilibrium still holds. That is, mass flowing into state i should balance mass flowing out of i.

Taking $i = 1, 2, 3$, we get

(8) $$\beta_3\lambda^{a_3}(t)\cdot p_{31}\lambda^3(t) \approx \beta_1\lambda^{a_1}(t)\cdot p_{12}\lambda(t) \approx \beta_2\lambda^{a_2}(t)\cdot p_{21}\lambda(t).$$

Comparing order of λ in (8) we have $a_3+3 = a_1+1 = a_2+1$. By conservation of mass $\Sigma\beta_i\lambda^{a_i}(t) \approx 1$. Thus $\min a_i = 0$ and then $a_3 = 0$, $a_1=a_2=3-1=2$. Hence $\underline{S} = \{3\}$. Similarly it can be shown that $\beta_3 = 1$, $\beta_1 = p_{31}/p_{12}$ and $\beta_2 = p_{31}/p_{21}$.

Example 2. $S = \{1, 2, 3, 4\}$. As before, a boldface arrow in Figure 2 means a cost with $U(i,j) = V(i)$. By the same argument above, states 1, 2 and states 3, 4 can be merged respectively with

(9) $$P(X(t) = 2) \approx \lambda(t)\cdot P(X(t) = 1), \quad P(X(t) = 4) \approx \lambda^7(t)\cdot P(X(t) = 3).$$

For brevity we consider here only the order estimates and omit the limiting constants β_i. Those four plain arrows, which played no role so far, are now used to define the costs between the newly formed states. See Figure 3. In other words, they come into stage only after equilibrium relations (9) have been reached. Then states (1+2) and (3+4) can be merged with

(10) $$P(X(t) = 3) + P(X(t) = 4) \approx \lambda^3(t)\cdot[P(X(t) = 1) + P(X(t) = 2)].$$

Since $\Sigma P(X(t) = i) = 1$, it follows from (9) and (10) that $P(X(t) = 1) \approx 1$, $P(X(t) = 2) \approx \lambda(t)$, $P(X(t) = 3) \approx \lambda^7(t)$ and $P(X(t) = 4) \approx \lambda^{10}(t)$.

Example 3. Same S as in example 1. Regarding $\underline{S} = \{3\}$ as an absorbing state, we delete in Figure 1 the arrows outgoing from state 3. Since no more cycle can be found the constant M in Theorem 3 is $\max V(i) = 1$. Let $F_i(t) = P(X(t) = i, \tau > t)$ for $i = 1, 2$. The PF property requires $p_{12} \neq p_{23}$.

Case 1. $p_{12} > p_{23}$. Write $F_1(t)/F_2(t) = [F_1(t)\exp\int_0^t p_{12}\lambda]/[F_2(t)\exp\int_0^t p_{12}\lambda]$. By using (5) and L'Hospital's rule it is not hard to show first $F_1(t) = O(F_2(t))$ and then $F_1(t) \approx p_{21}(p_{12}-p_{23})^{-1}\lambda(t)F_2(t)$.

Case 2. $p_{12} < p_{23}$. $F_2(t) = p_{12}(p_{23}-p_{12})^{-1}F_1(t)$ by a similar argument.

Case 3. $p_{12} = p_{23}$. The PF property fails. In this simple case, however, we can show $F_1(t) \approx c\lambda^{1/2}(t)F_2(t)$, $c = (p_{21}/p_{12})^{1/2}$, by using another common factor $\exp[\int_0^t p_{12}\lambda + c\lambda^{3/2}]$ before applying the L'Hospital's rule. Note that fractional power of λ appears.

References

1. E. Aarts and J. Korst, Simulated Annealing and Boltzmann Machines, Wiley, New York, 1989.
2. Y. Amit and U. Grenander, Compare sweeping strategies for stochastic relaxation, Preprint, Brown Univ, 1989.

3. R. Azencott, Simulated Annealing, Seminaire Bourbaki, 1987–88, no. 697.
4. P. Barone and A. Frigessi, Improving stochastic relaxation for Gaussian random fields, Preprint, IAC, Roma, 1988.
5. O. Catoni, Applications of sharp large deviations estimates to optimal cooling schedules, Preprint, Ecole Normale Supérieure, 1990.
6. V. Cerny, Thermodynamical approach to the traveling salesman problem: an efficient simulation algorithm, J. Opt. Theory Appl. 45(1985), 41–51.
7. T. S. Chiang and Y. Chow, On the convergence rate of annealing processes, SIAM J. Control Optim. 26(1988), 1455–1470.
8. T. S. Chiang and Y. Chow, A limit theorem for a class of inhomogeneous Markov processes, Ann. Probab. 17(1989), 1483–1502.
9. T. S. Chiang and Y. Chow, The asymptotic behavior of simulated annealing process with absorption, Preprint, 1989.
10. T. S. Chiang, C. R. Hwang and S. J. Sheu, Diffusion for global optimization in R^n, SIAM J. Control Optim. 25(1987), 737–753.
11. T. S. Chiang and Y. Chow, Asymptotic behavior of eigenvalues and random updating schemes, Preprint, 1990.
12. Y. Chow and J. Hsieh, On occupation times of annealing processes, Preprint, 1990.
13. D. P. Connors and P. R. Kumar, Simulated annealing type Markov chains and their order balance equations, SIAM J. Control Optim. 27(1989), 1440–1461.
14. L. Davis, Genetic Algorithm and Simulated Annealing, Pitman. London, 1987.
15. M. I. Freidlin and A. D. Wentzell, Random Perturbations of Dynamical Systems, Springer, Berlin, 1984.
16 A. Frigessi, C.R. Hwang, S.J. Sheu, P. Stefano, On the Gibbs sampler, the Metropolis algorithm and other single–site updating dynamics, Preprint, 1990.
17. A. Frigessi, C.R. Hwang and L. Younis, Optimal spectral structure of reversible stochastic matrices, Monte Carlo methods and the simulation of Markov random fields, Preprint, 1990.
18. N. Gantert, Laws of large numbers for the annealing algorithm, Preprint, Universitat Bonn, 1989.
19. D. Geman, Random fields and inverse problems in imaging, in Lecture Notes in Mathematics 1427, Springer–Verlag, 1990.
20. S. Geman and D. Geman, Stochastic relaxation, Gibbs distribution and the Bayesian restoration of images, IEEE PAMI 6(1984), 721–741.
21. S. Geman and C. R. Hwang, Diffusions for global optimization, SIAM J. Control Optim. 24(1986), 1031–1043.
22. B. Gidas, Global optimization via the Langevin equation, Proc. 24th IEEE Conf. Decision and Control, Fort Lauderdale, Florida, 1985, 774–778.
23. B. Hajek, Cooling schedules for optimal annealing, Math. Oper. Res. 13(1988), 311–329.
24. R. Holley and D. Stroock , Simulated annealing via Sobolev inequalities, Commun. Math. Phys. 115(1988), 553–569.
25. C.R. Hwang and S.J. Sheu, Large time behaviors of perturbed diffusion Markov processes with applications III: Simulated annealing, Preprint, 1986.
26. C.R. Hwang and S.J. Sheu, Singular perturbed Markov chains and exact behaviors of simulated annealing process, Preprint, 1988.
27. S. Kirkpatrick, C. Gebatt and M. Vecchi, Optimizations by simulated annealing, Science 220(1983), 671–680.
28. P. J. M. van Laarhoven and E. Aarts, Simulated Annealing: Theory and Applications, Reidel, Dordrecht, 1987.
29. D. Mitra, F. Romeo and A. Sangiovanni–Vincentelli, Convergence and finite time behavior of simulated annealing, Adv. Appl. Prob. 18(1986), 747–771.
30. A. D. Sokal, Monte Carlo Methods in Statistical Mechanics: Foundations and New Algorithms, Preprint, 1989.
31. J. N. Tsitsiklis, Markov chains with rare transitions and simulated annealing, Math. Oper. Res. 14(1989), 70–90.
32. A. D. Ventcel, On the asymptotics of eigenvalues of matrices with elements of order $\exp(-V_{ij}/2\epsilon^2)$, Soviet Math. Dokl. 13 (1972), 65–68.
33. G. Winkler, An ergodic L^2–theorem for simulated annealing in Bayesian image reconstruction, J. Appl. Prob. 27(1990), 779–791.

Statistical analysis of Markov random fields

using large deviation estimates

F. COMETS[1]

ABSTRACT

We discuss a probabilistic large deviation estimate and some applications to parametric estimation for Markov random fields. The major interest is that this estimate holds independently of phase transition. It yields general consistency results, covering for instance both maximum likehood estimator and pseudo-likelihood estimator for complete observations ; we also present other applications.

I. INTRODUCTION

Markov random fields (abrev. MRF) provide convenient statistical models for spatial data, since important features of the dependence structure can be captured in the energy function. D. et S. Geman noticed that, using a MRF as prior distribution for the "ideal" image together with a degradation model, one recovers the traditional reconstruction criteria used in non stochastic image processing methods ; see [GEM] for a survey. The parameters of the MRF correspond to smoothing parameters and the interest of stochastic methods lies in estimation of these parameters from the observations, instead of an arbitrary choice in the non stochastic set-up.

This paper is concerned with (asymptotical) statistical quality of estimators, not with computational aspects. In order to check if a given estimator is consistent, we will use a large deviation estimate (given in § 3) which describes the behaviour of samples from a MRF ; this approach works even in the case of long-range dependence and of translation invariance breaking. In particular, we give in § 4 a new consistency criterium for maximum objective estimators. In § 5, we

1 URA CNRS 756 et 1321

Address: Centre de Mathématiques Appliquées
Ecole Polytechnique
91128 PALAISEAU Cedex

will discuss other applications of this type of estimate (partial observations, optimality theory for estimators, coding theory for MRF sources).

II. MARKOV RANDOM FIELDS ON $\mathbb{Z}^2$

Let X_i, $i \in \mathbb{Z}^2$ be the pixel variables, with values in some space Ω_0 that we will assume to be finite[2] here for the sake of simplicity. We let $\Omega = \Omega_0^{\mathbb{Z}^2}$ be the set of all pixel configurations $x = (x_i)_{i \in \mathbb{Z}^2}$.

The dependence structure is expressed by an interaction function $\Phi(x_B)$, with values in $\mathbb{R}^p$ and depending only[3] on the restriction x_B of the configuration x to some fixed ball B centered at the origin ; here, p denotes the dimensionality of the parameters vector θ which ranges over some subset Ⓗ of $\mathbb{R}^p$. Given Φ and a window $\Lambda \subset \mathbb{Z}^2$, one defines the energy of the pixel configuration inside Λ as the sum of self-interaction and of the interaction with the boundary configuration (outside Λ)

$$\mathcal{U}_\Lambda(x) = \sum_{i : (i+B) \cap \Lambda \neq \varnothing} \Phi(\tau^i x)$$

where τ^i is the space translation of vector i ($\tau^i : \Omega \to \Omega$ with $(\tau^i x)_j = x_{i+j}$, $\forall j \in \mathbb{Z}^2$).

Let $\mathcal{P}(\Omega)$ be the set of all probability distributions on Ω. We introduce an exponential family of MRF in the following

<u>*DEFINITION*</u>

An element Π_θ of $\mathcal{P}(\Omega)$ is called a <u>MRF</u> (with parameter θ) if, for all finite subset Λ in $\mathbb{Z}^2$, the conditional distribution inside Λ given the configuration[4] ${}_\Lambda x$ outside Λ is equal to

$$\Pi_\theta\left(X_\Lambda = x_\Lambda \,/\, {}_\Lambda X = {}_\Lambda x\right) = \left[Z_\Lambda(\theta, {}_\Lambda x)\right]^{-1} \exp\left\{\theta . \mathcal{U}_\Lambda(x)\right\} \qquad (1)$$

<u>Example</u> (nearest neighbour Ising model).

$\Omega_0 = \{+1,-1\}$, $p = 2$, $\theta = (\beta, h)$, $\Phi(x) = \left[\frac{1}{2} x_0(x_{e_1} + x_{-e_1} + x_{e_2} + x_{-e_2}), x_0\right]$ with $e_1 = (1,0)$ and

2 The following results are true for very general spaces Ω_0 - polish spaces -, and in any dimension d - $i \in \mathbb{Z}^d$ - .

3 For general Ω_0, one must assume that Φ is a bounded continuous function of x_B.

4 *similar to $x_\Lambda = (x_i)_{i \in \Lambda}$, ${}_\Lambda x$ denotes the configuration outside Λ : ${}_\Lambda x = (x_i)_{i \notin \Lambda}$.*

$e_2 = (0,1)$. The energy is then

$$\theta.\mathcal{U}_\Lambda(x) = \beta \sum_{\langle i,j\rangle ;\, i,j \in \Lambda} x_i x_j + h \sum_{i \in \Lambda} x_i + \text{boundary terms}$$

where $\langle i,j\rangle$ denotes nearest neighbours.

A MRF possesses the spatial Markov property ; they are the extension of Markov chains with non-zero transition probabilities to higher dimension. The notion of Gibbs distribution is a slight generalization of MRF in the case of arbitrary large balls of interaction, and therefore it is not of practical interest here. The above definition, due to Dobrushin, Lanford and Ruelle, is concerned with a distribution on the whole lattice. To describe finite size images, one considers finite volume MRF on Λ, defined on the set of configurations x_Λ inside Λ by the formula (1) with some arbitrary choice of the boundary configuration ${}_\Lambda x$. MRF in the above definition are limit points of finite volume ones as $\Lambda \to \mathbb{Z}^2$ (thermodynamic limit), and therefore approximate the finite volume ones for large domains Λ of observation.

In our framework, the set $G(\theta)$ of MRF with parameter θ is non empty. Phase transition occurs if $G(\theta)$ contains more than one element. Then, some of the Π_θ are not ergodic (ergodicity occurs when any average in space converges to its expected value as $\Lambda \to \mathbb{Z}^2$) ; even worst, some of them may be non translation invariant. When using real data, we cannot ignore such a surprising behaviour.

III. PARAMETRIC ESTIMATION

We start with complete observations : we discuss the estimation of θ_0 from the observation of X_Λ, where X is a fixed sample from some $\Pi_{\theta_0} \in G(\theta_0)$[5]. Notice that there is no hope in estimating correctly Π_{θ_0} itself when $G(\theta_0)$ contains more than one element : since Π_{θ_0} is a convex combination of extremal elements in $G(\theta_0)$, drawing X may be seen as choosing first an extremal according to its weight in the convex combination, then sampling X from this extremal ; therefore, only the extremal may be correctly estimated, but not Π_{θ_0} itself.

Many classical estimators are obtained in maximizing with respect to θ some objective function. Popular examples are : maximum likelihood estimator (MLE)[6] :

5 There exist other ways to produce data [GID], which are covered too by the following results [COM].

$$\text{MLE} = \underset{\theta}{\text{Arg max}} \frac{1}{|\Lambda|} \text{Log}\, \Pi_\theta \left\{X_\Lambda / {}_\Lambda x\right\} \tag{2}$$

Besag's maximum pseudo-likelihood estimator (PMLE)[6] :

$$\text{PMLE} = \underset{\theta}{\text{Arg max}} \frac{1}{|\Lambda|} \sum_{i \in \Lambda} \text{Log}\, \Pi_\theta \left\{X_i / {}_i X\right\} \tag{3}$$

as well as least square estimators [DEL], the variational estimator [ALG] and the logistic estimator. These objective functions have a common feature : they are nearly space averages of some test function f, i.e., they are of the form

$$\frac{1}{|\Lambda|} \sum_{i \in \Lambda} f\left(\tau^i X\right) + \varepsilon_\Lambda$$

where the remainder term ε_Λ (coming from boundary effects) tends to 0. In (2), we have $f(x) = \theta.\Phi(x) - p(\theta)$ with $p(\theta) = \lim_{\Lambda \to Z^2} \frac{1}{|\Lambda|} \text{Log}\, Z_\Lambda(\theta, {}_\Lambda x)$, while $f(x) = \text{Log}\, \Pi_\theta(x_0 / {}_0 x)$ in (3). The natural tool to study space averages is the empirical field $R_{\Lambda,X}$ based on the observations, given by

$$R_{\Lambda,X} = \frac{1}{|\Lambda|} \sum_{i \in \Lambda} \delta_{\tau^i X} \ .$$

This is a random element in $\mathcal{P}(\Omega)$ which satisfies

$$\frac{1}{|\Lambda|} \sum_{i \in \Lambda} f(\tau^i X) = \int_\Omega f(x)\, R_{\Lambda,X}(dx) \ .$$

It is a convenient to use the following large deviation upper bound. Let $G_S(\theta)$ be the set of translation invariant elements in $G(\theta)$.

$\forall\, K : \mathcal{P}(\Omega) \to \mathbb{R}$ continuous, $\forall\, \varepsilon > 0, \exists\, \delta > 0, \exists\, c > 0$:

$$\Pi_{\theta_0} \left\{ \text{dist}\left[K(R_{\Lambda,X}) ; \left\{K(R) ; R \in G_S(\theta_0)\right\}\right] \geq \varepsilon \right\} \leq C \exp\left\{-|\Lambda|\delta\right\} \tag{4}$$

The exponential rate δ is explicit in terms of relative entropy of stationary fields and of ε (see [COM], [FOR], [OLL]). This estimate states that, with probability going to 1 exponentially fast, a

6 One has to specify a boundary configuration.

space average is not worse behaved than its expected value for the worst translation invariant MRF. It should be pointed out that the estimate holds for any $\Pi_{\theta_0} \in G(\theta_0)$, independently of phase transition.

How to use : for heuristics, assume that we are given a space average (like a moment estimator), $F_\Lambda(X_\Lambda) = \frac{1}{|\Lambda|} \sum_{i \in \Lambda} f(\tau^i X)$, to estimate some function of the parameter. If there is no phase transition at θ_0, Π_{θ_0} is ergodic and

$$F_\Lambda(X_\Lambda) \xrightarrow[\Lambda \to Z^2]{a.s.} F(\theta_0) = \int f \, d\Pi_{\theta_0} \tag{5}$$

Let's rephrase this in a slightly weaker form

$$\Pi_{\theta_0}\{\text{dist}[F_\Lambda(X_\Lambda); F(\theta_0)] \geq \varepsilon\} \longrightarrow 0 \tag{6}$$

In the case of phase transition, the ergodic theorem (5) fails in general, but (6) remains valid since (4) does ; now F is a multivalued function, and $F(\theta_0)$ is the set $\{\mathbb{E}^R f ; R \in G_S(\theta_0)\}$[7]. With a clever choice of f, there will exist some continuous function F^{-1} such that $F^{-1}(F(\theta)) = \theta$ for all $\theta \in (H)$. Then, $F^{-1}(F_\Lambda)$ is a consistent estimator of θ.

IV. CONSISTENCY CRITERIA

Let us consider first concave objective functions $k_\Lambda(\theta;X_\Lambda)$. Assume that there exists a continuous function $K : (H) \times \mathcal{P}(\Omega) \to \mathbb{R}$ with

A) $k_\Lambda(\theta;x_\Lambda) = K(\theta;R_{\Lambda,x}) + \varepsilon_\Lambda$, $\lim_{\Lambda \to Z^2} \varepsilon_\Lambda = 0$ uniformly in x and for θ in bounded sets.

B) $\forall \theta, \forall \Pi_\theta \in G_S(\theta)$, θ is the unique maximizer of $K(.;\Pi_\theta)$

C) $k_\Lambda(\theta;x_\Lambda)$ is concave in θ for all x.

---

7 In the example, $F(\theta)$ is an interval with endpoints $\int f d\Pi^+_{\theta_0}, \int f d\Pi^-_{\theta_0}$ for $\beta > \beta_c = 0{,}44 \ldots$ and $h = 0$.

<u>*THEOREM 1*</u> *: (th. 3.1 in* [COM]*). Assume* A,B,C. *Then, any measurable maximizer* $\widehat{\theta}_\Lambda$ *of* $k_\Lambda(\theta,X_\Lambda)$ *is a.s. consistent.*

From the estimate (4), it is clear that, for this consistency result as well as the one below, the gross error probability $\Pi_{\theta_0}\left\{\left|\widehat{\theta}_\Lambda - \theta_0\right| > \varepsilon\right\}$ goes to zero exponentially fast in $|\Lambda|$. In the case of complete observations, both MLE and PMLE satisfy these conditions, provided that

$$\theta \neq \theta' \Rightarrow G(\theta) \cap G(\theta') = \varnothing \tag{7}$$

This natural identifiability condition amounts to the uniqueness requirement in the assumption B. For the proofs, see [COM], th. 2.1, 3.1. We give now a new result.

<u>*THEOREM 2*</u> *: Assume conditions* A,B. *If* Ⓗ *is compact, any maximizer* $\widehat{\theta}_\Lambda$ *of* $k_\Lambda(\theta,X_\Lambda)$ *is a.s. consistent.*

□ Let $\delta,\varepsilon > 0$. From the definition of $\widehat{\theta}_\Lambda$, we have :

$$\left\{\left|\widehat{\theta}_\Lambda - \theta_0\right| \geq \delta\right\} \subset \left\{\sup_{|\theta - \theta_0| > \delta} k_\Lambda(\theta,X_\Lambda) \geq k_\Lambda(\theta_0,X_\Lambda)\right\}$$
$$\subset \left\{\sup_{|\theta - \theta_0| \geq \delta} K(\theta,R_{\Lambda,x}) - K(\theta_0,R_{\Lambda,x}) \geq -2\varepsilon\right\}.$$

using the assumption A) for large Λ. Since

$\{\theta \in$ Ⓗ ; $|\theta - \theta_0| \geq \delta\}$ is compact, and since K is continuous, the function $R \to \sup_{|\theta - \theta_0| \geq \delta} K(\theta,R) - K(\theta_0,R)$ is continuous.

It follows from B) that $K(\theta,R) - K(\theta_0,R) < 0$ for all $R \in G_S(\theta_0)$ and all θ with $|\theta - \theta_0| \geq \delta$. This, together with compactness of $G_S(\theta_0)$ implies that

$$-\delta_0 = \sup\{K(\theta,R) - K(\theta_0,R) \,;\, R \in G_S(\theta_0) ,\, |\theta - \theta_0| \geq \delta\} < 0.$$

Choosing $\varepsilon = \delta_0/4$, we obtain for large Λ

$$\Pi_{\theta_0}\left\{\left|\widehat{\theta}_\Lambda - \theta_0\right| > \delta\right\} \leq \Pi_{\theta_0}\left\{\sup_{|\theta - \theta_0| \geq \delta} K(\theta,R_{\Lambda,x}) - K(\theta_0,R_{\Lambda,x}) > -\frac{\delta_0}{2}\right\}$$
$$\leq C\, e^{-\delta_1|\Lambda|} \text{ for some } \delta_1 > 0$$

using the estimate (4) and the definition of δ_0. ■

This result does not require convexity in θ , hence it is useful for imperfect observations and for moment estimators : in the heuristics at the end of the previous section, we would maximize the (non concave) function - dist $(F_\Lambda(X_\Lambda) ; F(\theta))$. This yields exponential rate for convergence for the moment estimator studied in [FPI] when β is restricted to a bounded set.

To end this section, let's consider like in § 3 of [KUN], the case when the true distribution P of X does not belong to the model - i.e., $P \notin \bigcup_\theta G(\theta)$ - . Indeed, one does not really pretend that the true image is a realization of a MRF, but that some important features can be faithfully captured by a MRF model [AZE]. Let's assume for simplicity that P is ergodic. Then, the MLE converges to the unique value θ_1 of θ such that $\mathbb{E}^P \Phi \in \left\{\mathbb{E}^{\Pi_\theta} \Phi ; \Pi_\theta \in G_S(\theta)\right\}$; hence, the estimated model will be, in the limit, the unique MRF in the model with the same expected energy as P. On the other hand, the PMLE concentrates on the value θ_2 (unique in general) maximizing $\mathbb{E}^P \operatorname{Log} \Pi_\theta(X_0/_0X) = C(P) - \mathbb{E}^P\,\mathbb{E}^P \left\{\operatorname{Log} \frac{P(X_0/_0X)}{\Pi_\theta(X_0/_0X)} \Big/ {}_0X\right\}$: the PMLE fits the true conditional distribution with the markovian model. In general, $\theta_1 \neq \theta_2$, and for large Λ, MLE and PMLE will differ significatively. Therefore, the estimator $\hat{\theta}$ should be chosen in a coherent manner with the forthcoming tasks ; for instance, if these tasks involve simulation of whole configurations with $\Pi_{\hat{\theta}}$, MLE should be preferable.

V. FURTHER APPLICATIONS

We report in this section other applications of the estimate (4) in statistical analysis of MRF ; they require in general the expression of δ in (4) in terms of relative entropy.

5.a. Partial observations (noisy data, bluring, ...)

In image analysis, we do not observe directly the "original image" X, but some data Y obtained from X through some degradation mechanism, that we will assume to be known for simplicity.

The MLE involves lengthy computations via stochastic E.M. algorithm, but it may be implemented together with usual restoration tasks [YOU].

In [CGI], we study MLE for the model $Y_i = f[(M * X)_i , \eta_i]$ with a bluring matrix M, a non linear degradation f and a white noise η, under mild assumptions. The joint field (X,Y) is a MRF, and then the observation Y satisfies an estimate like (4), though it is not itself a MRF. We show consistency of MLE for compact (H) combining this estimate and (many) other steps.

A major question is to find (consistent) estimators, which are easier to compute than MLE. In some particular cases, moment estimators[8] have been proposed : one is based on analytic computations of this moment in the case of Ising model corrupted by a multiplicative Bernoulli white noise [FPI], another is based on the knowledge of the degradation (how the observation reflects the smoothness of the prior) [GMC] ; in this last case, the moment are computed by Monte-Carlo simulation for a few values of the parameter, and then interpolated between these values.

In the case of finite state space Ω_0, an estimator reminiscent of PMLE is proposed in [PIE], which amounts to estimating - in a "non parametric" way[8] - the marginal distribution of X itself. Derivations of more practical estimators are investigated in [VEI].

5.b. Optimality

If there is no phase transition at θ_0, MLE is in general asymptotically normal and efficient ([GID], [GUY], [JAN], [KUN], [PIC]). But when phase transition occurs, very little is known about second order asymptotics for the MLE, and Fisher information does not exist as $\Lambda \to \mathbb{Z}^2$. Therefore, one cannot follow Fisher's program, and it is tempting to define optimality in a somewhat relaxed sense. In the spirit of estimate (4), one will look at

$$c(\theta_0,\varepsilon) = - \lim_{\Lambda \to \mathbb{Z}^2} \frac{1}{|\Lambda|} \operatorname{Log} \Pi_{\theta_0} \left\{ \left| \hat{\theta}_\Lambda - \theta_0 \right| > \varepsilon \right\}$$

if the limit exists, and one will look for consistent estimators $\hat{\theta}_\Lambda$ depending on X_Λ which maximizes $c(\theta_0,\varepsilon)$ (optimal estimators). This approach to optimality in the i.i.d. case is well-known, and is related to the work of Bahadur. It is shown in [COM], sect. 4, that for MRF, the MLE is optimal in the above sense. A theoretical analysis of hypothesis testing can also be developped from the estimate (4) following Bahadur's ideas in the independent case [BAH] ; optimility amounts in optimizing the exact slope of test statistics.

On the other hand, a recent progress indicates that estimators other than MLE might be nicely behaved even at phase transitions : for the Ising model, Guyon and Künsch prove in this volume that these estimators are asymptotically normal provided that the underlying distribution is stationary and ergodic.

---

8 The difficulty is that the marginal distribution of a MRF are not explicit in term of θ, except in the gaussian case.

5.c. Coding Markov images

In information theory (Shannon's theory), the connection between large deviation estimates (like (4)) and coding for time-dependent signals has been extensively used. In [AMI], various fixed blocks coding schemes for signal from a MRF source are proposed, which asymptotically yields the best rate for coding ; they also prove that coding the signal with one-dimensional techniques leads to strictly worse rates.

REFERENCES

[ALG] M. ALMEIDA, B. GIDAS, *A variational method for estimating parameters for MRF from complete or incomplete data*, Preprint Brown Univ (1989).

[AMI] Y. AMIT, M. MILLER, *Large deviations for coding Markov chains and Gibbs random fields*, Preprint Brown Univ (1990).

[AZE] R. AZENCOTT, *Image analysis and Markov fields*, Proceed. congress ICIAM in Paris (1987).

[BAH] R.R. BAHADUR, *Some limit theorems in statistics*, SIAM, Philadelphia (1971).

[CGI] F. COMETS, B. GIDAS, *Parameter estimation for Gibbs distributions from partially observed data*, (1988). To appear in Ann. Appl. Prob.

[COM] F. COMETS, *On consistency of estimators for exponential families of Markov random field on the lattice* (1989). To appear in Ann. Stat.. For the proof of (4), see *Grandes déviations pour des mesures de Gibbs sur* $\mathbb{Z}^d$, C.R. Acad. Sci. Paris, Série 1, 303 n° 11, p.511-514 (1986).

[DEL] H. DERIN, H. ELLIOTT, *Modeling and segmentation of noisy and textured images using Gibbs random fields*, IEEE Trans. Pattern Anal. Machine Intell., vol. PAMI-9, p. 39-55 (1987).

[FOR] H. FOLLMER, S. OREY, *Large deviation for the empirical field of a Gibbs measure*, Ann. Prob. 16, p. 961-977 (1988).

[FPI] A. FRIGESSI, M. PICCIONI, *Parameter estimation for two-dimensional Ising fields corrupted by noise*, Stoch. Proc. and their Appl. 34, p. 297-311 (1990).

[GEM] D. GEMAN, Course in St-Flour 1988. Lect. Notes Math. 1427 (1989).

[GE2] D. GEMAN, S. GEMAN, *Stochastic relaxation, Gibbs distribrutions and the Bayesian restoration of images*, IEEE TRans. Pattern Anal. Machine Intell., vol. PAMI-6, p.

721-741 (1984).

[GMC] S. GEMAN, D.E. Mc CLURE, *Statistical methods for tomographic image reconstruction*, in Bull. ISI (Proc. 46th Session Int. Statistical Institute) vol. 52 (1987).

[GID] B. GIDAS, *Parametric estimation for Gibbs distributions, I : fully observed data*, to appear in "Markov random fields : theory and applications", Academic press.

[GK] X. GUYON, H. KÜNSCH, *Asymptotic comparison of estimators in the Ising model*, In this volume.

[GUY] X. GUYON, *Estimation d'un champ par pseudo-vraisemblance conditionnelle : étude asymptotique et application au cas Markovien*, in Spatial processes and spatial time series analysis, Proc. Franco-Belgian Stat. Meet., Ed. F. Droesbeke, Publ. Univ. St-Louis, Bruxelles (1987).

[JAN] M. JANZURA, *Statistical analysis of Gibbs random fields*, 10th Prague Conf. 1986 "Inform. th., Stat. decis. funct., random processes", Reidel Publishing comp., p. 429-438 (1988).

[KUN] H. KÜNSCH, *Thermodynamics and statistical analysis of Gaussian random fields*, Z. Wahrsch. Geb. 58, p. 407-421 (1981).

[OLL] S. OLLA, *Large deviation for Gibbs random fields*, Prob. th. Rel. Fields 77, p. 343-357 (1988).

[PIC] D. PICKARD, *Inference for discrete Markov field : the simplest non-trivial case*, JASA 82, p. 90-96 (1987).

[PIE] W. PIECZYNSKI, *Estimation of context in random fields*, J. Applied Stat., vol. 16 special issue "Stat. meth. in Image Anal.", n° 2, p.283-290 (1989).

[VEI] A. VEIJANEN, *An estimator for imperfectly observed Markov random fields*, preprint Univ. Helsinki (1990).

[YOU] L. YOUNES, *Parametric inference for imperfectly observed Gibbsian fields*, PTRF 82, n° 4; p. 625-645 (1989). See also "Estimation and Annealing for Gibbsian Fields", Ann. Inst. H. Poincaré 24, p. 269-294 (1988).

METROPOLIS METHODS, GAUSSIAN PROPOSALS AND ANTITHETIC VARIABLES

Peter J. Green and Xiao-liang Han

University of Bristol,
Department of Mathematics,
Bristol, BS8 1TW, UK.

ABSTRACT

We investigate various aspects of a class of dynamic Monte Carlo methods, that generalises the Metropolis algorithm and includes the Gibbs sampler as a special case. These can be used to estimate expectations of marginal distributions in stochastic systems. A distinction is drawn between speed of weak convergence and precision of estimation. For continuously distributed processes, a particular gaussian proposal distribution is suggested: this incorporates a parameter that may be varied to improve the performance of the sampling method, by adjusting the magnitude of an "antithetic" element introduced into the sampling. The suggestion is examined in detail in some experiments based on an image analysis problem.

Keywords: *autocorrelation time, convergence rate, dynamic Monte Carlo, Gibbs sampler, statistical image analysis.*

1. Introduction

Complex stochastic systems, large collections of random variables with non-trivial dependence structure, arise in probability modelling in many contexts. Examples include statistical mechanics, geographical epidemiology, pedigrees in genetics, statistical image analysis, and general multi-parameter Bayesian inference. In practice, distributions of variables in such systems are usually not amenable either to exact numerical calculation or to direct simulation. The dynamic Monte Carlo approach to the computation of probabilities and expectations in these systems is receiving much attention. The basic idea in such methods is to turn the given static problem into a dynamic one by constructing what is usually an artificial temporal stochastic process, that is known to converge weakly to the distribution of the original system, yet is easy to simulate. This temporal process is generally taken to be a time-homogeneous Markov chain, so that apart from the matter of checking the aperiodicity and irreducibility of the constructed process (and some additional regularity conditions when the state space is uncountable), analysis of the original problem is reduced to the construction of a Markov chain with a specified equilibrium distribution.

There will always be very many such chains, and the art of dynamic Monte Carlo is to choose one that uses computational resources effectively by striking the right balance between simplicity and speed. This paper is a contribution to the discussion of how to strike this balance. It is stimulated by our interest in a class of problems arising in a Bayesian approach to low-level image analysis, but in fact our observations and conclusions will be quite generally applicable.

After a brief survey of the statistical literature on dynamic Monte Carlo, the remainder of the paper is divided into three sections. In Section 2, we discuss how to define speed of convergence appropriately when estimating properties of the equilibrium distribution, and how to monitor and quantify convergence from a sample realisation. Section 3 contains a new proposal for a class of Markov chains with a given equilibrium, and an analysis of some properties of this class in certain simple situations. In Section 4, we present the results of some experiments making use of this class of chains in an idealised Bayesian image analysis context.

Dynamic Monte Carlo methods originated in computational physics research; the paper by Metropolis *et al.* (1953) is usually taken as starting the subject. An enormous literature has developed: the current state of the subject, with particular reference to applications in statistical mechanics and quantum field theory, is summarised in the excellent lecture notes of Sokal (1989). The monograph of Hammersley and Handscomb (1964) introduced the ideas to the statistical community, but only with the development of stochastic models for image analysis, and Geman and Geman's proposal (1984) to use the Gibbs sampler in image reconstruction, did interest really develop. The Metropolis algorithm and Gibbs sampler are described in Ripley (1987, p.113), and will be further discussed in Section 3. Many other applications have now been addressed by such approaches, including Monte Carlo testing, for example in the Rasch model, (Besag and Clifford, 1989), marginal distributions in Bayesian inference (Gelfand and Smith, 1990), and geographical epidemiology (Besag, York and Mollié, 1991).

There are connections with the development of simulated annealing as an approach to combinatorial optimisation, proposed by Kirkpatrick *et al.* (1983) as an analogue of annealing in physical systems, and also utilised by Geman and Geman (1984) in image reconstruction. Here a time-inhomogeneous Markov chain is constructed; instantaneously the transition mechanism has an equilibrium distribution that is a renormalised power of the distribution of interest. As the process evolves, this power is gradually increased: in the limit the distribution is concentrated on the value(s) of maximum probability, and if the power is increased sufficiently slowly, the process can be shown to converge to this maximum. We do not consider annealing in this paper.

2. Speed of convergence

We are concerned with a random vector $\mathbf{x}$ with components x_i indexed by $i \in S$, a finite set of sites or pixels. The set of possible values for $\mathbf{x}$ will be denoted by Ω which usually has the form C^S where C might be finite, countable, or an interval in $\mathbf{R}$ or $\mathbf{R}^d$ depending on context. For the most part, in this paper we will use notation appropriate to the countable case. In addition, in most applications there is a vector of observables $\mathbf{y}$. The object of interest is the distribution of $\mathbf{x}$ given $\mathbf{y}$: we denote its density with respect to an appropriate measure as $p(\mathbf{x}|\mathbf{y})$. In the context of image analysis, $\mathbf{y}$ represents an observed pixellated degraded digital image, and $\mathbf{x}$ an unobservable true image representing the "state of nature": study of the posterior distribution $p(\mathbf{x}|\mathbf{y})$ is our route to drawing inference about this truth. Note that in this and other examples, $\mathbf{x}$ and $\mathbf{y}$ are vectors of high dimensionality.

Our discussion is also applicable to the case where there are no observables: just regard $\mathbf{y}$ as null. We will use the symbol $\pi(\mathbf{x})$ to denote whichever of $p(\mathbf{x})$ or $p(\mathbf{x}|\mathbf{y})$ is of interest. The same computational ideas apply to both cases: thus Grenander's perceptive observation (1983, p.83) that "pattern analysis equals pattern synthesis".

Attention usually focusses on one or more functionals of the distribution $\pi(\mathbf{x})$: suppose that we wish to estimate the expectation $E_\pi(f)=\sum f(\mathbf{x})\pi(\mathbf{x})$. This is very general, for example $f(\mathbf{x})$ might be $\sum_{i\in A} x_i$ (total truth in a region $A\subset S$) or $I_{[x_k=0]}$ (to estimate the probability of a zero; $I_{[]}$ is the indicator function). If $V(\mathbf{x})$ is a sufficient statistic for a parameter β in the model $\pi(\mathbf{x})$, then $f(\mathbf{x})$ might be $V(\mathbf{x})$ or $I_{[V(\mathbf{x})\le t]}$, in order to construct procedures for inference about β.

Let P be a Markov transition function on the state space Ω, that is irreducible and aperiodic, and has π as equilibrium distribution, so that

$$\sum_{\mathbf{x}\in\Omega} \pi(\mathbf{x})P(\mathbf{x},\mathbf{x}') = \pi(\mathbf{x}')$$

for all $\mathbf{x}'\in\Omega$. Suppose we have a partial realisation $\{\mathbf{x}^{(t)}:t=0,1,2,\ldots,N\}$ from this Markov chain. Then our estimator of $E_\pi(f)$ will be the empirical average

$$\bar{f}_N = \frac{1}{N}\sum_{t=1}^{N} f(\mathbf{x}^{(t)}).$$

Motivated by the observation in Gelfand and Smith (1990) that Rao-Blackwellisation can be used to reduce mean squared error, in the case of replicated independent Monte Carlo runs, we are investigating the performance of modified estimators exploiting conditioning. For example, if $f(\mathbf{x})$ is actually $f_1(x_1)$, a function of a single component of $\mathbf{x}$, then $g(\mathbf{x})=E_\pi(f_1(x_1)|\mathbf{x}_{S\setminus 1})$ may sometimes be cheaply computed. It has the same expectation $E_\pi(f)=E_\pi(g)$, but $\bar{g}_N$ will have smaller mean squared error than $\bar{f}_N$. Since these have the same general form, the ensuing treatment continues to apply; we will not discuss this modification specifically here, but further details will be reported elsewhere.

How good are such estimators? This seems to depend on how fast the Markov chain $\{\mathbf{x}^{(t)}\}$ converges weakly to π. Under the irreducibility condition, P has only the single eigenvalue 1 on the unit circle, with the constant vectors as the corresponding right eigenvectors, so that the rate of convergence is given by R, the spectral radius of P acting on the orthogonal complement of the constant vectors, and we find

$$\sup_{\mathbf{x},\mathbf{x}'\in\Omega} |P^t(\mathbf{x},\mathbf{x}')-\pi(\mathbf{x}')| \sim cR^t. \qquad (1)$$

Then for any bounded, continuous function f,

$$|E(f(\mathbf{x}^{(t)}))-E_\pi(f)| \le c(f)R^t. \qquad (2)$$

Using R, we can define the exponential autocorrelation time $(-1/\log R)$: the number of steps of the Markov chain needed to reduce the "errors" $|P^t(\mathbf{x},\mathbf{x}')-\pi(\mathbf{x}')|$ by a factor of e asymptotically. Small R (small autocorrelation time) indicates rapid convergence, but will actually be a pessimistic measure for any particular f, for which the chain may achieve faster convergence of $E(f(\mathbf{x}^{(t)}))$.

But this discussion does not address performance of the estimator $\bar{f}_N$, obtained from the sample path of the process by integration over time, not over realisation. This estimator has bias and variance, whose asymptotic forms are:

$$E(\bar{f}_N)-E_\pi(f) = \frac{1}{N}\sum_{t=1}^{N}\{E(f(\mathbf{x}^{(t)}))-E_\pi(f)\}$$

$$\sim \frac{1}{N}\sum_{t=1}^{\infty}\{E(f(\mathbf{x}^{(t)}))-E_\pi(f)\}; \tag{3}$$

$$var(\bar{f}_N) = \frac{1}{N^2}\sum_{s=1}^{N}\sum_{t=1}^{N}cov(f(\mathbf{x}^{(s)}),f(\mathbf{x}^{(t)}))$$

$$\sim \frac{\sigma^2}{N}\sum_{t=-(N-1)}^{N-1}(1-\frac{|t|}{N})\rho_t(f)$$

$$\sim \frac{\sigma^2}{N}\sum_{t=-\infty}^{\infty}\rho_t(f) \tag{4}$$

where $\rho_t(f)$ is the autocorrelation function of the process $\{f(\mathbf{x}^{(t)})\}$, calculated under the equilibrium distribution π, and σ^2 is the equilibrium variance of $f(\mathbf{x})$. The asymptotic variance is a factor

$$\tau(f) = \sum_{t=-\infty}^{\infty}\rho_t(f)$$

times what would be obtained if independent random sampling of $\mathbf{x}$ from $\pi(\mathbf{x})$ could be achieved: we call $\tau(f)$ the integrated autocorrelation time (differing from Sokal's definition (1989) by a factor of 2). From (2), (3) and (4), it is evident that the asymptotic mean squared error of $\bar{f}_N$ as an estimator of $E_\pi(f)$ is determined by the variance, which is of order N^{-1} while that of the squared bias is N^{-2}.

For clarification of the distinction between rapid convergence (small R in (1) and (2)) and good estimation performance (small $\tau(f)$), it is helpful to study the finite reversible case, where explicit expressions can be given.

Suppose the Markov chain P is finite, reversible, irreducible and aperiodic, and that B is the diagonal matrix with entries $(\pi(\mathbf{x}),\mathbf{x}\in\Omega)$, the equilibrium probabilities for P. Reversibility means that $\pi(\mathbf{x})P(\mathbf{x},\mathbf{x}')=\pi(\mathbf{x}')P(\mathbf{x}',\mathbf{x})$ for all $\mathbf{x},\mathbf{x}'\in\Omega$, so that BP is a symmetric matrix. We then have the spectral representation

$$P = E\Lambda E^{\mathrm{T}}B$$

where $\Lambda=\mathrm{diag}(\lambda_1,\lambda_2,\ldots)$ is a diagonal matrix of eigenvalues of P, which are real, and E is the matrix whose columns are the corresponding right eigenvectors, normalised to be orthogonal with respect to B, so that $E^{\mathrm{T}}BE=I$. We take λ_1 to be the unique unit eigenvalue, so that $E_{\mathbf{x}1}=1$ for all $\mathbf{x}\in\Omega$. Then $P^n=E\Lambda^nE^{\mathrm{T}}B\rightarrow E\,\mathrm{diag}(1,0,\ldots,0)E^{\mathrm{T}}B=\mathbf{1}\pi^{\mathrm{T}}$ as expected.

If $\mathbf{f}$ is the vector with components $(f(\mathbf{x}),\mathbf{x}\in\Omega)$ and $\mathbf{a}$ the vector of initial probabilities for the chain, then

$$E(f(\mathbf{x}^{(t)})) = \mathbf{a}^{\mathrm{T}}P^t\mathbf{f}$$

$$= \sum_k\lambda_k^t(E^{\mathrm{T}}\mathbf{a})_k(E^{\mathrm{T}}B\mathbf{f})_k.$$

So

$$E(f(\mathbf{x}^{(t)}))-E_\pi(f) = E(f(\mathbf{x}^{(t)}))-\sum_{\mathbf{x}}\pi(\mathbf{x})f(\mathbf{x})$$
$$= \sum_{k\geq 2}\lambda_k^t(E^{\mathrm{T}}\mathbf{a})_k(E^{\mathrm{T}}B\mathbf{f})_k.$$

If λ_2 is an eigenvalue second largest in absolute value, then $|\lambda_2|=R$ and

$$|E(f(\mathbf{x}^{(t)}))-E_\pi(f)| = O(R^t), \tag{5}$$

where the multiplier depends on the initial distribution, the particular functional of interest, and on the transition matrix P.

Turning to the empirical average $\bar{f}_N$, we have

$$E(\bar{f}_N) = \frac{1}{N}\sum_{t=1}^{N}\sum_k\lambda_k^t(E^{\mathrm{T}}\mathbf{a})_k(E^{\mathrm{T}}B\mathbf{f})_k,$$

whence the bias is

$$E(\bar{f}_N)-E_\pi(f) = \sum_{k\geq 2}\{\frac{1}{N}\sum_{t=1}^{N}\lambda_k^t\}(E^{\mathrm{T}}\mathbf{a})_k(E^{\mathrm{T}}B\mathbf{f})_k$$
$$\sim \frac{1}{N}\sum_{k\geq 2}\frac{\lambda_k}{1-\lambda_k}(E^{\mathrm{T}}\mathbf{a})_k(E^{\mathrm{T}}B\mathbf{f})_k. \tag{6}$$

As for the variance, the equilibrium autocovariance is

$$\sigma^2\rho_t(f) = \sum_{\mathbf{x}}\sum_{\mathbf{x}'}f(\mathbf{x})f(\mathbf{x}')\pi(\mathbf{x})(P^t(\mathbf{x},\mathbf{x}')-\pi(\mathbf{x}'))$$
$$= \mathbf{f}^{\mathrm{T}}BP^t\mathbf{f}-\mathbf{f}^{\mathrm{T}}\pi\pi^{\mathrm{T}}\mathbf{f}$$
$$= \mathbf{f}B(P^t-\mathbf{1}\pi^{\mathrm{T}})\mathbf{f}.$$

Now $(P^t-\mathbf{1}\pi^{\mathrm{T}})=(P-\mathbf{1}\pi^{\mathrm{T}})^t$ for $t\geq 1$ so

$$\sum_{t=0}^{\infty}(P^t-\mathbf{1}\pi^{\mathrm{T}}) = (I-P+\mathbf{1}\pi^{\mathrm{T}})^{-1}-\mathbf{1}\pi^{\mathrm{T}}.$$

But from (4)

$$Nvar(\bar{f}_N) \sim \sigma^2\tau(f)$$
$$= \sigma^2\sum_{t=-\infty}^{\infty}\rho_t(f) = 2\sigma^2\sum_{t=0}^{\infty}\rho_t(f)-\sigma^2$$
$$= \mathbf{f}^{\mathrm{T}}\{2B(I-P+\mathbf{1}\pi^{\mathrm{T}})^{-1}-2B\mathbf{1}\pi^{\mathrm{T}}-B(P^0-\mathbf{1}\pi^{\mathrm{T}})\}\mathbf{f}$$
$$= \mathbf{f}^{\mathrm{T}}BE\,\mathrm{diag}\{2(1-\lambda_k+\delta_{k1})^{-1}-2\delta_{k1}-1+\delta_{k1}\}E^{\mathrm{T}}B\mathbf{f}$$
$$= \sum_{k\geq 2}\frac{1+\lambda_k}{1-\lambda_k}(E^{\mathrm{T}}B\mathbf{f})_k^2. \tag{7}$$

The matrix expression is noted by Peskun (1973), and the spectral expansion by Sokal (1989) and Frigessi, Hwang and Younes (1990).

Contrasting (5) with (7), we see that rapid weak convergence to equilibrium is obtained by having all eigenvalues λ_k other than $\lambda_1 = 1$ small in absolute value, whilst good asymptotic mean squared error of estimation is suggested by having $(1+\lambda_k)/(1-\lambda_k)$ small: "negative eigenvalues help". The rôle played by the eigenvectors in (7) should not be neglected, however, as two alternative transition matrices P will in general differ not only in their eigenvalues.

In practice, with a finite Monte Carlo sample size N, both of these aspects of convergence are relevant. The very complexity of the distribution π which suggested consideration of Monte Carlo simulation in the first place inhibits explicit calculation of eigen-decompositions, of course, and we need diagnostics for studying the rate of weak convergence and methods for estimating the integrated autocorrelation time. Such tools will be used both in studies aimed at making general recommendations, such as the present one, and routinely in the actual use of dynamic Monte Carlo methods (not least in order to attach standard errors to estimates of $E_\pi(f)$). Blind application of Gibbs or Metropolis samplers, with no examination of these issues, can produce completely meaningless results.

The conflicting demands of small $\sup_{k\geq 2}|\lambda_k|$ and small $(1+\lambda_k)/(1-\lambda_k)$ suggest a revised strategy of switching between different transition mechanisms as the simulation proceeds, producing a time-inhomogeneous Markov chain. In its simplest form, which seems to be commonly used in the physics literature, the idea would be to use an initial process P_0 for the first N_0 iterations, then to switch to P for another N updates. P_0 would be chosen to give rapid convergence to equilibrium and P for a small $\tau(f)$. The switch would take place when diagnostics suggested that the process was effectively in equilibrium, and the first N_0 iterations discarded for estimation purposes, so that the estimator is

$$\frac{1}{N}\sum_{t=N_0+1}^{N_0+N} f(\mathbf{x}^{(t)}).$$

More complicated variants of this, perhaps involving continuous alteration of the transition mechanism, and/or weighted averages of $\{f(\mathbf{x}^{(t)})\}$ may be worth exploring; another factor that can influence this discussion arises when the cost of computing f is high relative to that of the Markov transition, which will support sub-sampling the chain at equally spaced times at which $f(\mathbf{x}^{(t)})$ is computed, with a corresponding modification to the definition of autocorrelation time.

Rather than sample repeatedly from a single run of the process, some authors, for example Gelfand and Smith (1990), propose evaluating $f(\mathbf{x}^{(t)})$ only once (so that $N=1$, although N_0 is large), but then repeatedly restarting the whole process, so as to be able to average completely independent values of $f(\mathbf{x})$. But this seems to us to be inefficient, at least in the situations of our experience, where $\tau(f)$ is much less than $(-1/\log R)$ and $var_\pi(f)$ is sufficiently large that say 100 or 1000 effectively independent observations will be needed to estimate $E_\pi(f)$ to adequate precision. In this situation, more computing effort would be used to achieve the same precision if the chain were restarted.

We have only tentative recommendations to make regarding the diagnostic monitoring of convergence. Practical considerations limit attention to a few scalar-valued functionals $f_1(\mathbf{x}^{(t)}), f_2(\mathbf{x}^{(t)}), \ldots$, although there may be merit in also measuring aspects of several $\mathbf{x}^{(t)}$ jointly, such as a summary of the magnitude of the difference between

$\mathbf{x}^{(t-1)}$ and $\mathbf{x}^{(t)}$. Each functional f_j will have a characteristic spectral radius R_j governing the rate of convergence of its distribution to equilibrium: none of these can exceed R (see equation(2)). The hope in selecting a range of such functionals for study is that at least one R_j is close to R, so that we are not misled into an over-optimistic impression of the rate of convergence of the process as a whole. Our approach is then simply to plot the values of these functionals against iteration number: the clarity of the visual impression given about convergence depends on the equilibrium variance of the functional. See the examples in Section 4.

Estimating the integrated autocorrelation time $\tau(f)$, for any particular functional f, is a standard problem from the analysis of stationary time series. (We suppose that we only address this question after discarding the initial N_0 iterations according to the criteria just described, so that we can regard the process as in equilibrium). The integrated autocorrelation time is simply 2π times the normalised power spectral density function of the process evaluated at frequency 0 (Priestley, 1981, p.225), so we are dealing with a special case of spectral density estimation. The difficulties are well-known, the naive estimator $\sum_{t=-\infty}^{+\infty}\hat{\rho}_t(f)$ using the sample autocorrelations of the observed process being inconsistent as the length of the observed series increases (Priestley, p.429). The conventional solution is to apply a spectral window, that is to use a weighted estimator $\sum w_t\hat{\rho}_t(f)$, where the lag window function w_t decreases to 0 as $t\to\pm\infty$. In particular, Sokal (1989) recommends the truncated periodogram estimator $\sum_{|t|\le M}\hat{\rho}_t(f)$, with the window width M chosen adaptively as the minimum integer with $M\ge 3\hat{\tau}(f)$. In our implementation of this we estimate the autocorrelations from the Fourier transform of the process, thus wrapping the time axis onto a circle; this approximation is acceptable if $\tau(f)$ is small relative to the length of the series.

An alternative non-parametric estimator of $\tau(f)$ which is also appealing turns out to be related to the spectral density estimator using the Bartlett window (Priestley, p.439), and is recommended by Hastings (1970). If $N=bk$ and the series is broken into b non-overlapping blocks of k consecutive observations, then the between-blocks mean square

$$\frac{k}{b-1}\sum_{j=1}^{b}\left[\{\frac{1}{k}\sum_{t=(j-1)k+1}^{jk} f(\mathbf{x}^{(t)})\}-\bar{f}_N\right]^2 \tag{8}$$

is an approximately unbiased estimator of $\sigma^2\tau(f)$ as b and $k\to\infty$.

3. Gaussian proposals in the Metropolis method

Motivated by our interest in image analysis problems, we consider here a new class of samplers appropriate to the continuous case, where $\Omega=\mathbf{R}^S$, with particular emphasis on designing Markov chain methods with small integrated autocorrelation time.

The best known dynamic Monte Carlo method is the Metropolis algorithm (Metropolis, *et al.*, 1953). Here we describe it in the interesting variant due to Hastings (1970). Recall that we wish to construct a Markov chain with a prescribed equilibrium distribution $\pi(\mathbf{x})$. Let $q(\mathbf{x},\mathbf{x}')$ be an arbitrary irreducible aperiodic transition function on $\Omega\times\Omega$: how can this be modified to achieve the required equilibrium? Given $\mathbf{x}^{(t)}=\mathbf{x}$, a proposal $\mathbf{x}'$ is drawn from $q(\mathbf{x},\mathbf{x}')$, but not immediately taken as the new state of the chain. Rather, it is only accepted, and $\mathbf{x}^{(t+1)}$ set equal to $\mathbf{x}'$, with probability $\alpha(\mathbf{x},\mathbf{x}')$; otherwise it is rejected, and no move is made, so that $\mathbf{x}^{(t+1)}=\mathbf{x}$. The acceptance probability can always be chosen so that detailed balance is obtained:

$$\pi(\mathbf{x})P(\mathbf{x},\mathbf{x}') = \pi(\mathbf{x}')P(\mathbf{x}',\mathbf{x}) \tag{9}$$

for all $\mathbf{x},\mathbf{x}'\in\Omega$. One possibility for α is

$$\alpha(\mathbf{x},\mathbf{x}') = \min\left\{1, \frac{\pi(\mathbf{x}')q(\mathbf{x}',\mathbf{x})}{\pi(\mathbf{x})q(\mathbf{x},\mathbf{x}')}\right\}, \tag{10}$$

for which (9) is easily verified, the corresponding transition function being

$$\begin{aligned} P(\mathbf{x},\mathbf{x}') &= q(\mathbf{x},\mathbf{x}')\alpha(\mathbf{x},\mathbf{x}') \qquad \mathbf{x}'\neq\mathbf{x} \\ &= 1-\sum_{\mathbf{x}'\neq\mathbf{x}} q(\mathbf{x},\mathbf{x}')\alpha(\mathbf{x},\mathbf{x}') \qquad \mathbf{x}'=\mathbf{x} \end{aligned}$$

Among all possible α achieving detailed balance for a given q, the particular choice in (10) is shown by Peskun (1973) to give minimum integrated autocorrelation time.

This prescription is very general, and can be used to generate a wide variety of Markov chain simulation methods for different problems. The process $\mathbf{x}^{(t)}$ is usually highly multivariate, and in practice we usually concentrate on algorithms which only change one component of $\mathbf{x}$ at a time. (There are notable exceptions in special cases, for example the algorithm of Swendsen and Wang (1987)). This does not affect the validity of (10), but merely facilitates its computation. There are various valid ways to choose which component, i, of $\mathbf{x}^{(t)}$ is to be updated in the transition to $\mathbf{x}^{(t+1)}$: the common ones being a systematic choice, cycling through $i\in S$ in some fixed order, or a random choice, drawing i at random each time. The choice is reflected in q.

In the original application of this idea, Metropolis *et al.* (1953) considered a finite set of "colours": $\Omega=\{0,1,\ldots,L-1\}^S$, and the proposal that takes a uniformly distributed choice from among the $L-1$ colours different from the current one. In this case, and whenever there is symmetry of the proposal distribution, $q(\mathbf{x},\mathbf{x}')=q(\mathbf{x}',\mathbf{x})$, the expression for $\alpha(\mathbf{x},\mathbf{x}')$ simplifies to $\min\{1,\pi(\mathbf{x}')/\pi(\mathbf{x})\}$. But we shall see there is something to be gained by the slightly greater generality.

Two other points might be made about this prescription. One is the "distribution-free" nature of the simulation step: the transition function q is quite arbitrary (provided that $q(\mathbf{x},\mathbf{x}')$ and $q(\mathbf{x}',\mathbf{x})$ are either both zero or both positive). So use of the method is not restricted to those $\pi(\mathbf{x})$ which are convenient for simulation: the model π only enters the algorithm through the calculation of $\pi(\mathbf{x}')/\pi(\mathbf{x})$ in the definition of α. The second point is that although the whole procedure has the flavour of the conventional rejection methods for static Monte Carlo simulation, there is no requirement, as there, for the density that is used for simulation to envelope (a submultiple of) the density of interest.

One particular algorithm in this class has received a good deal of attention in the recent statistical literature: the Gibbs sampler. The proposal distribution q is defined as follows: a pixel i is chosen from S uniformly at random, the current value $x_i^{(t)}$ deleted, and the proposed new value drawn from the conditional distribution, under π, of x_i given the values of all other pixels: thus

$$q(\mathbf{x},\mathbf{x}') = \frac{1}{|S|}\sum_{i\in S}\pi(x'_i|x_{S\setminus i})$$

(Systematic scanning over the pixels is also commonly used). It is trivial to see that for $\mathbf{x}\neq\mathbf{x}'\in\Omega$, $q(\mathbf{x},\mathbf{x}')$ and $q(\mathbf{x}',\mathbf{x})$ are each positive if and only if $\mathbf{x}$ and $\mathbf{x}'$ differ in at most one coordinate, and that in that case

$$\frac{q(\mathbf{x},\mathbf{x}')}{q(\mathbf{x}',\mathbf{x})} = \frac{\pi(\mathbf{x}')}{\pi(\mathbf{x})}$$

so that the corresponding acceptance probability $\alpha(\mathbf{x},\mathbf{x}')$ is identically 1. From this point of view, the Gibbs sampler, or heat-bath method as it is known in the physical literature, is but an extreme form of the Metropolis method, with a highly model-dependent choice of proposal and zero probability of rejection.

Although informal heuristics suggest that eliminating rejection should reduce the integrated autocorrelation time, the computational price paid may be high. For the Gibbs sampler requires simulation from $\pi(x'_i|x_{S\setminus i})$, which may be quite unwieldy. Except when C consists of a small number of discrete colours, or when $\pi(\mathbf{x})$ is Gaussian, even normalisation of $\pi(x'_i|x_{S\setminus i})$ may be expensive. On the other hand, if some model-independent choice of proposal is made, we only need to be able to compute the ratio $\pi(x'_i|x_{S\setminus i})/\pi(x_i|x_{S\setminus i})$ of the posterior probabilities of the proposed and current values.

Thus if we know that the Gibbs sampler does yield good convergence properties, it may nevertheless be preferable in terms of computational cost to choose a proposal distribution that is merely reasonably close to $\pi(x'_i|x_{S\setminus i})$ from which it is easy to simulate, and to tolerate the consequent small probability of rejection.

The Gibbs sampler is not the only Metropolis method that gives zero rejection probability. In their study of stochastic relaxation in gaussian processes, Barone and Frigessi (1989) derived a class of samplers that include the Gibbs sampler as a special case. Suppose that μ_i and σ_i^2 are the expectation and variance of the conditional distribution $\pi(x'_i|x_{S\setminus i})$. The Gibbs sampler proceeds by drawing the new value x'_i from $N(\mu_i,\sigma_i^2)$. Barone and Frigessi's ω-stochastic relaxation (ω-*SR*) approach draws instead from $N((1+\theta)\mu_i-\theta x_i,(1-\theta^2)\sigma_i^2)$. (We use θ in place of their $\omega-1$). Validity of this method is most easily checked in the present context by noting that $q(\mathbf{x},\mathbf{x}')/q(\mathbf{x}',\mathbf{x})$ does not depend on θ. Barone and Frigessi prove that in the case of entirely positive association between the variables (all non-diagonal entries in the inverse of the variance matrix non-positive), the spectral radius R of the corresponding Markov chain is a decreasing function of θ at $\theta=0$. An intuitive explanation for this advantage of using $\theta>0$ in the case of positive association comes from noting that then the current value x_i is positively correlated with the values of its neighbours. If x_i is, say, in the lower tail of its marginal distribution under π, then the whole local conditional distribution $\pi(x'_i|x_{S\setminus i})$ will be biased towards this lower tail: hence the advantage in modifying the Gibbs sampler to improve convergence by "over-correcting" this bias.

A simpler yet stronger result holds for the asymptotic variance: for any linear function of $\mathbf{x}$, the asymptotic variance when using Barone and Frigessi's modified sampler, with systematic scanning of pixels, is proportional to $(1-\theta)/(1+\theta)$. Without loss of generality, we assume that the process has zero expectation and is in equilibrium.

Theorem.
Suppose that $\pi(\mathbf{x})$ is the gaussian distribution $N(0,V)$ where V is non-singular, and that the pixels are indexed by $i\in S=\{1,2,\ldots,n\}$. Let a stationary process $\{\mathbf{x}^{(t)},t\in \mathbf{Z}\}$ with marginal distribution $\pi(\mathbf{x})$ be defined by updating x_i cyclically for $i=1,2,\ldots,n,1,2,\ldots$ by resampling x_i from

$$N((1+\theta)\mu_i-\theta x_i,(1-\theta^2)\sigma_i^2) \tag{11}$$

where μ_i and σ_i^2 are the mean and variance of the distribution $\pi(x_i|x_{S\setminus i})$. Then for any vector of constants $\mathbf{c}$,

$$N\,var(\mathbf{c}^{\mathrm{T}}\frac{1}{N}\sum_{t=1}^{N}\mathbf{x}^{(t)}) \;\to\; \frac{1-\theta}{1+\theta}\mathbf{c}^{\mathrm{T}}V\mathrm{diag}(V^{-1})V\mathbf{c}.$$

Proof. We first consider the stationary first-order matrix autoregression defined by $\mathbf{x}^{(t+1)}=A\mathbf{x}^{(t)}+\mathbf{z}^{(t+1)}$, where $\{\mathbf{z}^{(t)}:t\in \mathbf{Z}\}$ are independent and identically distributed gaussian random vectors with zero mean. (Since the process is stationary, and is to have marginal distribution π, it follows that we must have $var(\mathbf{z}^{(t)})=V-AVA^{\mathrm{T}}$). Now for any $t\ge 0$, $E(\mathbf{x}^{(t)}\mathbf{x}^{(0)\mathrm{T}})=E((A^t\mathbf{x}^{(0)}+\sum_{r=1}^{t}A^{t-r}\mathbf{z}^{(r)})\mathbf{x}^{(0)\mathrm{T}})=A^tE(\mathbf{x}^{(0)}\mathbf{x}^{(0)\mathrm{T}})+0 \;=A^tV$. Thus

$$\begin{aligned}\sum_{t=-\infty}^{\infty}E(\mathbf{x}^{(t)}\mathbf{x}^{(0)\mathrm{T}}) &= \sum_{t=0}^{\infty}A^tV+\sum_{t=0}^{\infty}(A^tV)^{\mathrm{T}}-V\\ &= (I-A)^{-1}V+V(I-A)^{-1}-V\end{aligned}$$

Now

$$\begin{aligned}var(\mathbf{c}^{\mathrm{T}}\frac{1}{N}\sum_{t=1}^{N}\mathbf{x}^{(t)}) &= \frac{1}{N^2}\mathbf{c}^{\mathrm{T}}\left[\sum_{s=1}^{N}\sum_{t=1}^{N}E(\mathbf{x}^{(s)}\mathbf{x}^{(t)\mathrm{T}})\right]\mathbf{c}\\ &\sim \frac{1}{N}\mathbf{c}^{\mathrm{T}}[(I-A)^{-1}V+V(I-A)^{-1}-V]\mathbf{c}\\ &= N^{-1}\mathbf{c}^{\mathrm{T}}(I-A)^{-1}(I+A)V\mathbf{c}.\end{aligned}$$

We now have to write Barone and Frigessi's sampler in the matrix autoregressive form. But

$$\begin{aligned}\mu_i &= E(x_i|x_{S\setminus i}) = -g_{ii}^{-1}\sum_{j\ne i}g_{ij}x_j,\\ \sigma_i^2 &= var(x_i|x_{S\setminus i}) = g_{ii}^{-1},\end{aligned}$$

where $G=(g_{ij})=V^{-1}$. Thus the sampler (11) can be written

$$x_i^{(t+1)} = \sum_{j=1}^{i-1}b_{ij}x_j^{(t+1)} + \sum_{j=i}^{n}b_{ij}x_j^{(t)} + z_i^{(t+1)},$$

where $B=(b_{ij})=I-(1+\theta)\Gamma G$ and $var(z_i^{(t+1)})=(1-\theta^2)\Gamma_{ii}$ where $\Gamma=(\mathrm{diag}(G))^{-1}$. Let L denote the lower triangle of B. Then in matrix form we have

$$\mathbf{x}^{(t+1)} = L\mathbf{x}^{(t+1)} + (B-L)\mathbf{x}^{(t)} + \mathbf{z}^{(t+1)}$$

or

$$\mathbf{x}^{(t+1)} = (I-L)^{-1}(B-L)\mathbf{x}^{(t)} + (I-L)^{-1}\mathbf{z}^{(t+1)}.$$

This is a matrix autoregression with $A=(I-L)^{-1}(B-L)$. Thus

$$(I-A)^{-1}(I+A)V = (I-B)^{-1}(I+B-2L)V$$
$$= (1+\theta)^{-1}G^{-1}\Gamma^{-1}\{2I-(1+\theta)\Gamma G+2(1+\theta)\Gamma H\}V$$

where H is the lower triangle of G

$$= (1+\theta)^{-1}V\{2\Gamma^{-1}-(1+\theta)G+2(1+\theta)H\}V$$
$$= (1+\theta)^{-1}V\{(1-\theta)\Gamma^{-1}+(1+\theta)(H-H^{\mathrm{T}})\}V$$

since G is symmetric. On pre- and post-multiplying by the same vector $\mathbf{c}$, the anti-symmetric term vanishes, and we obtain the required result.

The implication of this result is that, for linear functionals in the gaussian case, and considering only the asymptotic variance, best performance in this class of procedures is obtained by letting $\theta \to +1$. This is a dynamic analogue of the conventional idea of using antithetic variables to reduce Monte Carlo variance. It is interesting to note that this effect is anticipated, without explanation, in a simple example in Hastings (1970, p.101).

All of this applies only to gaussian distributions $\pi(\mathbf{x})$, and our real interest is in other cases with continuously distributed $\mathbf{x}$. Only in rather special cases could we expect to find a family of samplers analogous to that of Barone and Frigessi, indexed by an "antithetic parameter" θ and including the Gibbs sampler, yet convenient for simulation. As a general procedure, however, we suggest using a gaussian proposal of the form

$$x_i' \sim N((1+\theta)\mu-\theta x_i, (1-\theta^2)\sigma^2) \tag{12}$$

in the Metropolis/Hastings algorithm, with appropriately chosen μ, σ^2 and θ (these can depend on all variables in the model except x_i). The acceptance probability is

$$\alpha(\mathbf{x},\mathbf{x}') = \min\{1, \frac{\pi(\mathbf{x}')q(\mathbf{x}',\mathbf{x})}{\pi(\mathbf{x})q(\mathbf{x},\mathbf{x}')}\}$$

which simplifies to

$$= \exp[\min\{0, g(x_i')-g(x_i)\}],$$

where

$$g(x_i') = \log \pi(x_i'|x_{S\setminus i}) + \frac{1}{2\sigma^2}(x_i'-\mu)^2. \tag{13}$$

Note that α does not depend on the antithetic parameter θ. We can now choose μ and σ, depending on $x_{S\setminus i}$, to ensure that $g(x_i')$ is approximately constant over a range of arguments including x_i and the most probable values from the proposal distribution, so that $\alpha(\mathbf{x},\mathbf{x}')$ is close to 1 with high probability. Such a choice of μ and σ can be made by expanding $g(x_i')$ to second order about an appropriate approximate centre: for example, we have used the mean of neighbouring x_j when simulating from Gibbs distributions.

The nub of the idea is to use a gaussian approximation to the Gibbs sampler, but

- this need only be a good approximation in the centre of the conditional distribution,
- exact detailed balance is restored by the acceptance/rejection decisions, and
- we still have the parameter θ free to help improve asymptotic variance.

A full analysis of the spectrum of such a Markov chain seems to be a challenging problem, but intuitively one might be concerned that as θ increases towards 1, the spectral radius may approach or even attain the value 1. There may therefore be less freedom of choice in general than in Barone and Frigessi's gaussian case. This underlines the need to monitor convergence carefully as the simulation proceeds.

We are not aware of any classes of distributions for use in generating proposals, other than the gaussian family (12), into which it is possible to introduce an antithetic parameter θ that cancels on forming the ratio $q(\mathbf{x},\mathbf{x}')/q(\mathbf{x}',\mathbf{x})$. Thus if our procedure were modified to use a non-gaussian proposal distribution, the details would be a little more complicated. The simplest way to extend the idea to state spaces other than $\Omega = \mathbf{R}^S$ would be by transformation, for example, by replacing x throughout by $\log(x)$ if x takes only non-negative values.

Further insight into these sampling methods can be gained by considering a toy example. Suppose we have just two sites, and three possible values $\{1,2,3\}$ at each. There are then only nine possible states of the system, $\{11,12,\ldots,33\}$, and spectral decompositions are easily computed numerically. Updating is by uniform random choice of site, and when visiting site 1, the probabilities of some possible transitions are

$$\begin{array}{lc} 12 \to 12 & \alpha \\ 12 \to 22 & 1-\alpha-\gamma \\ 12 \to 32 & \gamma \\ 11 \to 21 & \delta \\ 11 \to 11 & 1-2\delta \end{array}$$

All other probabilities are determined from these by symmetry over permutations of sites and values. It is easily verified that the unique equilibrium distribution of the chain has $\pi(11) = \kappa/(6+3\kappa) = \pi(22) = \pi(33)$, and $\pi(12) = 1/(6+3\kappa) = \pi(13) = \ldots$etc., where $\kappa = (1-\alpha-\gamma)/\delta$. This is a symmetric Potts model on 3 colours (Potts (1952)). For any fixed value of $\kappa > 1$, the simple Metropolis method, in which the proposal is an equally likely choice among the colours different from the current one, is the case $\alpha = 0, \gamma = 0.5$. The Gibbs sampler is the case $\alpha = \gamma = \delta$, where the new value is (conditionally) independent of the old. A crude analogue of the sampler (12) for $\theta > 0$ is obtained by reducing α and γ, and increasing δ accordingly to preserve the equilibrium, thus increasing the (equilibrium) probability of change at a transition, which is proportional to $(1-\alpha-\gamma/2)$.

Table 1 displays some values of $\tau(f)$ and R for selected parameter values, including the simple Metropolis method, the Gibbs sampler, those achieving minimum $\tau(f)$ or R, and for contrast, an extreme case with very poor performance. The values for $\tau(f)$ apply to any function f of x_1 alone: the invariance follows from the symmetry in this example. These figures confirm that minimum $\tau(f)$ and R are not the same thing, that Gibbs sampling achieves neither, and that Metropolis methods *can* be very poor.

Table 1. $\tau(f)$ and R in the toy example.

$\kappa = 1.5$

	α	γ	$\tau(f)$	R
min $\tau(f)$	0	0.25	1.7701	0.7500
min R	0	0.4	1.9762	0.4000
simple	0	0.5	2.2063	0.5000
Gibbs	0.2857	0.2857	3.1667	0.5714
poor	0.99	0	265.67	0.9943

$\kappa = 3$

	α	γ	$\tau(f)$	R
min $\tau(f)$	0	0	3.0667	0.6667
min R	0	0.0659	3.1111	0.6052
simple	0	0.5	4.5111	0.7101
Gibbs	0.2	0.2	4.5238	0.7000
poor	0.99	0	405.67	0.9961

There is an analysis of the spectral radius R for various samplers in the finite state space case in Frigessi, Hwang, Sheu, and di Stefano (1990), including some numerical comparisons for the Ising model.

4. Experiments with the new sampler

In this last section, we present a few of the results from some fairly extensive experimentation with the Metropolis algorithm with the gaussian proposal distribution suggested in the previous section. Kirkland (1989) performed a thorough study of a number of samplers for the case of binary Markov random fields; here, of course, we are considering only the continuous case.

The context is of an idealised image analysis problem based on artificial data. In all of the experiments to be described, both the true and observed images, $\mathbf{x}$ and $\mathbf{y}$ respectively, consist of 64×64 pixels. The model to be assumed in the analysis for $\mathbf{x}$ is gaussian:

$$p(\mathbf{x}) \propto \exp\{-\beta \sum_{[i,j]} (x_i - x_j)^2\} \tag{14}$$

where the sum is over orthogonal neighbours only. Each pixel has four neighbours, except for those on the boundary of the array which have three or two. The true $\mathbf{x}$ images from which our artificial data are generated are drawn from the same model except that (a) a possibly different parameter value β_0 is used, and (b) the overall average x value is adjusted to the level 25 by adding a constant to all x_i (under (14), the average has an improper distribution).

Two different models for $p(\mathbf{y}|\mathbf{x})$ will be used, in each case both for simulating and analysing the data. Under each of the models, the $\{y_i\}$ are conditionally independent, given $\mathbf{x}$, and we have respectively:

$$y_i \sim N(x_i, 25) \qquad \text{(gaussian)}$$

$$y_i \sim Poisson(25\exp(x_i/25-1)) \qquad \text{(Poisson)}$$

Note that these models have been devised so that they are comparable in terms of mean and variance; the second allows us to study Monte Carlo methods in the presence of Poisson variation with mean of similar order to that found in much of the medical imagery we see.

For the gaussian model, we use Barone and Frigessi's sampler; this is straightforward. A little work is needed to set up the corresponding Metropolis algorithm for the Poisson case, however. The function g defined in (13) is given by

$$\begin{aligned} g(x_i') &= \log p(y_i|x_i') + \log p(x_i'|x_{S\setminus i}) + \frac{1}{2\sigma^2}(x_i'-\mu)^2 + \text{constant} \\ &= y_i \log(me^{x_i'/m-1}) - me^{x_i'/m-1} - \beta\nu_i(x_i'-\bar{x}_i)^2 \\ &\quad + \frac{1}{2\sigma^2}(x_i'-\mu)^2 + \text{constant} \\ &= \frac{y_i x_i'}{m} - me^{x_i'/m-1} - \beta\nu_i(x_i'-\bar{x}_i)^2 \\ &\quad + \frac{1}{2\sigma^2}(x_i'-\mu)^2 + \text{constant} \\ &= x_i'\{\frac{y_i}{m} - me^{\bar{x}_i/m-1}(\frac{1}{m} - \frac{\bar{x}_i}{m^2}) + 2\beta\nu_i\bar{x}_i - \frac{\mu}{\sigma^2}\} \\ &\quad + x_i'^2\{-\frac{me^{\bar{x}_i/m-1}}{2m^2} - \beta\nu_i + \frac{1}{2\sigma^2}\} \\ &\quad + O((x_i'-\bar{x}_i)^3) + \text{constant} \end{aligned} \tag{15}$$

where $m = 25$ is the overall level assumed in the model, and ν_i and $\bar{x}_i$ are the number of neighbouring x values and their mean respectively. Thus if we choose

$$\mu = \bar{x}_i + \frac{\sigma^2}{m}(y_i - me^{\bar{x}_i/m-1})$$

and

$$\sigma^2 = \{2\beta\nu_i + \frac{e^{\bar{x}_i/m-1}}{m}\}^{-1}$$

the first and second order terms vanish, and there is a prospect that g will be nearly constant in the range of interest. These are the values used in the experiments we report. There may be merit in examining alternative quadratic approximations to the exponential function in (15) in the hope of obtaining values for μ and σ^2 that give higher average acceptance probability by making g closer to constant over a wider range, but we do not pursue that here.

Our experiments consider three different functionals f, chosen to reflect different aspects of the distribution π, but in no sense claimed to be thoroughly exploring the eigenspace of P. The functionals are

- Mean: the overall mean
- 8-Co: the lag-(8,0) spatial autocorrelation
- PL: the statistic that would give the maximum pseudo-likelihood estimate of β for directly observed $\mathbf{x}$ from the model (14), namely $N/(2\sum \nu_i (x_i - \bar{x}_i)^2)$.

We first present, in Table 2, estimates of the integrated autocorrelation time, for all three functionals, for three values of β, for both the gaussian and Poisson cases, and for three independent replicates of each. Four different samplers are compared: three of these are the $\theta = 0.5, \theta = 0$ and $\theta = -0.5$ versions of our proposed method. The other is a simple Metropolis method using a proposal drawn from a gaussian distribution centred at the current value, and with standard deviation 3: thus $x_i' \sim N(x_i, 3^2)$. This corresponds formally to (12), in the limit as $\theta \to -1$ and $\sigma^2 \to \infty$ while $(1-\theta^2)\sigma^2 \to 9$. We use Sokal's estimator $\hat{\tau}(f)$ (see section 2). Each estimate is based on the last 4096 sweeps of a run of 5000, starting from $\mathbf{x} = \mathbf{y}$. Our experience has been that Sokal's estimator is somewhat more stable than the between-blocks mean square (8) with $N = bk = 4000$ and $k = 50$ or 100; most of the exceptions to this pattern being with the simple Metropolis sampler.

Table 2. Estimates of integrated autocorrelation time.

(a) Gaussian

		Mean			8-Co			PL		
$\beta = 0.001$	θ=0.5	0.09	0.15	0.10	0.14	0.15	0.16	0.14	0.16	0.15
	θ=0.0	1.15	1.21	1.12	1.12	1.16	1.18	0.92	0.98	0.98
	θ=−0.5	3.42	3.31	3.25	4.29	3.28	3.24	2.50	2.36	2.62
	simple	14.96	17.14	14.87	22.47	11.34	14.39	12.14	11.09	10.66
$\beta = 0.01$	θ=0.5	0.97	0.97	0.96	0.94	0.90	0.88	1.48	1.38	1.47
	θ=0.0	2.72	3.25	2.89	2.39	2.51	3.03	0.98	0.98	1.04
	θ=−0.5	9.39	6.53	6.63	7.41	6.66	6.85	1.79	1.73	1.65
	simple	19.75	17.60	20.05	19.88	16.61	19.64	5.33	6.08	6.15
$\beta = 0.1$	θ=0.5	6.55	6.65	8.52	5.39	4.58	6.13	2.21	1.94	2.09
	θ=0.0	21.24	14.91	18.53	14.11	10.47	16.03	1.02	1.08	1.20
	θ=−0.5	55.59	56.84	108.8	28.03	26.65	27.51	1.59	1.37	1.59
	simple	93.77	104.0	68.84	58.62	62.36	70.64	4.41	3.85	3.77

(b) Poisson

		Mean			8-Co			PL		
$\beta = 0.001$	θ=0.5	1.29	1.22	1.12	1.10	1.03	1.11	3.52	3.31	3.16
($\beta_0 = 0.1$)	θ=0.0	2.16	1.84	1.84	1.75	1.49	1.40	2.58	3.61	2.64
	θ=−0.5	5.48	5.25	4.73	3.48	3.52	3.36	5.29	5.65	5.55
	simple	22.00	20.37	13.17	12.66	14.51	11.43	9.83	13.80	12.62
$\beta = 0.01$	θ=0.5	1.18	1.25	1.34	1.14	1.17	1.18	1.44	1.45	1.53
	θ=0.0	3.41	3.50	3.34	3.23	2.98	3.05	1.13	1.13	1.13
	θ=−0.5	9.14	10.87	10.23	8.58	10.15	8.10	2.08	2.27	2.26
	simple	18.28	27.69	21.28	11.72	22.75	19.88	3.73	5.63	4.86
$\beta = 0.1$	θ=0.5	8.33	6.55	7.85	4.41	5.75	5.75	2.45	2.13	2.15
	θ=0.0	19.75	21.61	19.64	12.28	12.61	14.98	1.24	1.15	1.27
	θ=−0.5	43.63	81.21	80.97	49.27	34.53	70.60	1.58	1.52	1.62
	simple	50.99	56.13	40.62	46.57	46.55	81.15	3.95	4.43	3.63

In earlier experiments, we found that in the case of low interaction parameter, $\beta = 0.001$, the consequent wide range in values in the generated true **x** led to very unstable results. Such wide variation in **x** does not occur in most real image analysis problems, and so our studies in this case have used simulations using $\beta_0 = 0.1$ instead.

It is clear from Table 2 that the different samplers have very different behaviour as measured by autocorrelation time. For two of the functionals, the $\theta = 0.5$ sampler is always the best, often giving asymptotic variance as small as would arise from independent random sampling. This is remarkably good performance, and very encouraging. It confirms the heuristic interpretation given earlier of the antithetic properties of the sampler when $\theta > 0$. In contrast, the simple Metropolis method performs very badly, suggesting in some cases that a run 100 times as long as for independent sampling is needed to give the same asymptotic variance. The pattern for the third functional, namely the pseudo-likelihood statistic, is somewhat different: in most of the cases considered, the best performance is obtained with $\theta = 0$, corresponding exactly or approximately to the Gibbs sampler. Of the three, this functional depends most directly on local conditional distributions, so it is intuitively reasonable that resampling directly from these distributions should be close to optimal.

Other features of the Table are that there is apparently little difference in performance in the gaussian and Poisson cases, and that, as would be expected, the autocorrelation time increases with β.

It is of interest to compare these numerical estimates with the conclusions of the Theorem in the previous section. The only case to which the Theorem applies exactly is that of the mean functional in the gaussian case, for which it is apparent from Table 2 that the estimated autocorrelation times are indeed approximately proportional to $(1-\theta)/(1+\theta)$. The Table also suggests that the conclusions of the Theorem hold more widely, to a rough approximation.

In Table 3, we present another property of the same four samplers: the empirical acceptance rates, expressed as percentages, and computed only after equilibrium is reached. For the Barone-Frigessi samplers in the gaussian case, of course there is

100% acceptance, but we see that the rate is about 90% or better even in the Poisson case with $\beta = .001$ (the situation among those considered where the quadratic approximation to the exponential function is (15) is least adequate).

Table 3. Empirical Metropolis acceptance rates in equilibrium.

(a) Gaussian

	β = 0.001	β = 0.01	β = 0.1
θ=0.5	100%	100%	100%
θ=0.0	100%	100%	100%
θ=−0.5	100%	100%	100%
simple	79.82%	69.64%	40.38%

(b) Poisson

	β = 0.001 (β_0 = 0.1)	β = 0.01	β =0.1
θ=0.5	89.73%	98.18%	99.93%
θ=0.0	91.27%	98.45%	99.94%
θ=−0.5	93.58%	98.83%	99.95%
simple	79.89%	69.50%	40.35%

Before coming to a general conclusion that these Metropolis methods all perform well in the circumstances of this example, we should seek some reassurance that the Markov chains we are simulating do actually converge in a reasonable number of steps.

In Figure 1, we display the values of our three functionals, for a single realisation of the chain, plotted against "time" measured in units of complete sweeps through the image; this Figure is for the gaussian model, with $\beta = 0.01$. In order to make the initial transient more visible, we have deliberately chosen a poor starting value for the run, namely $x_i \sim U(0,10)$. Each panel of the Figure displays four trajectories, one for each of the four sampling procedures represented in Tables 2 and 3. It is evident that in this case, the three Barone-Frigessi methods all converge quickly (as judged by these functionals): equilibrium is effectively reached by time 25. The simple Metropolis method takes somewhat longer, until approximately time 70. Of course, there are visible differences in character between the three sets of trajectories, reflecting the differing equilibrium variances of the three functionals.

Figure 2 reveals a dramatically different picture for the Poisson model: all other details are the same as for Figure 1. With a poor starting value, the $\theta = +0.5$ sampler converges extremely slowly, and has not reached equilibrium even by time 1000. For the PL functional, this is also true for $\theta = 0.0$. However, the remaining two samplers apparently converge by time 80. The unacceptably slow convergence when $\theta \geq 0.0$ is apparently due to a very low average acceptance probability when the process is far from equilibrium: recall that the values presented in Table 3 applied to equilibrium only.

Recommendation of choice of sampler using the criteria of convergence speed is therefore in stark contrast to that suggested by our discussion of the autocorrelation

Figure 1. Gaussian Case (β =0.01) (a) Mean

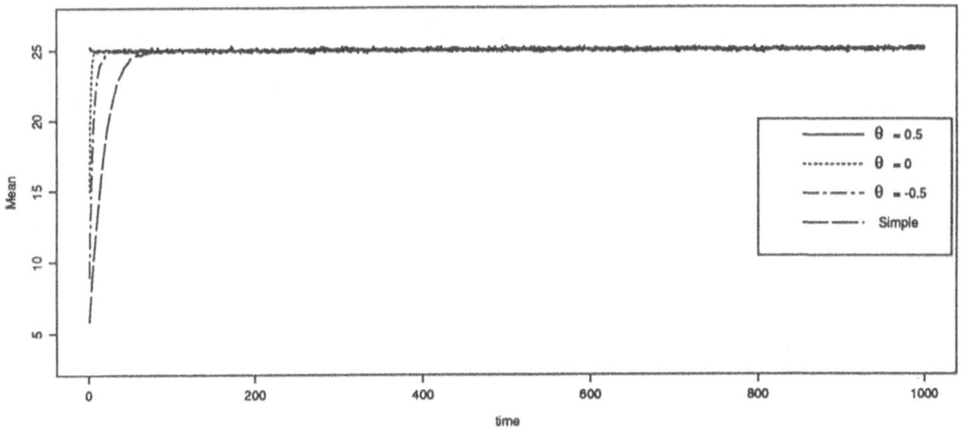

(b) 8-Co

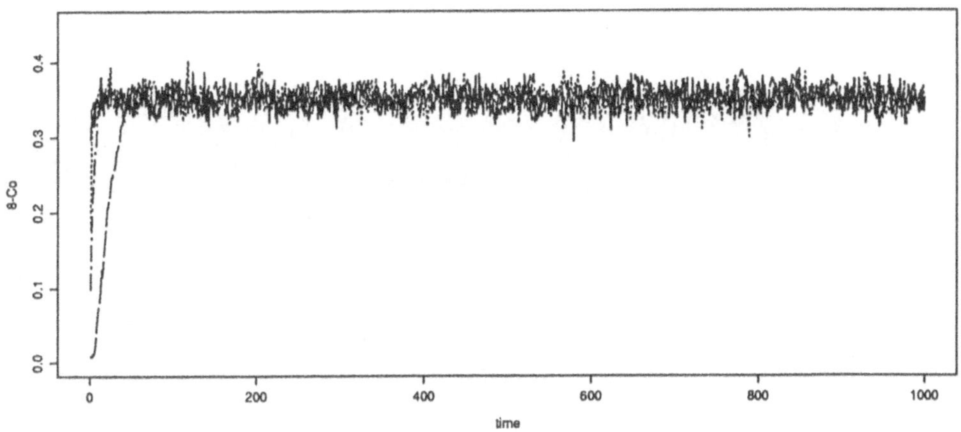

(c) PL

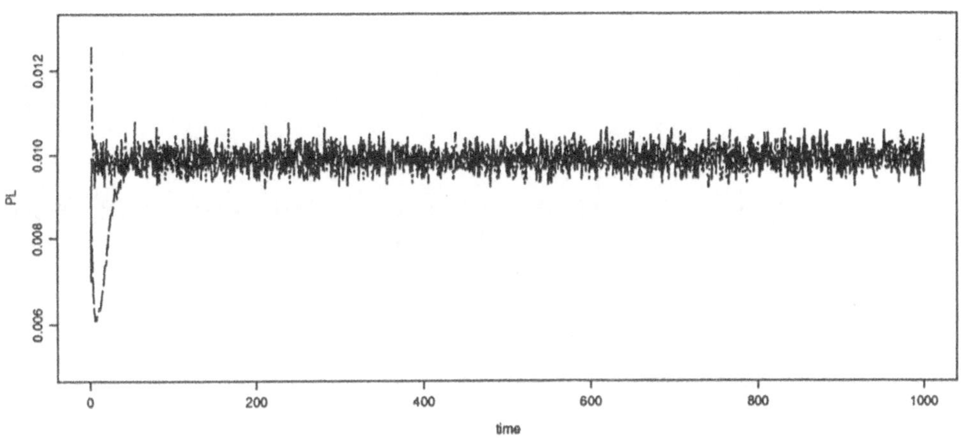

Figure 2. Poisson Case (β=0.01) (a) Mean

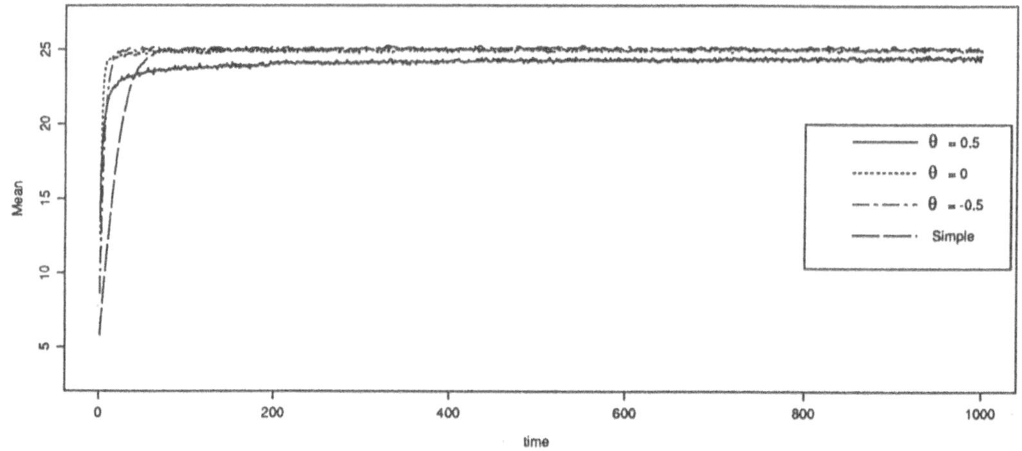

(b) 8-Co

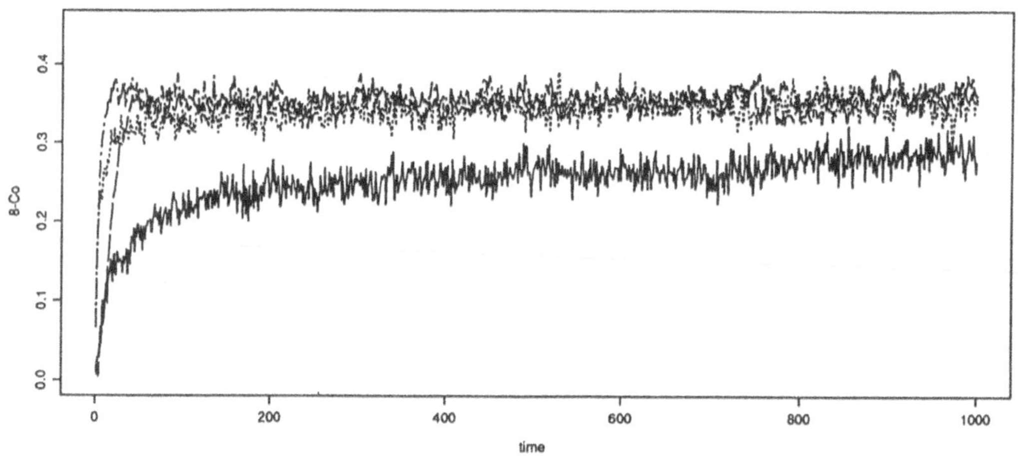

(c) PL

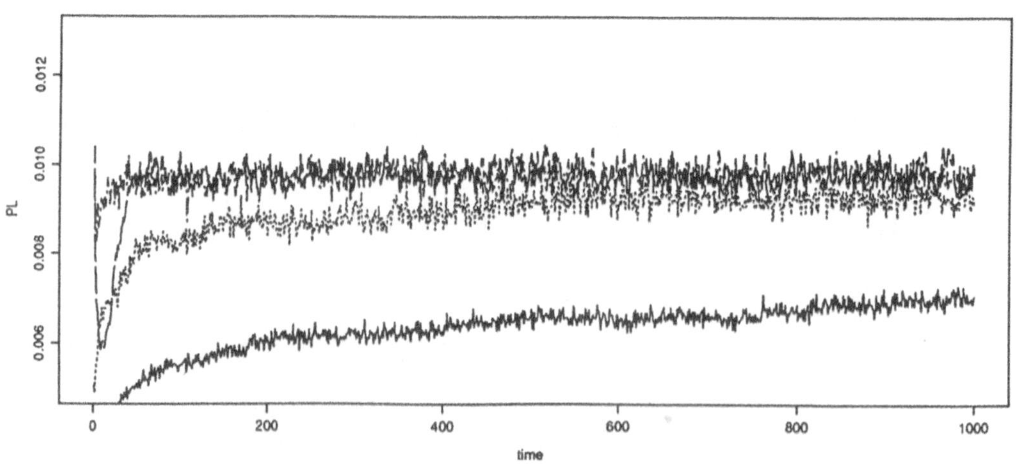

Figure 3. Poisson Case (β =0.1) (a) Mean

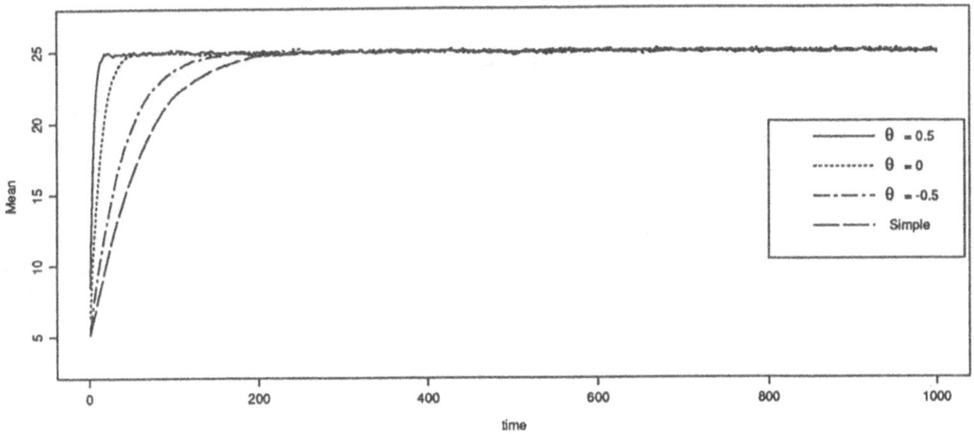

(b) 8-Co

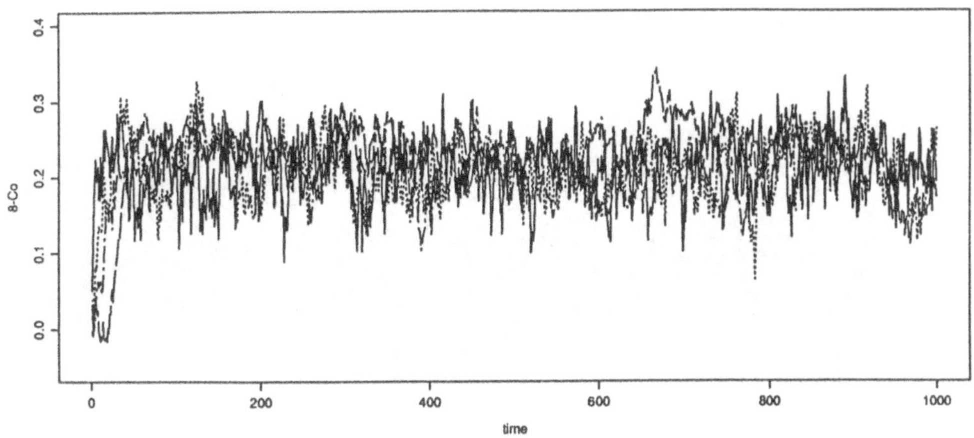

(c) PL

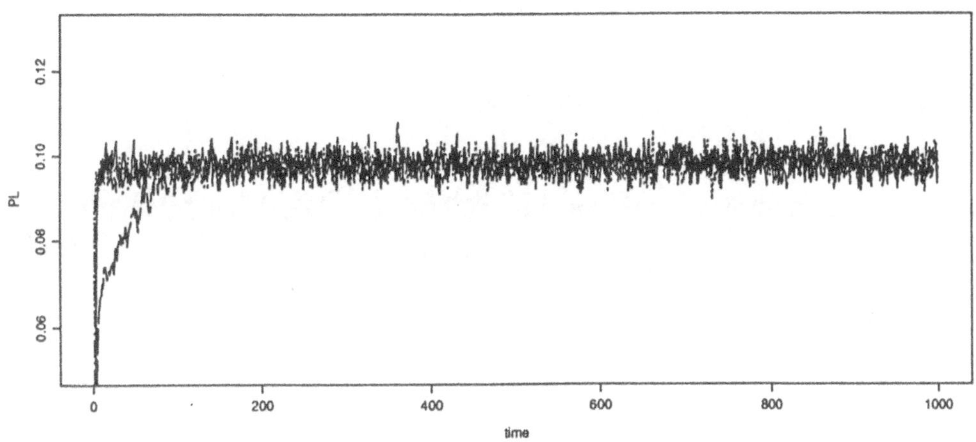

Figure 4. Poisson Case (β =0.001 [βo=0.1]) (a) Mean

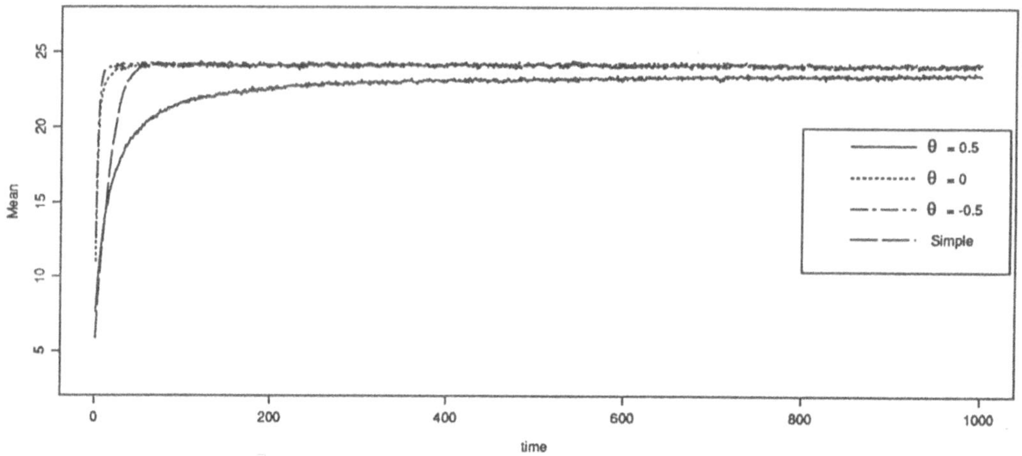

(b) 8-Co

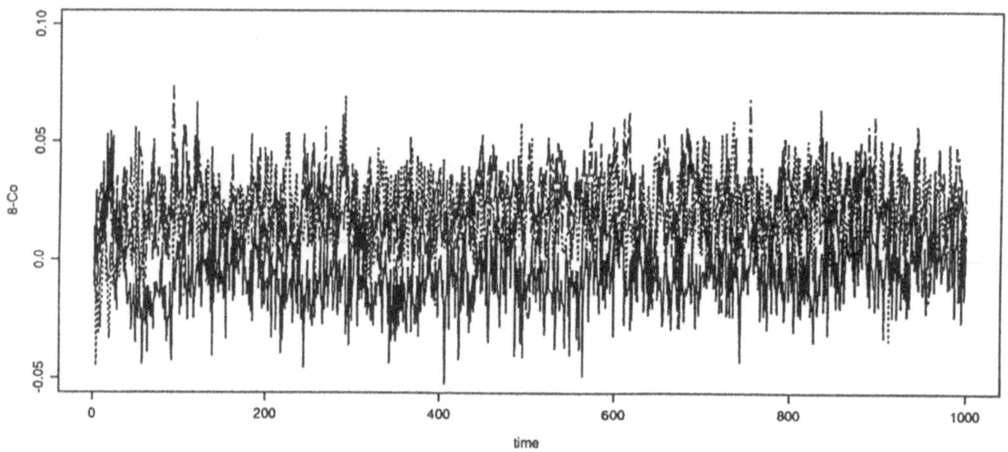

(c) PL

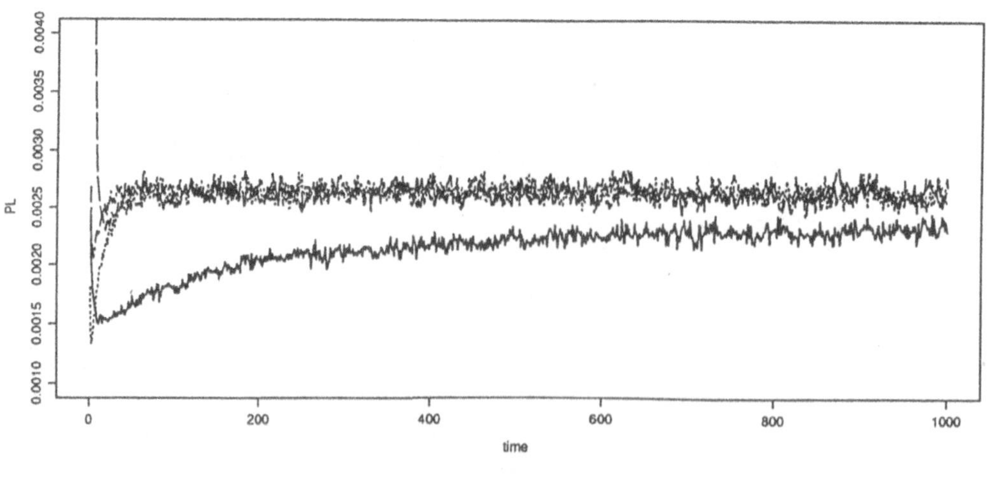

times in Table 2. We feel these observations strongly support the strategy, mentioned in Section 2, of switching from one type of sampler to another, as convergence is achieved. Indeed, one might envisage working entirely with our gaussian proposals, and adaptively varying the value of θ according to the behaviour of the sample trajectories of functionals such as those in Figure 2. We have not investigated this suggestion, but it seems a promising line for future enquiry.

In Figure 3, we display similar information for the Poisson case, with the prior interaction parameter β increased to 0.1. As we observed earlier, increased β means increased $\tau(f)$, but we see that the performance in terms of converge speed is improved, presumably because the quadratic approximation to (15) is better for larger β.

Finally, we turn in Figure 4 to the case of low prior interaction, $\beta = 0.001$, in the Poisson case. As with Tables 2 and 3, we simulated the true $\mathbf{x}$ using $\beta_0 = 0.1$. The results are consistent with the pattern observed in the other examples, but perhaps convey a suggestion that the poor performance in Figure 2 is partly due to the wide intensity range in $\mathbf{x}$ in consequence of simulating from (14) with $\beta = 0.01$.

In conclusion, we underline the points made in Section 2. Successful use of dynamic Monte Carlo methods demands attention to two different aspects of the constructed Markov chains: speed of weak convergence and integrated autocorrelation time. Both are determined by the spectral structure of the transition mechanism of the chain, but are computationally inaccessible except in trivial examples. However, careful monitoring of sample paths of the process can give useful information. Without such monitoring, conclusions of Monte Carlo calculations on stochastic systems can be very misleading.

The performance of Metropolis methods using the gaussian proposal (12) can be very good, providing that care is taken in selecting the antithetic parameter θ.

Acknowledgements

We are grateful for the insightful comments of Julian Besag, Arnoldo Frigessi and Brian Ripley in discussing this work, and acknowledge the financial support of the Complex Stochastic Systems initiative of the Science and Engineering Research Council.

References

Barone, P. and Frigessi, A. (1989) Improving stochastic relaxation for gaussian random fields. Probability in the Engineering and Informational sciences, 4, 369-389.

Besag, J. and Clifford, P. (1989) Generalized Monte Carlo significance tests. Biometrika, 76, 633-642.

Besag, J., York, J. C., and Mollié, A. (1991) Bayesian image restoration, with two applications in spatial statistics (with discussion). Ann. Inst. Statist. Math., 43, 1-59.

Frigessi, A., Hwang, C-R., Sheu, S-J. and di Stefano, P. (1990) Convergence rates of the Gibbs sampler, the Metropolis algorithm, and other single-site updating dynamics. IAC quaderno 6/90, Rome.

Frigessi, A., Hwang, C-R. and Younes, L. (1990) Optimal spectral structure of reversible stochastic matrices, Monte Carlo methods and the simulation of Markov random fields. Submitted to Ann. Appl. Probab.

Gelfand, A. E. and Smith, A. F. M. (1990) Sampling based approaches to calculating marginal densities. J. Amer. Statist. Assoc., 85, 398-409.

Geman, S. and Geman, D. (1984) Stochastic relaxation, Gibbs distributions and the Bayesian restoration of images. IEEE Trans. Pattern Anal. Machine Intell., 12, 609-628.

Grenander, U. (1983) Tutorial in pattern theory, Brown University, Division of Applied Mathematics, Providence, RI.

Hastings, W. K. (1970) Monte Carlo simulation methods using Markov chains and their applications. Biometrika, 57, 97-109.

Kirkland, M. (1989) Simulating Markov random fields. Ph. D. thesis, University of Strathclyde.

Kirkpatrick, S., Gellatt, C. D. and Vecchi, M. P. (1983) Optimization by simulated annealing. Science, 220, 671-680.

Metropolis, N., Rosenbluth, A. W., Rosenbluth, M. N., Teller, A. H. and Teller, E. (1953) Equations of state calculations by fast computing machines. J. Chem. Phys., 21, 1087-1092.

Peskun, P. H. (1973) Optimum Monte-Carlo sampling using Markov chains. Biometrika, 60, 607-612.

Potts, R. B. (1952) Some generalised order-disorder transformations. Proc. Camb. Phil. Soc., 48, 106-109.

Priestley, M. (1981) Spectral analysis and time series. Academic Press, London.

Ripley, B. D. (1987) Stochastic simulation. Wiley, New York.

Sokal, A. D. (1989) Monte Carlo methods in statistical mechanics: foundations and new algorithms. Troisième cycle de la Physique en Suisse Romande lecture notes.

Swendsen, R. H. and Wang, J-S. (1987) Non-universal critical dynamics in Monte Carlo simulations. Phys. Rev. Lett., 58, 86-88.

THE CHI-SQUARE CODING TEST FOR NESTED MARKOV RANDOM FIELD HYPOTHESES

X. Guyon , C.Hardouin

University Paris 1, UFR Maths-Informatique
12 place du Panthéon, 75005 Paris, France.

Summary : Let Y be a Markov random field on S parametrized via its local conditional specifications. We study consistency of coding and pseudo-likelihood estimators. Then we obtain conditional asymptotic normality for the coding estimator and deduce that the difference of coding statistic for two nested hypotheses is, unconditionally, a chi-square. For these results, we do not need regularity of the lattice, translation invariance for the specification or weak dependence for the field.

Résumé : Soit Y un champ de Markov sur S, paramétrisé via ses spécifications conditionnelles locales. On étudie la consistance des estimateurs de codage et de pseudo-vraisemblance. Nous obtenons alors la normalité asymptotique conditionnelle pour l'estimateur de codage et en déduisons que le test asymptotique de différence de codage pour deux hypothèses emboîtées est une statistique du chi 2. Pour établir ces résultats, il n'est pas nécessaire de supposer le lattice S régulier ni la spécification invariante par translation ni le champ Y faiblement dépendant.

Key words : Conditional specification, Markov random field, coding and pseudo-likelihood estimators, difference of coding test.

AMS Classification : 62.M.02, 62.M.05

I- Introduction and notations

The main result of this work is the following : if Y is a Markov random field (M.R.F) on S described by a p-dimensional parametric hypothesis (H_p) on its conditional specifications and if we want to test a regular sub-hypothesis (H_q) of (H_p), then we construct a difference of coding contrasts statistic whith as an asymptotic law a $\chi^2(p-q)$ under (H_q). This result is general in the sense that regularity of the lattice, translation invariance of the specification or weak dependency of the field are not required.

We first describe the framework of the problem. S is an infinite countable set of sites equipped with a symetric graph of vicinity $\mathcal{S}$: i and j are neighbour-sites if and only if $(i,j) \in \mathcal{S}$. The neighbourhood of A will be noted ∂A :

$$\partial A = \{j \in S : j \notin A \text{ and } \exists i \in A \text{ s.t. } (i,j) \in \mathcal{S}\}$$

We shall use ∂i for $\partial\{i\}$.

Let Y be a random field with its values in a measurable state space $\Omega = \prod_{i \in S} \Omega_i$. If $A \subset S$, we shall write $Y_A = (Y_i, i \in A)$. The random field is **markovian** if the law of Y_A, given $y_{\bar{A}}$ $(\bar{A} = S \setminus A)$ depends only on $y_{\partial A}$. If for each i, λ_i is a reference measure on Ω_i, we shall use $\pi_A(y_A/y_{\partial A})$ for the density of the conditional law : $\{\pi_A,\ A \subset S,\ A \text{ finite}\}$ is called the **conditional specification** of the M.R.F. We take as an hypothesis that π has a parametric form π_θ, θ belonging to the interior of Θ, a compact subset of $\mathbb{R}^p$.

If (D_n) is a strictly increasing sequence of domains of observations for Y, the **conditional pseudo-likelihood** of Y on D_n is (see J.Besag [1]) :

$$U_n(\alpha) = -d_n^{-1} \sum_{\mathring{D}_n} \log \pi_i(y_i/y_{\partial i}, \alpha), \quad \alpha \in \Theta \tag{1}$$

where $\mathring{D}_n = \{i \in D_n \text{ s.t. } \partial i \subset D_n\}$, d_n is the cardinal of $\mathring{D}_n$.

$C \subset S$ is a **coding subset** if for any $i,j \in C$, $i \neq j$, i and j are not neighbour-sites. If C is an infinite coding set, we define the coding contrast :

$$U_n^C(\alpha) = -c_n^{-1} \sum_{C_n} \log \pi_i(y_i/y_{\partial i}, \alpha) \tag{2}$$

$C_n = C \cap \mathring{D}_n$, c_n is the cardinal of C_n.

The associated estimators of θ, the true unknown parameter value, are the **pseudo maximum likelihood** (PML) **estimator** :

$$\hat{\theta}_n = \operatorname{Arg}\min_{\Theta} U_n(\alpha) \tag{1'}$$

and the **coding estimator** :

$$\hat{\theta}_n^C = \operatorname{Arg}\min_{\Theta} U_n^C(\alpha) \tag{2'}$$

We shall study first the consistency of both estimators (§ III) : this consistency is a consequence of a preliminary result of convergence for general minimum contrast estimator (§ II) generalizing to non ergodic situations the one given by D.Dacunha-Castelle and M.Duflo ([6]). Then we shall prove that the statistic of difference of coding for nested hypotheses is asymptotically a chi-square (§ IV). The particular case of translation invariant specification will be examined in § V.

Statistical results for M.R.F are studied in various works, see for example [4], [8], [9], [10], [11], [12], [13], [14] in the case of fully observed data, [5], [7], [17] in the case of partially observed data. The sub-ergodicity result of lemma 4 appears first in [8] for translation invariant models and a large deviation approach is developped in [4]. Note the important generalization of conditional pseudo-likelihood given in [14] for point processes or marked point processes. All the results given here take their origin in [11].

II- Consistency of minimum contrast estimator : the non ergodic case

Let $\{\Omega, \mathcal{A}, (P_\alpha), \alpha \in \Theta\}$ be a parametric model, θ, the true unknown parameter value belonging to the interior of Θ, a compact subset of $\mathbb{R}^p$. Here we generalize notions and result on minimum contrast estimator given in [6].

A **contrast function** relative to θ is a (non stochastic) function $\alpha \to K(\theta, \alpha)$ from Θ to $\mathbb{R}^+$ with a unique minimum in $\alpha = \theta$.

A **contrast process** relative to θ and K is a process (U_n) from Θ to $\mathbb{R}$ depending on step n of the observation $Y(n) = \{Y_j,\ j \in D_n\}$ and such that :

$$\liminf_n \left(U_n(\alpha) - U_n(\theta)\right) \geq K(\theta, \alpha) \quad \text{in } P_\theta - \text{probability} \tag{3}$$

A random variable such that : $$\hat{\theta}_n = \operatorname{Arg}\min_{\alpha \in \Theta} U_n(\alpha) \tag{4}$$

is called a minimum contrast estimator.

<u>Theorem 1 :</u> ([6], [11])

Suppose that the two following hypotheses are satisfied :

1) $\alpha \to U_n(\alpha, Y(n))$ *and* $\alpha \to K(\theta, \alpha)$ *are continuous*

2) *Define* $W_n(\eta) = \sup\{ |U_n(\alpha) - U_n(\theta)| \; ; \; |\alpha - \beta| \leq \eta, \; \alpha, \; \beta \in \Theta\}$
There exists $\varepsilon_k, \; \varepsilon_k > 0, \; \varepsilon_k \to 0$ *such that for each* k :

$$\lim_n P_\theta\big(W_n(1/k) \geq \varepsilon_k\big) = 0$$

Then $\lim_n \hat{\theta}_n = \theta$ *in* P_θ*-probability.*

The result is given in [6] where, in place of (3), we have the "ergodic" condition :

$$\lim_n \big(U_n(\alpha) - U_n(\theta)\big) = K(\theta, \alpha) \quad \text{in } P_\theta - \text{probability} \tag{3'}$$

Proof :

If follows [6]. Take B a non vacuous open ball centered in θ. Then for some $\varepsilon > 0$, $K(\theta, .)$ is greater than 2ε on $\Theta \setminus B$. Take k such that $\varepsilon_k < \varepsilon$ and a finite covering of $\Theta \setminus B$ with a finite number N of balls centered in θ_i and of radius $\leq 1/k$. Then :

$$(\hat{\theta}_n \notin B) \subset \{\inf_{\Theta \setminus B} U_n(\alpha) < U_n(\theta)\} \subset \{\inf_{i=1,N} \big(U_n(\theta_i) - U_n(\theta)\big) \leq W_n(1/k)\}$$

$$(\hat{\theta}_n \notin B) \subset \{\inf_{i=1,N} \big(U_n(\theta_i) - U_n(\theta)\big) < \varepsilon\} \cup \{W_n(1/k) > \varepsilon\}$$

From (3) we can deduce :

$$\begin{aligned} \liminf_n \big(\inf_{i=1,N} (U_n(\theta_i) - U_n(\theta))\big) &= \inf_{i=1,N} \big(\liminf_n (U_n(\theta_i) - U_n(\theta))\big) \\ &\geq \inf_{i=1,N} K(\theta, \theta_i) \; \geq \; 2\varepsilon \quad \text{in } P_\theta - \text{probability} \end{aligned}$$

and then, using the condition on W_n, we deduce that $\hat{\theta}_n$ is consistent. ∎

Corollary 2 :

Suppose that the process U_n *is given by a mixture of contrast processes* $U_{n,k}$, $k = 1, K$:

$$U_n = \sum_{k=1,K} a_{n,k} U_{n,k} \, , \quad a_{n,k} \geq 0$$

such that :

1) $U_{n,k}$, $k = 1, K$, *satisfy the assumptions of theorem 1*

2) *For one* k_0, $1 \leq k_0 \leq K$, U_{n,k_0} *fullfills* (3) *for some continuous contrast function* $K_{k_0}(\theta, \alpha)$

3) $\liminf a_{n,k_0} > 0$

Then, the minimum U_n*-contrast estimator is consistent.*

Proof : Straightforward.

As a consequence of the corollary, we shall see in the next paragraph that the consistency of the P.M.L estimator ensures from the consistency of the coding estimator.

Remark : If in (3) the inequality is P_θ-almost surely, and if the condition on W_n is strengthened in :

$$P_\theta(\limsup_n W_n(1/k) \geq \varepsilon_k) = 0$$

then $\hat{\theta}_n$ is strongly consistent (see [11]).

III- Consistency of coding and P.M.L estimators

Let us define the three following conditions ; the first one is on the graph structure, the two others on the specification.

(M_1) C is the disjoint union of C_k, $k = 1, K$ such that :

1) Each C_k is a strongly coding set in the following sense : for all distinct points $i, j \in C_k$, ∂i and ∂j are not neighbours.

2) For each $i \in C_1$, $\Omega_{\partial i}$ is a fix set $\mathcal{X}$ and if $c_{n,1}$ is the cardinal of $C_{n,1} = C_1 \cap \overset{\circ}{D}_n$, we have : $\liminf(c_{n,1} c_n^{-1}) = \tau > 0$

3) $\liminf(c_n d_n^{-1}) > 0$

(M_2) There exist $c > 0$ and $\varepsilon(\eta) > 0$, $\varepsilon(\eta) \to 0$ as $\eta \to 0$ such that for all $i, y_i, y_{\partial i}, y_{\partial\partial i}$, we have :

1) $\pi_i(y_i/y_{\partial i}, \alpha) \geq c$

2) $|\pi_i(y_i/y_{\partial i}, \alpha) - \pi_i(y_i/y_{\partial i}, \beta)| \leq \varepsilon(\eta)$ if $\alpha, \beta \in \Theta$, $|\alpha - \beta| \leq \eta$

3) For $i \in C_1$, and at the true parameter value θ : $\pi_{\partial i}(y_{\partial i}/y_{\partial\partial i}, \theta) \geq c$

Let $m_i(\theta, \alpha; y_{\partial i}) = -E_\theta\left[\log \dfrac{\pi_i(Y_i/y_{\partial i}, \alpha)}{\pi_i(Y_i/y_{\partial i}, \theta)}\right] \geq 0$, and for $i \in C_1$ let λ be the reference measure on $\mathcal{X}$.

(M_3) There exists a function $m(\theta, \alpha; y) \geq 0$ for $\alpha, y \in \Theta \times \mathcal{X}$, continuous in α and measurable in y, λ-integrable in y for each α, such that :

1) $m_i(\theta, \alpha; y) \geq m(\theta, \alpha; y)$ if $i \in C_1$

2) $\alpha \to K(\theta, \alpha) = \int_{\mathcal{X}} m(\theta, \alpha; y)\ \lambda(dy)$ is continuous with a unique minimum in $\alpha = \theta$.

In spite of their apparent complexity, conditions $M_1 - M_3$ are easy to manipulate. In the case of a translation invariant specification, continuity for π_0 and $\pi_{\partial 0}$ and the condition of proper representation of $\theta \to \pi_0(./., \theta)$ will imply $M_1 - M_3$. (see §V);

Theorem 3 :
Under conditions $M_1 - M_3$ for (Y, C), both coding estimator and PML estimator are consistent.

Proof :

Here as in the following results, the basic property is the conditional independence of the Y_i, $i \in C$, given $y_{\bar{C}}$. We first prove the sub-ergodicity lemma :

Lemma 4 : ([8], [11])

Let A be a measurable subset of $\mathcal{X}$, $\quad F_n(C_1, A) = c_{n,1}^{-1} \sum_{C_{n,1}} 1_{(Y_{\partial i} \in A)}$

Then, $P_\theta - a.s. : \quad \liminf_n F_n(C_1, A) \geq \frac{c}{2}\lambda(A)$

Proof of the lemma :

The variables $1_{(Y_{\partial i} \in A)}$, $i \in C_1$, have a mean greater than $c\lambda(A)$ $(M_2 - 3)$, bounded variance, and are, conditionally to $y_{\overline{\partial C}_1}$, independent. Then, from the strong law of large numbers for L^2-independent variables (see for example Breiman, theorem 3-27), we have :

$$\liminf_n F_n(C_1, A) \geq \frac{c}{2}\lambda(A) \quad P_{\theta, y_{\overline{\partial C}_1}} - a.s.$$

As both sides are independent of $y_{\overline{\partial C}_1}$, the same inequality is valid P_θ-a.s. ■

First, let us consider the C_1-coding contrast :

$$U_n^{C_1}(\alpha) = -c_{n,1}^{-1} \sum_{C_{n,1}} \log \pi_i(Y_i / y_{\partial i}, \alpha)$$

and write : $U_n^{C_1}(\alpha) - U_n^{C_1}(\theta) = A_n + B_n$, with :

$$A_n = -c_{n,1}^{-1} \sum_{C_{n,1}} \left(\log \frac{\pi_i(Y_i / y_{\partial i}, \alpha)}{\pi_i(Y_i / y_{\partial i}, \theta)} + m_i(\theta, \alpha; y_{\partial i}) \right)$$

A_n is a sum of zero mean variables of bounded variance $(M_2 - 1)$, conditionally independent given $y_{\bar{C}_1}$. Then, in the same way as in lemma 4, we have :

$$\lim_n A_n = 0 \quad P_\theta - a.s.$$

and the following relations P_θ-a.s. :

$$\begin{aligned} \liminf_n (U_n^{C_1}(\alpha) - U_n^{C_1}(\theta)) &= \liminf_n B_n \\ &\geq \liminf_n \int_{\mathcal{X}} m(\theta, \alpha; x)\ F_n(C_1, dx) \quad (M_3 - 1) \\ &\geq \int_{\mathcal{X}} m(\theta, \alpha; x)\ \liminf_n F_n(C_1, dx) \quad (m \geq 0) \\ &\geq \frac{c}{2} \int_{\mathcal{X}} m(\theta, \alpha; x)\ \lambda(dx) \quad := \frac{c}{2} K(\theta, \alpha) \quad \text{(from lemma 4)} \end{aligned} \tag{5}$$

Then condition (3) is satisfied for $U_n^{C_1}$ respectively to $\frac{c}{2}K(\theta, \alpha)$.

Consistency of the coding estimator $\hat{\theta}_n^C$:

We have :

$$U_n^C(\alpha) = \sum_{k=1,K} c_{n,k} c_n^{-1} U_n^{C_k}(\alpha)$$

where $U_n^{C_k}$ is the C_k-contrast defined in the same way as for C_1, and $c_{n,k}$ is the cardinal of $C_k \cap \overset{\circ}{D}_n$. The consistency of $\hat{\theta}_n^C$ ensures from (5), $M_1 - 2$, M_2 and corollary 2 with $k_0 = 1$.

Consistency of the pseudo-maximum likelihood estimator $\hat{\theta}_n$:

As we have $M_1 - 3$ and the following decomposition of U_n:

$$U_n = c_n d_n^{-1} U_n^C + (d_n - c_n) d_n^{-1} U_n^{\bar{C}}$$

the same conclusion applies for $\hat{\theta}_n$. ∎

IV- Conditional normality for $\hat{\theta}_n^C$ and unconditional chi-2 coding test for nested markovian hypotheses

$\hat{\theta}_n^C = \hat{\theta}_n^C(y_C, y_{\bar{C}})$ is a random variable depending on both stochastic realisations of Y on C and $\bar{C}$. We show that the asymptotic law of $\hat{\theta}_n^C$ conditionally to $y_{\bar{C}}$ is normal. Then we prove that the difference of coding contrasts test for nested hypotheses is a chi-2 test, unconditionally to $y_{\bar{C}}$.

Let $\pi^{(k)}$ be used for the k-derivative of π in the parameter.

(N_1) $\pi_i \ : \ \Omega_i \times \Omega_{\partial i} \times \Theta \to \mathbb{R}^+$ are of class C^3 and $\{\pi_i^{-1},\ \pi_i^{(k)},\ k = 1, 2, 3\}$ are uniformly in i bounded.

(N_2) There exists $V : \mathcal{X} \to \mathbb{R}^p$ such that for $i \in C_1$:

1) $E_{\theta, y_{\partial i}} [\pi_i^{(1)}(Y_i/y_{\partial i}, \theta)\ \pi_i^{(1)}(Y_i/y_{\partial i}, \theta)^t] \geq VV^t(y_{\partial i})$

2) $\int_{\mathcal{X}} VV^t(x)\ \lambda(dx)$ is regular

Define :

$Z_i = -\frac{\partial}{\partial \theta} \log\ \pi_i(Y_i/y_{\partial i}, \theta)$, $\quad I_i(y_{\partial i}) = \mathrm{Var}_{\theta, y_{\partial i}} Z_i$, and the conditional coding information :

$$I_n(\theta) = I_n(\theta, y_{\bar{C}}) = c_n^{-1} \sum_{C_n} I_i(y_{\partial i}) \tag{6}$$

Theorem 5 :

Suppose that $M_1 - M_3$, N_1, N_2 are satisfied ; then, conditionally to $y_{\bar{C}}$, we have :

$$I_n(\theta, y_{\bar{C}})^{1/2}\ \sqrt{c_n}\ (\hat{\theta}_n^C - \theta) \xrightarrow{w} N_p(0, I)$$

Proof :

In all this proof, $y_{\bar{C}}$ is a fixed configuration of Y on $\bar{C}$ and we work under the conditional probability $P_{\theta,y_{\bar{C}}}$. For simplicity we use U_n for U_n^C, $\hat{\theta}_n$ for $\hat{\theta}_n^C$.

For some $\theta_n^\star \in [\theta, \hat{\theta}_n]$, we have :

$$0 = \sqrt{c_n}U_n^{(1)}(\theta) + U_n^{(2)}(\theta_n^\star)\sqrt{c_n}(\hat{\theta}_n - \theta) \tag{7}$$

(i) Asymptotic normality of $\sqrt{c_n}U_n^{(1)}(\theta)$:

$\sqrt{c_n}U_n^{(1)}(\theta) = \dfrac{1}{\sqrt{c_n}}\sum_{C_n} Z_i$ where $\{Z_i,\ i \in C\}$ are zero mean independent variables, with bounded third moments. Then we can apply a CLT for independent non identically distributed variables (see Breiman theorem 9.2), provided the sufficient condition :

$$\liminf_n \Big(c_n^{-1}\ \mathrm{var}(\sum_{C_n} Z_i)\Big) \geq I_0 > 0$$

for I_0 a symmetric positive definite matrix. But this condition ensures from N_2 and lemma 4 :

$$\begin{aligned}
\liminf_n \Big(c_n^{-1}\mathrm{Var}(\sum_{C_n} Z_i)\Big) &\geq \liminf_n \Big(c_n^{-1}\mathrm{Var}(\sum_{C_{n,1}} Z_i)\Big) \\
&\geq \frac{\tau}{2}\liminf_n \Big(c_{n,1}^{-1}\mathrm{Var}(\sum_{C_{n,1}} Z_i)\Big) && (M_1 - 2) \\
&\geq \frac{\tau}{2c^2}\liminf_n (c_{n,1}^{-1}\sum_{C_{n,1}} E_\theta[\pi_i^{(1)}\pi_i^{(1)\ t}]) && (N_1 \text{ and } \pi_i^{-1} \geq c > 0) \\
&\geq \frac{\tau}{4c}\int_{\mathcal{X}} VV^t(v)\ \lambda(dv) \quad := I_0 && (\text{lemma 4 and } N_2)
\end{aligned}$$

(ii) On the other part, as $\theta_n^\star \to \theta$ in P_θ-probability, it is straightforward to obtain :

$$(U_n^{(2)}(\theta_n^\star) + I_n(\theta, y_{\bar{C}})) \longrightarrow 0 \quad \text{in } P_{\theta,y_{\bar{C}}} - \text{probability} \tag{8}$$

Theorem 5 ensures immediately from (7), (i), and (8). ■

Description of nested hypotheses $(H_p) \supset (H_q)$

(H_q) , $q < p$ is a q-dimensional sub-hypothesis of (H_p) defined as follows : let Λ be a compact set of $\mathbb{R}^q$, ϕ in the interior of Λ, $r : \Lambda \to \Theta$ with $\theta = r(\phi)$ and r of class C^2. Suppose that $R = \dfrac{\partial r}{\partial \beta}(\phi)$ is of full rank q. Then, $\{P_\alpha : \alpha = r(\beta),\ \beta \in \Lambda\}$ defines (H_q).

Let C **be a coding set for** (H_p), $\hat{\theta}_n$ (resp. $\hat{\phi}_n$) the coding estimator under (H_p) (resp. (H_q)), and write $U_n^C(\hat{\phi}_n) = U_n^C(r(\hat{\phi}_n))$. Then we have the following **unconditional** chi-2 asymptotic result :

Theorem 6 :

Suppose that $M_1 - M_3, N_1, N_2$ are satisfied ; then if (H_q) is satisfied :

$$\Delta_n = 2c_n\left(U_n^C(\hat{\phi}_n) - U_n^C(\hat{\theta}_n)\right) \xrightarrow{w} \chi^2_{p-q}$$

Proof :

Given a fixed realisation $y_{\bar{C}}$, we are going to show :

$$\Delta_n(y_{\bar{c}}) \xrightarrow{w} \chi^2_{p-q} \quad \text{under } \left(P_{r(\phi),y_C}\right) \tag{9}$$

As the limit law does not depend on $y_{\bar{c}}$, we get the same result unconditionally for $y_{\bar{c}}$, and this is the result of theorem 6.

Then let us fix $y_{\bar{C}}$, and work under $P_{r(\phi),y_C}$. To simplify, let us write U_n for U_n^C. When $\theta = r(\phi)$, we have :

$$U_n(\theta) = U_n(\phi)\ , \quad I_n(\phi) = R^t I_n(\theta) R \tag{10}$$

and $U_n^{(2)}(\theta_n^\star) + I_n(\theta)$ and $U_n^{(2)}(\phi_n^\star) + I_n(\phi) \longrightarrow 0$ in probability.

We obtain, in a standard way :

$$\Delta_n = \varepsilon_n^t B_n \varepsilon_n\ , \text{ with } \varepsilon_n = I_n(\theta)^{-1/2}\sqrt{c_n}U_n^{(1)}(\theta) \xrightarrow{w} N_p(0,I) \tag{11}$$

$B_n = I_n(\theta)^{1/2}F_n I_n(\theta)^{1/2}$, with

$F_n = U_n^{(2)}(\tilde{\theta}_n)^{-1}U_n^{(2)}(\theta_n^\star)U_n^{(2)}(\tilde{\theta}_n)^{-1} \ - \ RU_n^{(2)}(\tilde{\phi}_n)^{-1}U_n^{(2)}(\phi_n^\star)U_n^{(2)}(\tilde{\phi}_n)^{-1}R^t$

and $\tilde{\theta}_n, \theta_n^\star \in [\theta, \hat{\theta}_n^C]$, $\quad \tilde{\phi}_n, \phi_n^\star \in [\phi, \hat{\phi}_n^C]$

Let us define :

$$A_n(y_{\bar{C}}) = I - I_n(\theta)^{1/2} R I_n(\phi)^{-1} R^t I_n(\theta)^{1/2} \tag{12}$$

Then, as a consequence of (10), A_n is idempotent of rank $p-q$ and $(A_n - F_n)$ goes to zéro in probability. As ε_n converges weakly, (ε_n) is tighted so $\varepsilon_n^t(A_n - F_n)\varepsilon_n$ goes to zéro in probability and then we get (9). ∎

Remarks :

1) For each coding set C we have a coding test. But as these tests are correlated for two distinct coding sets, it is not easy in general to obtain a global statistic from different coding tests.

2) The chi-2 asymptotic test is unconditional when asymptotic normality of $\hat{\theta}_n^C$ considered in theorem 5 is conditional to $y_{\bar{C}}$. There are situations where $\hat{\theta}_n^C$ is unconditionally normal as $\hat{\theta}_n$:

- Y is ergodic ; then put $I(\theta) = \lim I_n(\theta, y_{\bar{C}})$ in place of $I_n(\theta, y_{\bar{C}})$ (see [12] in the case of the Ising model).

- Y is weakly dependent ; then put $\bar{I}_n(\theta) = E_\theta\big[I_n(\theta, Y_{\bar{C}})\big]$ in place of $I_n(\theta, y_{\bar{C}})$ (see [10]).
3) If Y is weakly dependent, it is possible to construct an explicit log pseudo-likelihood ratio test : the asymptotic law is a mixture of $p-q$ independent χ_1^2 weighted by the eigen values of a (non idempotent) matrix equivalent to A_n (12) expressed in the terms of both hypotheses, see [11]. In the context of Whittle gaussian contrast for a second order stationary random field, see [16].

V- The case of a translation invariant specification

Suppose that Y is a MRF on $S = \mathbb{Z}^d$ with a translation invariant specification π, $\Omega_i = \Omega,\ |\partial i| = m,\ |\partial\partial i| = M,\ D_n = [1, n]^d$.

(M) : (M_1) For $a = (a_1, a_2, \cdots, a_d) \in \mathbb{N}^{*d}, \quad C \supseteq a\mathbb{Z}$

(M_2) Ω is compact and $\pi = \pi_0(y/v, \alpha),\ \pi_{\partial 0}(v/w, \alpha)$ are continuous in their three arguments and strictly positive.

(M_3) $\alpha \rightarrow \pi(./., \alpha)$ is proper : for each $\alpha \neq \beta$ there exists $(y, v) \in \Omega \times \Omega^m$ such that $\pi(y/v, \alpha) \neq \pi(y/v, \beta)$

(N) : (N_1) $\pi^{(k)},\ k = 1, 2, 3$ exist and are continuous in their three arguments

(N_2) $\displaystyle\int_{\Omega^m} E_\theta^v\left[\pi^{(1)}\pi^{(1)t}(Y/v, \theta)\right]\ \lambda(dv)$ is regular

If $(H_q) \subset (H_p)$ are nested hypotheses defined as in § IV, theorems 3 and 6 can be reformulated.

<u>Corollary 7 :</u>
Let Y be a MRF with translation invariant specification π and C a coding set for (H_p) :
(i) If condition (M) is satisfied, then both PML estimator and coding estimator are consistent.
(ii) If furthermore (N) is verified, then :

$$2c_n\left(U_n^C(\hat{\phi}_n^C) - U_n^C(\hat{\theta}_n^C)\right) \xrightarrow[(H_q)]{w} \chi_{p-q}^2$$

Let us consider two examples where π is given by a conditional energy :

$$\pi(y/v) = \Big(\sum_{z\in\Omega} \exp -U(z/v)\Big)^{-1} \exp -U(y/v)$$

Example 1 : Ising model ; $y \in \{-1,+1\}$, $U(y/v) = y\ \theta^t\ V(v)$, $\theta, V(v) \in \mathbb{R}^p$, $v \in \Omega^m$. Conditions (M) and (N) are satisfied if $\sum_{v\in\Omega^m} V(v)V(v)^t$ is regular.

Test of isotropy : If (H_3) is the four-nearest-neighbour's model, $\theta^t = (\alpha, \beta, \gamma)$, $V^t(v_{ij}) = (1,\ y_{i-1,j}+y_{i+1,j},\ y_{i,j-i}+y_{i,j+1})$, isotropy (H_2) $(\beta = \gamma)$ can be tested using coding estimators on $\mathbb{Z}^2_+ = \{(i,j),\ i+j \text{ is even}\}$ for both models. The test statistic is asymptotically a χ^2_1.

Example 2 : Models of K label pictures ([2], [15]) : $\Omega = \{1, 2, \cdots, K\}$ is a set of qualitative labels.

$$\begin{cases} U(Y_i = k/v_i, \theta) = \alpha_k - \sum_{l\neq k} \beta_{kl}\ n_i(l) \\ n_i(l) = \sum_{j\in\partial i} 1_{(Y_j=l)}\ ;\quad \beta_{kl} = \beta_{lk},\quad \alpha_K = 0,\quad k,l = 1,\cdots,K \end{cases}$$

$\theta = (\alpha_1, \alpha_2, \cdots, \alpha_{K-1}; \beta_{kl},\ 1 \le k < l \le K)$ is a proper parametrisation and conditions (M) and (N) follow. To test the exchangeability of the labels (that is $\alpha_1 = \cdots = \alpha_K = 0$, $\beta_{kl} = \beta$) we can use the $\chi^2(\frac{K^2+K-4}{2})$ coding test.

REFERENCES

[1] BESAG J. (1974) : Spatial interaction and the statistical analysis of lattice systems. *J.R.S.S. b*, *n* 36, *p*.192 – 236.

[2] BESAG J. (1986) : On the statistical analysis of dirty pictures. *J.R.S.S. b*, *n* 48, *p*.259–302.

[3] BREIMAN L. (1968) : Probability. *Addison Wesley.*

[4] COMETS F. (1989) : On consistency for exponential families of Markov random fields on the lattice. *Prepub. n 89-30, Université d'Orsay, to appear.*

[5] COMETS F. et GIDAS B. (1988) : Parameter estimation for Gibbs distributions from partially observed data. *to appear in Ann. Appl. Prob.*

[6] DACUNHA-CASTELLE D. et DUFLO M. (1983) : Probabilités et statistiques, Tome 2 : Problèmes à temps mobile. *Masson.*

[7] FRIGESSI A. and PICCIONI M. (1990) : Parameter estimation for two-dimensional Ising fields corrupted by noise. *Stoch. Proc. and their Appl. 34, p.297-311.*

[8] GEMAN D. et GRAFFIGNE C. (1986) : Markov random field image models and their application to computer vision. *Proceeding of the International congress of math., Ed. A. M. Gleason, A.M.S., Providence.*

[9] GIDAS B. : Parametric estimation for Gibbs distributions, I : fully observed data. *to appear in "Markov random fields : theory and applications", Acad. Press.*

[10] GUYON X. (1987) : Estimation d'un champ gaussien par pseudo-vraisemblance conditionnelle : étude asymptotique et application au cas markovien. *in Spatial processes and spatial time series analysis, proceeding of 6th Franco-Belgian meeting of statisticians. Ed. Droesbeke, Pub. Fac. Univ. St Louis, Bruxelles, p.15-62.*

[11] HARDOUIN C. (1991) : Processus de contraste fort et applications. *Thesis, in progress.*

[12] GUYON X. et KUNSCH H.R. (1990) : Asymptotic comparison of estimators in the Ising model. *Same volume.*

[13] JANZURA M; (1988) : Statistical analysis of Gibbs random fields. *10th Prague conference 1986 "Inform. th., Stat. decis. func. and random processes", Reidel publishing comp., p.429-438.*

[14] JENSEN J.L. and MØLLER J. (1989) : Pseudo-likelihood for exponential family models of spatial processes. Research report n 203, Dept. of th. Statistics, Univ. of Aarhus.

[15] STRAUSS D.J. (1975) : Analysing binary lattice data with nearest-neighbour property. *J.A.P. 12, p.702-715.*

[16] YAO J.F. (1990) : Méthodes bayésiennes en segmentation d'image et estimation par rabotage des modèles spatiaux. Thèse Univ. Orsay.

[17] YOUNES L. (1988) : Parametric inference for imperfectly observed Gibbsian fields. *Prob. Th. and Rel. Fields 82, n 4, p.625-645.*

ASYMPTOTIC COMPARISON OF ESTIMATORS IN THE ISING MODEL

Xavier GUYON
Université Paris I, 12 Place du Panthéon, 75005 Paris

Hans R. KÜNSCH
Seminar für Statistik, E.T.H. Zentrum, CH 8092 - Zurich.

1. Introduction

Because of their use as priors in image analysis, the interest in parameter estimation for Gibbs random fields has rosen recently. Gibbs fields form an exponential family, so maximum likelihood would be the estimator of first choice. Unfortunately it is extremly difficult to compute. Other estimators which are easier to compute have been proposed : the Coding and the pseudo-maximum likelihood estimator (Besag, 1974), a minimum chi-square estimator (Glötzl and Rauchenschwandtner, 1981 ; Possolo, 1986-a) and the conditional least squares estimator (Lele et Ord, 1986), cf the definitions below in section 2.2. - These estimators are all known to be consistent. Hence it is a natural question to compare efficiency among these simple estimators and with respect to the maximum likehood estimator. We do this here in the simplest non trivial case, the d-dimensional nearest neighbor isotropic Ising model with external field. We show that both the pseudo maximum likelihood and the conditional least squares estimator are asymptotically equivalent to a minimum chi-square estimator when the weight matrix for the latter is chosen appropriately (corollary 2). These weight matrices are different from the optimal matrix. Hence we expect also the resulting estimators to be different although in all our examples the maximum pseudo likelihood and the minimum chi-square estimator with optimal weight turned out to be asymptotically equivalent. In particular, our results do not confirm the superior behavior of minimum chi-square over pseudo maximun likehood reported in Possolo (1986a). By example, we show that conditional least squares and minimum chi-square with the identity matrix as weights can be worse than the optimal minimum chi-square estimator. Compared with the maximum likelihood, the easily computable estimators are not bad if the interaction is weak, but much worse if the interaction is strong. Our results suggest that their asymptotic efficiency tends to zero as one approaches the critical point. This is confirmed by considering the Gaussian analogue in section 5.

Estimation for Gibbs random fields is also of theoretical interest because it provides examples with nonstandard behavior of the maximum likelihood estimator : the speed of convergence may be faster than the square root of the number of observations, the limit law may be non-normal and the asymptotic variance not continuous (Pickard 1987 ; Comets and Gidas, 1991). These phenomena are due to phase transition and long range dependence. We show here that the easily computable estimators are much more nicely behaved : they are asymptoticaly normal with the usual speed of convergence for all parameter values at leat in the stationary and ergodic case.Finally we should like to mention that our comparison is being made under the assumption that the model holds. In imaging applications, this might not be a reasonable assumption because a realization of the Ising model is quite different from a real world image. When the underlying distribution is different from the model, the estimators will in general converge to different values. It seems to be an open problem how we then should compare the different estimator.

2. Definition of the model and the estimators

2.1. The Ising model. Let Ω be the configuration space of a binary random field on the d-dimensional lattice, i.e. $\Omega = \{-1, +1\}^{\mathbb{Z}^d}$. We equip Ω with the σ-field $\mathfrak{F}$ generated by the coordinate mappings $X_t : (x_s) \in \Omega \to x_t \in \{-1, +1\}$ $(t \in \mathbb{Z}^d)$. An Ising model is then a probability distribution P on $(\Omega, \mathfrak{F})$ such that the conditional distributions at a site t given the values outside t take the following form :

$$(2.1) \qquad P[X_t = x_t \mid X_s = x_s, s \neq t] = \frac{\exp(x_t(h + \beta v_t))}{2\cosh(h + \beta v_t)} =: \pi(x_t \mid v_t) \text{ where } v_t = \sum_{\| s-t \| = 1} x_s .$$

When $\beta > 0$, there is a tendency for values at neighboring sites to be alike. The parameter h regulates the preference of + 1's regardless of neighboring values.
The Ising model has been investigated extensively in statistical physics. We mention here some results without proof wich we are going to use later on.

A1. For $d = 1$, P is a stationary Markov chain with transition probabilities

$$(2.2) \qquad p := P[X_t = 1 \mid X_{t-1} = 1] = e^{-h} / (\cosh(h) + D(\beta, h))$$

$$(2.3) \qquad q := P[X_t = -1 \mid X_{t-1} = -1] = e^{h} / (\cosh(h) + D(\beta, h))$$

and marginal probability

$$(2.4) \qquad \alpha := P[X_t = 1] = \frac{1}{2}(1 + \sinh(h)/D(\beta,h)) \quad \text{where } D(\beta,h) = (e^{-4\beta} + \sinh(h)^2)^{1/2},$$

see Georgii (1988, section 3.2).

A2. For any finite $\Lambda \subset \mathbb{Z}^d$ the conditional distribution of $X_t, t \in \Lambda$, given $X_t, t \notin \Lambda$, takes the following form :

$$(2.5) \qquad P[X_t = x_t, t \in \Lambda \mid X_t = x_t, t \notin \Lambda] = \pi^\Lambda[x_\Lambda \mid x_{\partial\Lambda}] = (Z^\Lambda(x_{\partial\Lambda}))^{-1} \exp(-H^\Lambda(x_\Lambda \mid x_{\partial\Lambda}))$$

where Z^Λ is a normalizing constant and

$$(2.6) \qquad H^\Lambda(x_\Lambda \mid x_{\partial\Lambda}) = -h \sum_{t \in \Lambda} x_t - \frac{\beta}{2} \sum_{\substack{t, s \in \Lambda \\ \|t-s\|=1}} x_t x_s - \beta \sum_{\substack{t \in \Lambda, s \notin \Lambda \\ \|t-s\|=1}} x_t x_s .$$

$H^\Lambda(x_\Lambda \mid x_{\partial\Lambda})$ is called in physics the energy in Λ with boundary condition $x_{\partial\Lambda}$. For a proof, see e.g. Georgii (1988, chap. 1.2).

A3. For some parameter values (β,h) there is more than one probability distribution P satisfying (2.1). Let us denote the set of all these probabilities by $\mathcal{G}(\beta,h)$. If $|\mathcal{G}(\beta,h)| > 1$, one speaks of phase transition. The existence of phase transition makes the Ising model interesting for physics, but it is the cause of several difficulties when estimating β and h as we shall see later. The following is known about $\mathcal{G}(\beta,h)$, see e.g. Georgii (1988, Sections 6.2, 18.3.2 and the Bibliographical Notes) : If $d = 1$, then $|\mathcal{G}(\beta,h)| = 1$ for all β and h. If $d \geq 2$, and $\beta \geq 0$, $h \neq 0$, then $|\mathcal{G}(\beta,h)| = 1$. If $d \geq 2$, then $|\mathcal{G}(\beta,0)| = 1$ if $0 < \beta < \beta_c$ and $|\mathcal{G}(\beta,0)| > 1$ if $\beta_c < \beta$. If $d = 2$, then $\beta_c = \frac{1}{2} \log(1 + \sqrt{2})$ and $\mathcal{G}(\beta,0)$ consists of convex combinations of $P_+, P_- \in \mathcal{G}(\beta,0)$ with P_+ and P_- stationary, ergodic, $E_{P_+}[X_t] = -E_{P_-}[X_t] > 0$. If $d \geq 2$, then there are also nonstationary elements of $\mathcal{G}(\beta,0)$. Finally if $d \geq 2$ and $\beta < 0$, $|\mathcal{G}(\beta,0)| > 1$ occurs also for $h \neq 0$.

2.2. Estimators. Let Λ_n denote the cube $\{1, ..., n\}^d$ and $\bar{\Lambda}_n$ the set obtained by including all nearest neighbors of Λ_n. Assume that we have observed a single realization of an Ising model in $\bar{\Lambda}_n$ and that we want to estimate the parameters $\theta = (h,\beta)^T$ from it. We will investigate the following estimators.

B1) (conditional) Maximum likelihood (ML). This is defined by

$$\text{(2.7)} \qquad \hat{\theta}_n = \arg\min_{\theta} \left(\log Z_\theta^{\Lambda_n}(x_{\partial\Lambda_n}) + H_\theta^{\Lambda_n}(x_{\Lambda_n} \mid x_{\partial\Lambda_n}) \right)$$

At first this looks impossible to compute because there is no way to calculate $Z_\theta^{\Lambda_n}$ exactly. However Younes (1988) has shown that $\hat{\theta}_n$ can be computed by solving the normal equations with a stochastic gradient algorithm. As $n \to \infty$ it is known that

$$\text{(2.8)} \qquad p(\theta) = \lim |\Lambda_n|^{-1} \log Z_\theta^{\Lambda_n}(x_{\partial\Lambda_n})$$

exists and is independent of the boundary condition, see e.g. Georgii (1988, Theorem (15.30) a)). Still $p(\theta)$ is available in closed form only for $d = 1$ and $d = 2$, $h = 0$. For $d = 1$ the unconditional log likelihood can also be witten down, using (2.2) - (2.4).

B2) The Coding Estimator (Besag, 1974).
Let $\Lambda_{n,+} = \{t \in \Lambda_n \mid t_1 + ... + t_d \text{ is even}\}$ and $\Lambda_{n,-} = \{t \in \Lambda_n \mid t_1 + ... + t_d \text{ is odd}\}$ From (2.5) it follows that X_t, $t \in \Lambda_{n,+}$ are conditionally independent given X_t, $t \in \Lambda_{n,-} \cup \partial\Lambda_n$. The coding estimator is the conditional Maximum likelihood estimator of X_t, $t \in \Lambda_{n,+}$ which is equal to

(2.9) $$\hat{\theta}_n = \arg\min \sum_{t \in \Lambda_{n,+}} \{\log(\cosh(h+\beta v_t)) - hx_t - \beta\, x_t\, v_t\}$$

The minimization has to be done iteratively, but in contrast to B1) the function to be minimized is available in closed form.

B3) The Pseudo-Maximum likelihood Estimator (PML), (Besag 1974). B2) seems not to make full use of all data. If we let the sum in (2.9) run over all $t \in \Lambda_n$, the data are used more fully, but we are no longer maximizing a likelihood :

(2.10) $$\hat{\theta}_n = \arg\min \sum_{t \in \Lambda_n} \{ (\log(\cosh(h+\beta v_t)) - hx_t - \beta\, x_t\, v_t \}$$

Nevertheless this estimator has reasonable properties as we shall see.

B4) The Minimum-chisquare Estimator (MCh) (Glötzl-Rauchenschwandtner 1981, Possolo 1986a). It is derived from an analogy of the Ising model with logistic regression. It is defined as

(2.11) $$\hat{\theta}_n = \arg\min \sum_{v,v' \in V/} (y(v) - h - \beta\, v)\, A\, (v,v')\, (y(v') - h - \beta\, v').$$

where $V/$ is the set of possible values of $V_t = \sum_{\|s-t\|=1} X_s$, i.e. $V/ = \{ - 2d, -2d+2, ..., 2d\}$,

(2.12) $$y(v) = \frac{1}{2} \log (N(+, v) / N(- , v)),$$

(2.13) $$N(+,v) = |\Lambda_n|^{-1} \sum_{t \in \Lambda_n} 1_{[X_t = +1,\, V_t = v]}\,,\quad N(-,v) = |\Lambda_n|^{-1} \sum_{t \in \Lambda_n} 1_{[X_t = -1,\, V_t = v]}$$

The weight matrix A is symmetric and positive definite, otherwise arbitrary. The two most important cases are A = Id (ordinary least squares) and $A = [\mathrm{Cov}(y(v), y(v'))]^{-1}$ (generalized least squares) The latter is usually unknown and depends on θ, but it can be estimated using subsampling techniques, see Possolo (1986-b) and remark 1 below. In contrast to all previous proposals, this estimator is available in closed form :

(2.14) $$\hat{\theta}_n = \sum_{v \in V/} c(v)\, y(v)$$

where c(v) are the columns of the $2 \times (2d+1)$ matrix C :

(2.15) $$C = (B^T A\, B)^{-1}\, B^T A$$

and B is the $(2d+1) \times 2$ matrix with rows (1,v).

B5). Conditional Least Squares (CLS). (Lele and Ord, 1986). This is an analogy of a well known estimator in time series analysis. It is defined as

(2.16) $$\hat{\theta}_n = \arg\min \sum_{t \in \Lambda_n} (x_t - \mu(\theta, v_t))^2$$

where $\mu(\theta, v_t) = E_\theta [X_t \mid X_s, s \neq t] = \tanh(h + \beta v_t)$.

We can define B4) and B5) also on the coding lattice $\Lambda_{n,+}$. This is mainly of theoretical interest in order to see how much information is lost when passing from Λ_n to $\Lambda_{n,+}$.

All the above estimators have a common structure : They are obtained by minimizing a certain criterion $L(x_{\bar{\Lambda}_n}, \theta)$:

(2.17) $$\hat{\theta}_n = \arg\min L(x_{\bar{\Lambda}_n}, \theta).$$

This will be used in the next section when investigating their asymptotics.

3. Asymptotics of the estimators

3.1. Consistency. All the above estimators are consistent for all parameter values, irrespective of phase transitions. The three basic steps are the same as for i.i.d. or time series situations :

i) Show that $|\Lambda_n|^{-1} L(x_{\bar{\Lambda}_n}, \theta) \to \ell(P_{\theta_o}, \theta)$ a.s. under P_{θ_o}.

ii) Show that $\arg\min_\theta \ell(P_{\theta_o}, \theta) = \theta_o$

iii) Justify the exchange of letting n tend to infinity and of taking the minimum.

Step iii) can be done with the classical argument of Wald (1949). Step ii) is straightforward in cases B2) - B5). In case B1) it follows from the variational principle (see Georgii, 1988, chapter 15). Step i) requises some care because of phase transitions. Assume first that P_{θ_0} is stationary and ergotic. Then i) follows from the ergodic theorem (in case B2 we first condition on $\mathbb{Z}_+^d$ and apply the law of large numbers for independent non-identical random variables). An arbitrary stationary $P_{\theta_0} \in \mathcal{G}(h_o, \beta_o)$ is a mixture of stationary and ergodic Ising models (Georgii 1988, Theorem (14.15)). Hence strong consistency for the latter imply also strong consistency for the former. The case where $P_{\theta_0} \in \mathcal{G}(h_o, \beta_o)$ is not stationary is the most difficult. Comets (1989, see also the same issue) has solved this using the method of large deviations (see also Guyon et Hardouin, (1990) where consistency is proved on a weaker condition).

3.2. Asymptotic normality of the estimators B3) - B5)

In spatial situations it is usually more difficult to determine the speed of convergence and the limit distribution of $\hat{\theta}_n - \theta_o$. Standard proofs of asymptotic normality for an estimator obtained by minimizing a criterion as in (2.17) consist of the following two steps :

i) By a Taylor expansion of $L(x_{\bar{\Lambda}_n}, \theta)$ of order two around θ_o one approximates $(\hat{\theta}_n - \theta_o)$ by an arithmetic mean.

ii) One shows a Central Limit Theorem for the approximating arithmetic mean, e.g. by checking mixing conditions.

The problem with step ii) is that for the Ising model the mixing conditions have been established only for some parameter values : If $|\beta|$ is small or $|h|$ large, Dobrushin's (1968) uniqueness condition holds and mixing follows (see Künsch, 1982). If $d = 2$, $h = 0$ and $\beta > \log(3)$ then mixing holds for the ergodic Gibbs measures $P_{\theta_o,+}$ and $P_{\theta_o,-}$ (Martin-Löf, 1973). Surprisingly we are able to carry out the above two steps for the estimators B3) - B5) for any $\theta_o \in \mathbb{R}^2$. We begin with step i).

Theorem 1. Assume that P_{θ_o} is stationary and ergodic. Then

i) For the Pseudo-Maximum-likelihood Estimator :

$$\hat{\theta}_n - \theta_o = \left(E_{\theta_o}\left[\sigma^2(\theta_o, V_t)\begin{pmatrix} 1 & V_t \\ V_t & V_t^2 \end{pmatrix}\right]\right)^{-1} |\Lambda_n|^{-1} \sum_{t \in \Lambda_n} (X_t - \mu(\theta_o, V_t)) \begin{pmatrix} 1 \\ V_t \end{pmatrix} (1 + o_p(1)).$$

ii) For the minimum-chisquare Estimator :

$$\hat{\theta}_n - \theta_o = |\Lambda_n|^{-1} \sum_{t \in \Lambda_n} P_{\theta_o}[V_t]^{-1} \sigma^{-2}(\theta_o, V_t) \; (X_t - \mu(\theta_o, V_t)) \; c(V_t) \; (1 + o_p(1)).$$

iii) For the Conditional Leat Squares estimator :

$$\hat{\theta}_n - \theta_o = \left(E_{\theta_o}\left[\sigma^4(\theta_o, V_t)\begin{pmatrix} 1 & V_t \\ V_t & V_t^2 \end{pmatrix}\right]\right)^{-1} |\Lambda_n|^{-1} \sum_{t \in \Lambda_n} (X_t - \mu(\theta_o, V_t))$$

$$\sigma^2(\theta_o, V_t) \begin{pmatrix} 1 \\ V_t \end{pmatrix} (1 + o_p(1)).$$

Here $\sigma^2(\theta, v) = \mathrm{Var}_\theta [X_t \mid X_s, s \neq t] = 1 - \mu^2(\theta, v)$, $P_\theta[v] = P_\theta[V_t = v]$,

and $c(v)$ is given in (2.15).

Proof. : see Appendix A1

From this we obtain the following

Corollary 2 :

i) The pseudo-maximum likelihood Estimator is asymptotically equivalent to the minimum chisquare estimator with weight matrix $A = \mathrm{diag}(P_{\theta_0}[\,v\,]\,\sigma^2(\theta_o,v))$.

ii) The conditional Least Squares estimator is asymptotically equivalent to the Minimum chisquare estimator with weight matrix $A = \mathrm{diag}(P_{\theta_0}[\,v\,]\,\sigma^4(\theta_o,v))$.

Proof : This follows easily by inserting the definition of c (see (2.15)) into the right hand side of Theorem 1, ii) and comparing it with i) or iii) respectively. q.e.d.

Hence the class of minimum chisquare estimators is sufficiently rich to include the other simple estimators. However the weight matrices for B3), B4) with ordinary least squares, B4) with generalized least squares and B5) are all different. We expect that these four estimators will usually be not asymptotically equivalent. In specific situation they can still be equivalent, see Theorems 4 and 5 below.

Next we carry out step ii).

Theorem 3 : Assume that P_{θ_0} is stationary and ergodic. Then for all estimators B3) - B5) we have

$$|\Lambda_n|^{1/2}\,(\hat{\theta}_n - \theta_o) \xrightarrow{d} \mathcal{N}_2(0,\ \Gamma)$$

$\Gamma = \sum_{\|t\| \le 1} R(t)$ where $R(t)$ is the covariance matrix between two summands in the linearization given in Theorem 1 at lag t. (Note that $R(t) = 0$ if $\|t\| > 1$). Γ is strictly positive definite for all θ_o.

Proof : see Appendix A2.

This behavior has to be contrasted with the behavior of the maximum likelihood estimator where phase transitions can give rise to nonstandard phenomena, compare the discussion in Section 3.4 below.

Remark 1 : From Theorems 1 and 3 and their proofs we can deduce two more things. First we see that $|\Lambda_n|^{1/2}\,(y(v) - h_0 - \beta_0 v)_{v \in V/}$ is asymptotically normal and their asymptotic covariance matrix has elements

$$\sigma^{-2}(\theta_o,v)\, P_{\theta_o}[v]^{-1}\,\delta_{vv'} + \sigma^{-2}(\theta_o,v)\,\sigma^{-2}(\theta_o,v')\, P_{\theta_o}[v]^{-1} P_{\theta_o}[v']^{-1}$$

$$\sum_{\|t\|=1} E_{\theta_o}[\,(X_o-\mu(\theta_o,v))(X_t-\mu(\theta_o,v')]\, 1_{[V_o=v]}\, 1_{[V_t=v']}$$

Hence the optimal weight matrix A which is the inverse of the asymptotic covariance matrix is different from the weight matrices given in Corollary 2. So we expect the estimators not to be equivalent in general. The second fact we would like to note is that the results for the minimum chisquare estimator remain valid if A is replaced by a stochastic weight matrix A_n which converges to A as $n \to \infty$. Hence one can try to estimate the asymptotic covariances of (Y(v)). This can be done either parametrically by estimating all terms in the above formula or nonparametrically by a resampling procedures (Possolo, 1986b, Künsch, 1989).

3.3. Asymptotic normality of the estimator B2

In this case it is also possible to carry out Steps i) and ii) pour Section 3.2. For step ii) we first condition on $\mathbb{Z}_-^d$ so that we have an arithmetic mean of independant but not identically distributed random variables. Lindeberg's Theorem can be applied, and it turns out that the asymptotic covariances do not depend on the values we are conditioning on if P_{θ_o} is stationary and ergodic. For details see Guyon (1987).

3.4. Asymptotic behavior of the estimator B1

In this case both steps i) and ii) of Section 3.2. may fail. The asymptotic behaviour of the maximum likelihood estimator is still not fully known. Being a finite sum of C_∞ functions $Z_\theta^{\Lambda_n}(x_{\partial\Lambda_n})$ is also C_∞, but its limit $\lim_{n\to\infty} |\Lambda_n|^{-1} \log Z_\theta^{\Lambda_n}(x_{\partial\Lambda_n}) = p(\theta)$ (cf. (2.8)) is not differentiable for all parameter values : for instance if $\beta \geq 0$ then $p(\theta)$ is differentiable with respect to h iff $|\mathcal{G}(\theta)| = 1$ (Lebowitz and Martin-Löf, 1972). In order to carry out step i) we need that the first and second derivatives of $|\Lambda_n|^{-1} \log Z_G^{\Lambda_n}(x_{\partial\Lambda_n})$ converge and that the limits are independant of $(x_{\partial\Lambda_n})$. This holds for instance under Dobrushin's uniqueness conditions, see Künsch, 1982, section 5. We then get :

$$(\hat{\theta}_n-\theta_o)=I(\theta_o)^{-1}\,|\Lambda_n|^{-1}\begin{pmatrix} \sum_t (X_t - E_{\theta_o}[X_t]) \\ \frac{1}{2}\sum_{\substack{t,s\in\Lambda_n \\ \|t-s\|=1}} (X_tX_s - E_{\theta_o}[X_tX_s]) + \sum_{\substack{t\in\Lambda_n \\ s\notin\Lambda_n, \|t-s\|=1}} (X_tX_s - E_{\theta_o}[X_tX_s]) \end{pmatrix}\cdot(1+o_p(1))$$

where $I(\theta_0) = \lim\, |\Lambda_n|^{-1} \dfrac{\partial^2}{\partial\theta^2} \log Z^{\Lambda_n}(x_{\partial\Lambda_n}) =$

$$= \begin{pmatrix} \sum_{t\in\mathbf{Z}^d} \mathrm{Cov}_{\theta_o}(X_0, X_t) & * \\ \frac{1}{2}\sum_{t\in\mathbf{Z}^d} \mathrm{Cov}_{\theta_o}\Big(X_0, \sum_{\|t-s\|=1} X_tX_s\Big) & \frac{1}{4}\sum_{t\in\mathbf{Z}^d} \mathrm{Cov}_{\theta_o}\Big(\sum_{\|s\|=1} X_0X_s, \sum_{\|s-t\|=1} X_tX_s\Big) \end{pmatrix}$$

Under Dobrushin's uniqueness condition we can carry out also step ii) and thus find

$$|\Lambda_n|^{1/2}(\hat{\theta}_n - \theta_o) \xrightarrow{d} \mathcal{N}_2(0, I(\theta_o)^{-1}).$$

What can one say without Dobrushin's uniqueness conditions ? Efforts are concentrated on the case $d = 2$, $\beta > 0$. In Pickard (1987) the behavior of $\hat{\beta}_n$ if $h = 0$ is known is given. He shows that $a_n(\hat{\beta}_n - \beta_o)$ is asymptotically normal with $a_n = |\Lambda_n|^{1/2}$ for $\beta \neq \beta_c$, $a_n = |\Lambda_n|^{1/2} \log|\Lambda_n|$ for $\beta = \beta_c$. He does not consider exactly he same model as we do : instead of a stationary and ergodic model (i.e. either $P_{\theta_{o,+}}$ or $P_{\theta_{o,-}}$) he considers "free boundary conditions" which correspond to $\frac{1}{2}P_{\theta_{o,+}} + \frac{1}{2}P_{\theta_{o,-}}$. We expect however the same behavior of β_n under $P_{\theta_{o,+}}$ or $P_{\theta_{o,-}}$.

If also h is unknown we don't know any rigorous result published. At $\beta = \beta_c$ the variables (X_t) are strongly dependent and thus the sum defining $I(\theta_0)$ diverges, see Kindermann and Snell (1980). Because of this we expect a faster convergence rate than $|\Lambda_n|^{1/2}$. Comets and Gidas (1991) consider the so called Curie-Weiss model which corresponds to the limit case $d \to \infty$. There they find for instance non-normal limit distribution.

4. Some numerical examples

The estimators B2) - B5) are all relatively easy to calculate. Hence we would like to compare these amoung themselves and with the presumably optimal Maximum likelihood estimator. Unfortunately calculations are possible in closed form only in very special cases. Some of these are presented here.

4.1. The case $d = 2$ and $\theta_0 = 0$. We have

<u>*Theorem 4*</u> :

i) For $d = 2$ and $\theta_0 = 0$ the estimators B1), B3), B4) with optimal weights (generalized least squares) and B5) are all asymptotically equivalent and it holds

$$|\Lambda_n|^{1/2}(\hat{\theta}_n - \theta_o) \xrightarrow{d} \mathcal{N}_2\left[0, \begin{pmatrix} 1 & 0 \\ 0 & 0.5 \end{pmatrix}\right].$$

ii) In the same situation, the estimator B2) satisfies

$$|\Lambda_n|^{1/2}(\hat{\theta}_n - \theta_o) \xrightarrow{d} \mathcal{N}_2\left[0, \begin{pmatrix} 2 & 0 \\ 0 & 0.5 \end{pmatrix}\right].$$

iii) In the same situation, the estimator B4) with A = Id (ordinary lead squares) satisfies

$$|\Lambda_n|^{1/2}(\hat{\theta}_n - \theta_o) \xrightarrow{d} \mathcal{N}_2\left(0, \begin{pmatrix} 1.71 & 0 \\ 0 & 0.59 \end{pmatrix}\right).$$

Proof: see Appendix A3.

We see here that it is worthwhile to look for good weights when using the Minimum Chisquare estimator. The coding estimator loose efficiency for h but surprisingly not for β. Between the other estimators, there is no difference in this situation.

4.2. The one dimensional case

We make use of the Markov chain representation of (X_t) (cf. (2.2) - (2.4) and use the parameters of the transition matrix p and q instead of β and h. Note also that $|\Lambda_n| = n$.

<u>*Theorem 5*</u> : For $d = 1$ and the Maximum likelihood Estimator $\hat{\theta}_n$ we have

$$n^{1/2}(\hat{\theta}_n - \theta_o) \xrightarrow{d} \mathcal{N}_2(0, \Gamma) \quad \text{with}$$

$$\Gamma = \frac{2-p-q}{16\,p(1-p)q(1-q)}\begin{bmatrix} 4(p(1-q)^2 + q(1-p)^2) & 2(p-q) \\ 2(p-q) & p+q \end{bmatrix}$$

Proof: see Appendix A4.

For the orther estimator the calculations of the asymptotic covariances are not difficult in principle, but extremely time consuming. We therefore restrict ourselves to the case $h_0 = 0$.

Theorem 6 : If $d = 1$ and $h_0 = 0$, then the asymptotic covariance matrix Γ of $\hat{\theta}_n$ is given as follows

i) For the estimator B2), $\Gamma = \mathrm{diag}\left(\frac{p^2 + (1-p)^2}{p(1-p)}, \frac{p^2 + (1-p)^2}{8\,p^2(1-p)^2}\right)$.

ii) For the estimator B3) and B4) with generalized least squares, $\Gamma = \mathrm{diag}\left(\frac{1-p}{p}, \frac{1}{16\,p^2(1-p)^2}\right)$.

iii) For B5) we find

$$\Gamma = \mathrm{diag}\left(\frac{1}{2p(1-p)}\left\{\frac{1 + X(p)^5}{(1 + X(p)^3)^2} + 1 - 2p\right\}, \frac{1}{16\,p^2(1-p)^2}\right) \quad \text{where } X(p) = \frac{2p(1-p)}{p^2 + (1-p)^2}$$

p	0	0.1	0.2	0.3	0.4	0.5	0.6	0.7	0.8	0.9	1
B1)+B3)+B4)	∞	9	4	2.33	1.5	1	0.67	0.43	0.25	0.11	0
B5)	∞	10	6	3.1	1.5	1	0.67	0.55	1	1	1
B2)	∞	9.1	4.2	2.8	2.2	2	2.2	2.8	4.2	9.1	∞

Some numerical values for $\lim_{n \to \infty} n \operatorname{var} \hat{h}_n$

Proof: see Appendix A5

From theorems 4 and 5 we see that for estimating β, B1) is better than B3) - B5) and there are better than B2). These inequalities are strict except if $\beta_0 = 0$, i.e $p = 1/2$. Moreover as $\beta \to \pm\infty$ the relative efficiency of B3)- w.r. to B1) goes to zero, but the relative efficiency of B2) to B3) is bounded by 1/2. For estimating h , B1) is equivalent to B3) - B4) and these are better than B2). The relative efficiency of B2) w.r. to B1) tends to 1 as $\beta \to -\infty$ and to ∞ as $\beta \to \infty$.

Remark : As $\hat{h}_n$ and $\hat{\beta}_n$ are asymptotically independent (for $h_0 = 0$) for B1) - B5), we have the same asymptotic variance for $\hat{\beta}_n$ in the one parameter model with $h = 0$ known.

5. The Gaussian analogue

In 4.2 we have seen that the efficiency of B2) - B5) relative to the maximum likelihood estimator goes to zero if $\beta \to \infty$. In a certain sense ∞ is the critical point for d = 1 at which phase transition occurs. For $d \geq 2$ we have seen that B2) - B5) always converge with speed $|\Lambda_n|^{1/2}$ whereas the maximum likelihood estimator presumably converges faster at $\beta = \beta_c$. Here we investigate whether a similar phenomenon occurs in the Gaussian case. We thus assume that $(X_t, t \in \mathbb{Z}^d)$ is a centered stationary Gaussian field with the following local conditional distribution

$$(X_t \mid X_s, s \neq t) \sim \mathcal{N}\Big(\alpha \sum_{\|t-s\|=1} X_s, 1\Big).$$

The conditional density of X_t given X_s, $s \neq t$, is then

$$\frac{1}{\sqrt{2\pi}} \exp\Big(-\frac{1}{2}\big(x_t - \alpha \sum_{\|t-s\|=1} x_s\big)^2\Big) = \exp\Big(-\frac{1}{2} x_t^2 + \alpha x_t \sum_{\|t-s\|=1} x_s\Big) \cdot \frac{1}{Z(x_s, s \neq t)}$$

which shows the analogy to the Ising model with h = 0. This model is well known (see e.g. Besag (1976, 1977)). In particular it exists iff $d \leq 2$ and $\alpha < (2d)^{-1}$ or $d \geq 3$ and $\alpha \leq (2d)^{-1}$. For $d \geq 3$ $\alpha = (2d)^{-1}$ exhibits phase transition because (X_t) need not to have expectation zero any more. The estimators B1) - B3) can be defined easily. Besag (1976, 1977) gives the asymptotic variances of the estimators :

B1) $\lim_{n\to\infty} n \operatorname{Var} \widehat{\alpha}_n = 2 j^{-1}$ B2) $\lim_{n\to\infty} n \operatorname{Var} \widehat{\alpha}_n = \alpha(1 - 2 d\alpha\rho) / (d\rho)$

B3) $\lim_{n\to\infty} n \operatorname{Var} \widehat{\alpha}_n = \alpha^2(1 - 2 d\alpha\rho)^2 / (d\rho^2)$

where ρ is the correlation at lag one and

$$j = (2\pi)^{-d} \int_{[-\pi,\pi]^d} \Big(\sum_{i=1}^{d} \cos\lambda_i\Big)^2 \Big(1 - 2\alpha \sum_{i=1}^{d} \cos\lambda_i\Big)^{-2} d\lambda_1 \dots d\lambda_d$$

B1) is superior to B3), B3) is superior to B2) and the three estimators have the same efficiency in the situation of independance. The next result gives the rate of relative efficiency in terms of $\delta = 1-2\alpha d$ as $\delta \to 0$, $\delta > 0$, that is when α goes to the critical value from below.

Theorem 7 : As $\delta \searrow 0_+$, the rate of efficiency in δ for B1), B2), B3), are given by the following

Dimension d of the Lattice	B2	B3	B1
d = 1	$\delta^{1/2}$	δ	$\delta^{3/2}$
2	$(-\log\delta)^{-1}$	$(-\log\delta)^{-2}$	δ
3	0(1)	0(1)	$\delta^{1/2}$
4	0(1)	0(1)	$(-\log\delta)^{-1}$
≥ 5	0(1)	0(1)	0(1)

Proof : See Appendix A6.

When $d \geq 5$, the three methods have the same order of efficiency. B1) is strictly superior to B3) for $d \leq 4$ and B3) to B2) for $d \leq 2$.

Appendix

A.1. Proof of theorem 1

i) The normal equations for θ_n are as follows

$$\text{(A.1)} \qquad \sum_{t \in \Lambda_n} (x_t - \mu(\hat{\theta}_n, V_t)) \begin{pmatrix} 1 \\ V_t \end{pmatrix} = 0.$$

Because $\hat{\theta}_n$ is close to θ_0 by consistency we make a first order Taylor expansion of these equations.

Because $\frac{\partial}{\partial\theta}\mu(\theta, v) = \sigma^2(\theta, v)\begin{pmatrix} 1 \\ v \end{pmatrix}$, we obtain :

$$\text{(A.2)} \qquad 0 = \sum_{t \in \Lambda_n} (x_t - \mu(\theta_0, V_t)) \begin{pmatrix} 1 \\ V_t \end{pmatrix}^T \sum_{t \in \Lambda_n} \begin{bmatrix} \sigma^2(\tau_n^1, V_t) & \sigma^2(\tau_n^1, V_t)V_t \\ \sigma^2(\tau_n^2, V_t)V_t & \sigma^2(\tau_n^2, V_t)V_t^2 \end{bmatrix} - (\hat{\theta}_n - \theta_0)$$

where $\quad \| \tau_n^i - \theta_0 \| \leq \| \hat{\theta}_n - \theta_0 \| \ (i = 1, 2)$

Consider the first element of the matrix on the right hand side divided by $| \Lambda_n |^{-1}$:

$$|\Lambda_n|^{-1} \sum_{t \in \Lambda_n} \sigma^2(\tau_n^1, V_t) = E_{\theta_0}[\sigma^2(\theta_0, V_t)] + |\Lambda_n|^{-1} \sum_{t \in \Lambda_n} (\sigma^2(\theta_0, V_t) - E_{\theta_0}[\sigma^2(\theta_0, V_t)])$$

$$+ |\Lambda_n|^{-1} \sum_{t \in \Lambda_n} (\sigma^2(\tau_n^1, V_t) - \sigma^2(\theta_0, V_t))$$

The second term on the right hand side goes to zero by the ergodic theorem.The third term is bounded by $\sup_{v \in V/} \sup_{\| \theta - \theta_0 \| \leq \varepsilon} | \sigma^2(\theta, v) - \sigma^2(\theta_0, v) |$ on $\| \hat{\theta}_n - \theta \| \leq \varepsilon$.

As $\sigma^2(\theta,v) = 1 - \tanh(h+\beta v)$ and $V/$ is compact, this third term goes to 0 as ε goes to 0.

Arguing similarly for the other elements we find that the matrix on the right hand side of (A.2) converges in probability to $E_{\theta_0} \left[\sigma^2(\theta_0, V_t) \begin{bmatrix} 1 & V_t \\ V_t & V_t^2 \end{bmatrix} \right]$

This last matrix is strictly positive definite because

$$(a, b)\begin{bmatrix} 1 & v_t \\ v_t & v_t^2 \end{bmatrix}\begin{pmatrix} a \\ b \end{pmatrix} > 0 \text{ if } (a, b) \neq \text{const.}(v_t, -1).$$

This complete the proof of i)

ii) By a Taylor expansion of $\log(u)$ at $u = 1$

(A.3) $$y(v) = h_o + \beta_0 v + \frac{1}{2}\left[\frac{N(+, v)}{\pi_{\theta_0}(+ \mid v)P_{\theta_0}[v]} - \frac{N(-, v)}{\pi_{\theta_0}(- \mid v)P_{\theta_0}[v]}\right](1 + o_p(1))$$

The term in square brackets is

$$\left(\pi_{\theta_0}(+ \mid v)\,\pi_{\theta_0}(- \mid v)\,P_{\theta_0}[v]\right)^{-1} |\Lambda_n|^{-1} \sum_{t \in \Lambda_n} 1_{[V_t = v]}\left(\pi_{\theta_0}(- \mid v)\,1_{[X_t = +1]} - \pi_{\theta_0}(+ \mid v) 1_{[X_t = -1]}\right)$$

$$= \left(\pi_{\theta_0}(+ \mid v)\pi_{\theta_0}(- \mid v)P_{\theta_0}[v]\right)^{-1} \frac{1}{2} \mid \Lambda_n \mid^{-1} \sum_{t \in \Lambda_n} X_t - \mu(\theta_0, v))\, 1_{[v_t = v]} .$$

Now we plug (A.3) into (2.14). Because :

$$\sum_v c(v)\,(h_0 + \beta_0 v) = \theta_0 \quad \text{and} \quad \pi_{\theta_0}(+ \mid v)\pi_{\theta_0}(- \mid v) = \frac{1}{4}\,\sigma^2(\theta_0, v)$$

this gives :

(A.4) $$\hat{\theta}_n - \theta_0 = \mid \Lambda_n \mid^{-1} \sum_{t \in \Lambda_n} P_{\theta_0}[\,V_t\,]^{-1}\,\sigma(\theta_0, V_t)^{-2}\,c(V_t)\,(X_t - \mu(\theta_0, V_t))\,(1 + o_p(1))$$

iii) This is almost the same proof as in case i). We leave it to the reader. q.e.d.

A.2. Proof of theorem 3 :

Obviously it is sufficient to prove a Central Limit Theorem for,

$$S_n = \sum_{t \in \Lambda_n} Y_t$$

where $Y_t = (X_t - \mu(\theta_0, V_t))\; b(V_t)$, $b : V/ \to \mathbb{R}$ is arbitrary.

The crucial property is that :

(A.5) $$E_{\theta_0}[\,Y_t \mid X_s,\, s \neq t\,] = 0,$$

and Y_u is a function of X_s, $s \neq t$, if $\| t - u \| > 1$.

Our proof is a modification of the proof in Bolthausen (1982). At those places where he uses the mixing conditions, we use (A.5) above

Define : $$S_n = \sum_{t \in \Lambda_n} Y_t, \qquad S_{t,n} = \sum_{\substack{s \in \Lambda_n \\ \|t - s\| \leq 1}} Y_s, \qquad a_n = \sum_{t \in \Lambda_n} E_{\theta_0}[Y_t\, S_{t,n}]$$

$$S_n = a_n^{-1/2}\, S_n, \qquad S_{t,n} = a_n^{-1/2}\, S_{t,n}$$

This is the same as in Bolthausen (1982) except that we have a fixed $m = 1$ in the definition of $S_{t,n}$ instead of $m = m(n)$ increasing. Next we consider a decomposition

$$(i\lambda - S_n)\, e^{i\lambda S_n} = A_1 - A_2 - A_3$$

as in Bolthausen (1982). We have to show that $E_{\theta_0}[A_1], E_{\theta_0}[A_2], E_{\theta_0}[A_3]$ all tend to 0. But :

$$E_{\theta_0}[A_3] = a_n^{-1/2} \sum_{t \in \Lambda_n} E_{\theta_0} [Y_t \exp (i \lambda (S_n - S_{t,n}))] = 0$$

by first conditioning on X_s, $s \neq t$. Moreover

$$E_{\theta_0}[\mid A_2 \mid] \leq \text{const.}\ a_n^{-3/2} \sum_{t \in \Lambda_n} E_{\theta_0} [S_{t,n}^2] \rightarrow 0$$

because $S_{t,n}$ is bounded and $a_n \sim \mid \Lambda_n \mid \sum_{\| t \| \leq 1} E_{\theta_0} [Y_0 Y_t]$. Finally :

$$E_{\theta_0}[|A_1|] \leq \text{const.}\ E_{\theta_0} [\mid a_n^{-1} \sum_{t \in \Lambda_n} (X_t S_{t,n} - E_{\theta_0}[X_t S_{t,n}]) |]$$

$$= \text{const.}\ E_{\theta_0} [| \, |\Lambda_n|^{-1} \sum_{t \in \Lambda_n} (X_t R_t - E[X_t R_t] |] + 0 (|\Lambda_n|^{-1/d})$$

where $R_t = \sum_{\| s - t \| \leq 1} Y_s$ because $R_t = S_{t,n}$ except for t on the boundary. Hence $E_{\theta_0} [\mid A_1 \mid] \rightarrow 0$ by the L_1-version of the ergodic theorem.

To prove that Γ is strictly positive définite, we use a conditioning argument that has been suggested to us by Jens Jensen, Aarhus, in a private communication. Let :

(1) $Y_t = (X_t - \mu(\theta_0, V_t)) b(V_t)$.

It is then sufficient to prove that for any $b : V \rightarrow \mathbb{R}$, $b \not\equiv 0$, it holds that

(2) $\mathrm{Var}\left(\sum_{t \in \Lambda_n} Y_t \right) \geq \text{const.} \mid \Lambda_n \mid$ with const. > 0

Consider a sublattice $L_0 = \{ t \in \mathbb{Z}^d \mid t = 4s, s \in \mathbb{Z}^d \}$ containing every fourth point. Then

$$\mathrm{Var}\left(\sum_{t\in\Lambda_n} Y_t\right) = \mathrm{Var}\left(E\left[\sum_{t\in\Lambda_n} Y_t \middle| X_s, s\notin L_o\right]\right) +$$

$$E\left[\mathrm{Var}\left[\sum_{t\in\Lambda_n} Y_t \middle| X_s, s\notin L_o\right]\right] \geq E\left[\mathrm{Var}\left[\sum_{t\in\Lambda_n} Y_t \middle| X_s, s\notin L_o\right]\right]$$

$$\sim |\Lambda_n \cap L_o| \, E\left[\mathrm{Var}\left(Y_o + \sum_{\|t\|=1} Y_t \middle| X_s, \|s\| = 1 \text{ ou } 2\right)\right].$$

Here we have used that given X_s, $s \notin L_o$, the X_t, $t \in L_o$, are independent. Hence it is sufficient to show that

(3) $\mathrm{Var}\left(Y_o + \sum_{\|t\|=1} Y_t \middle| X_s, \|s\| = 1 \text{ or } 2\right) > 0$

for some boundary condition X_s, $\|s\| = 1$ or 2. We have - suppressing θ_o in μ -

$$Y_o + \sum_{\|t\|=1} Y_t = X_o b(V_o) + \sum_{\|t\|=1} X_t b(W_t + X_o) - \sum_{\|t\|=1} \mu(W_t + X_o) b(W_t + X_o)$$

$$- \mu(V_o) b(V_o)$$

where $W_t = V_t - X_o$ for $\|t\| = 1$.

Hence (3) is equivalent to

(4) $2b(V_o) + \sum_{\|t\|=1} X_t \left(b(W_t + 1) - b(W_t - 1)\right)$

$$- \sum_{\|t\|=1} \{\mu(W_t + 1) b(W_t + 1) - \mu(W_t - 1) b(W_t - 1)\} \neq 0$$

Now to any $j \in \{-2d+1, ..., 2d-1\}$ we can choose X_s, $\|s\| = 2$, such that $W_t \equiv j$ ($\|t\| = 1$)[1]. In this case (4) becauses

(5) $2b(V_o) + V_o(b(j+1) - b(j-1)) - 2d.\mu(j+1)b(j+1) + 2d\mu(j-1)\, b(j-1) \neq 0.$

Assume first that $b \not\equiv$ const. Then $b(j+1) - b(j-1) \neq 0$ for some j. Chosse X_t, $\|t\| = 1$, such that $V_o = j+1$ and $V_o = j-1$ respectively. If (5) would not hold in both cases, we obtain by subtraction

$$2(b(j+1) - b(j-1)) + 2(b(j+1) - b(j-1)) = 0$$

which is a contradiction. If $b \equiv$ const. $\neq 0$ then (5) is equivalent to

---

[1] Checked only for $d = 1,2,3$

(6) $\mu(j+1) - \mu(j-1) \neq \frac{1}{d}$.

It is easily seen that (6) holds for some $j \in \{-2d+1, \dots 2d-1\}$. q.e.d.

A.3 Proof of theorem 4

i) We drop the index θ_0. It is easily checked that $\mu(v) \equiv 0$, $\sigma^2(v) \equiv 1$,

$$E[X_t] = E[V_t] = 0\,, \qquad E[V_t^2] = 4$$

$$E[X_t\, X_s] = \delta_{ts} \qquad E[X_t\, X_s\, V_t] \equiv 0$$

$$E[X_t\, X_s\, V_t\, V_s] = \begin{cases} E[V_t^2] = 4 & \text{if} \quad t = s \\ E[X_t^2]^2 = 1 & \text{if} \quad \| t - s \| = 1 \\ 0 & \text{otherwise} \end{cases}$$

Hence we obtain the asymptotic covariance matrix Γ in the following cases

MLE:
$$\Gamma = \begin{bmatrix} 1 & 0 \\ 0 & \frac{1}{4}\,(4+4.1) \end{bmatrix}^{-1} = \begin{bmatrix} 1 & 0 \\ 0 & 0.5 \end{bmatrix}$$

PML and CLS :
$$\Gamma = \begin{bmatrix} 1 & 0 \\ 0 & 4 \end{bmatrix}^{-1} \begin{bmatrix} 1 & 0 \\ 0 & (4+4.1) \end{bmatrix} \begin{bmatrix} 1 & 0 \\ 0 & 4 \end{bmatrix}^{-1} = \begin{bmatrix} 1 & 0 \\ 0 & 0.5 \end{bmatrix}.$$

For the minimum chisquare estimator we first calculate the asymptotic covariance Σ between $Y(v)$ and $Y(v')$ as given in remark 1, section 3 :

$$\Sigma = (P[v]P[v'])^{-1} \left(E[X_t^2\, 1_{[V_t=v]}\, 1_{[V_t=v']}] + \sum_{\|t-s\|=1} E[X_t X_s\, 1_{[V_t=v]}\, 1_{[V_s=v']}] \right)$$

If $\| t\text{-}s \| = 1$, $E[X_t X_s\, 1_{[V_t=v]}\, 1_{[V_s=v']}] = E[X_t\, 1_{[V_t=v]}].E[X_s\, 1_{[V_s=v]}]$

and by a straightforward calculation one obtains

$$E[X_s\, 1_{[V_t=v]}] = \begin{cases} \pm\frac{1}{16} & \text{if} \quad v = \pm 4 \\ \pm\frac{1}{8} & \text{if} \quad v = \pm 2 \\ 0 & \text{if} \quad v = 0 \end{cases}$$

Hence the asymptotic covariance matrix of the $Y(v)$'s is

$$\Sigma = \mathrm{diag}\left(16, 4, \frac{8}{3}, 4, 16\right) + (2, 1, 0, -1, -2)^T (2, 1, 0, -1, -2).$$

Its inverse is by the matrix inversion Lemma

$$\Sigma^{-1} = \mathrm{diag}\left(\frac{1}{16}, \frac{1}{4}, \frac{3}{8}, \frac{1}{4}, \frac{1}{16}\right) - \frac{1}{2}\left(\frac{1}{8}, \frac{1}{4}, 0, -\frac{1}{4}, -\frac{1}{8}\right)^T \left(\frac{1}{8}, \frac{1}{4}, 0, -\frac{1}{4}, -\frac{1}{8}\right)$$

Hence for generalized least squares we find :

$$\Gamma = (B^T \Sigma^{-1} B)^{-1} = \left(\begin{pmatrix} \frac{1}{16} & \frac{1}{4} & \frac{3}{8} & \frac{1}{4} & \frac{1}{16} \\ \frac{1}{8} & \frac{1}{4} & 0 & -\frac{1}{4} & -\frac{1}{8} \end{pmatrix} \begin{pmatrix} 1 & 1 & 1 & 1 & 1 \\ 4 & 2 & 0 & -2 & -4 \end{pmatrix}^T \right) = \begin{pmatrix} 1 & 0 \\ 0 & 2 \end{pmatrix}^{-1}$$

ii) $X_+ = \{ X_i, i \in \mathbb{Z}^d_+\}$ being stationary and ergodic, we have for the coding estimator B2) formula i) of theorem 1 with $\Lambda_{n,+}$ in place of Λ_n.

As $(X_t - \mu(\theta_0, V_t)), t \in \Lambda_{n,+}$ are independent given X on $\mathbb{Z}^d_-$, we obtain for general θ_o :

$$\lim_{n \to \infty} |\Lambda_{n,+}| \operatorname{Var} \widehat{\theta}_n = \left[E_{\theta_0}\left(\sigma^2(\theta_0, V_t) \begin{pmatrix} 1 & V_t \\ V_t & V_t^2 \end{pmatrix} \right) \right]^{-1}$$

which is result ii) of theorem 4.

iii) Finally for B4) with ordinary least squares

$$\Gamma = (B^T B)^{-1} (B^T \Sigma B)(B^T B)^{-1} =$$

$$= \begin{pmatrix} 5 & 0 \\ 0 & 40 \end{pmatrix}^{-1} \left(\begin{pmatrix} 16 & 5 & \frac{8}{3} & 4 & 16 \\ 104 & 28 & 0 & -28 & -104 \end{pmatrix} \begin{pmatrix} 1 & 1 & 1 & 1 & 1 \\ 4 & 2 & 0 & -2 & -4 \end{pmatrix}^T \right) \begin{pmatrix} 5 & 0 \\ 0 & 40 \end{pmatrix}^{-1}$$

$$= \begin{pmatrix} \frac{1}{25} \cdot \frac{128}{3} & 0 \\ 0 & \frac{1}{1600} \cdot 944 \end{pmatrix}$$

A4. Proof of theorem 5

For $\Lambda_n = \{1,2,\ldots n\}$ and conditional on $X_0 = x_0$, the log likelihood is $[N(1,1) \log p + N(1,-1) \log(1-p) + N(-1,1). \log(1-q) + N(-1,-1) \log q]$ with :

$$N(a, b) = \sum_{i=1,n} 1(X_{i-1} = a, X_i = b)$$

Let $N(a) = N(a,1) + N(a,-1)$; then the maximum likehood estimator is

$\widehat{p}_n = N(1, 1) / N(1)$ if $N(1) \neq 0 = 0$ otherwise

$\widehat{q}_n = N(-1, -1) / N(-1)$ if $N(-1) \neq 0 = 0$ otherwise

As $N = (N(a, b), a, b \in \{-1, 1\})$ is a multinomial with parameters

$$\left(n ; \frac{p(1-q)}{2-p-q}, \frac{(1-p)(1-q)}{2-p-q}, \frac{(1-p)\,(1-q)}{2-p-q}, \frac{(1-p)q}{2-p-q} \right)$$

we easily obtain asymptotic normality for θ_n with asymptotic variance :

$$\lim_{n \to \infty} n \,.\, \mathrm{Var}\,\widehat{\theta}_n = \mathrm{Diag}\left(\frac{p(1-p)\,(2-p-q)}{1-q}, \frac{q(1-q)(2-p-q)}{1-p} \right)$$

From (2.2) - (2.4) it follows that

$$h = \frac{1}{2} \log(p/q), \qquad \beta = \frac{1}{4} \log\,(p.q\,/\,(1-p)(1-q))$$

Hence the change of parameters from (h,β) to (p,q) is regular, and the result is obtained easily, q.e.d.

A5. Proof of theorem 6

i) For the coding estimator, we have, as noted in appendix A3 :

$$\lim_{n \to \infty} n\ \mathrm{Var}\,\widehat{\theta}_n = 2 I_\theta^{-1}, \qquad \text{with } I_\theta = E_\theta\left(\sigma^2(\theta, V_t) \begin{pmatrix} 1 & V_t \\ V_t & V_t^2 \end{pmatrix} \right).$$

Here we can compute explicity the expectation on the right hand side for any value (h, β). A direct calculus gives :

$$I_\theta = \frac{4(1-p)^2 q^2 (1-q)}{(2-p-q)((1-p)(1-q)+q^2)} \begin{bmatrix} 1 & -2 \\ -2 & 4 \end{bmatrix} + \frac{4\,p(1-p)q(1-q)}{(2-p-q)\,(p+q)} \begin{bmatrix} 1 & 0 \\ 0 & 0 \end{bmatrix}$$

$$+ \frac{4\,p^2(1-p)(1-q)^2}{(2-p-q)((1-p)(1-q)+p^2)} \begin{bmatrix} 1 & 2 \\ 2 & 4 \end{bmatrix}$$

When $h = 0$, $p = q$, we find $I = \mathrm{diag}\left(\frac{2p(1-p)}{p^2 + (1-p)^2}, \frac{16p^2(1-p)^2}{p^2 + (1-p)^2} \right)$ which gives result i).

ii) From part i) of theorem 1, we have for B3) : $\lim_{n \to \infty} n\ \mathrm{Var}\,\theta_n = I_\theta^{-1} J_\theta I_\theta^{-1}$

with : $\quad J_\theta = I_\theta + 2\,\mathrm{Cov}\left((X_0 - \mu(\theta, V_0)) \begin{pmatrix} 1 \\ V_0 \end{pmatrix}, (X_1 - \mu(\theta, V_1)) \begin{pmatrix} 1 \\ V_1 \end{pmatrix} \right)$

For $p = q$, we find : $\quad J = \mathrm{Diag}\left(\frac{4p(1-p)^3}{(p^2 + (1-p)^2)^2}, \frac{16p^2(1-p)^2}{(p^2 + (1-p)^2)^2} \right)$

which give ii) for B3).

For B4) with generalized least-square, we use equality ii) of theorem 1 with c given by (2.15). $(y(v) - h - \beta v)$ is asymptoticaly equivalent to $\varepsilon(v)$, the term in square bracket in (A3)

$$\varepsilon(v) = \frac{1}{2} |\Lambda_n|^{-1} \left(\pi(+|v)\pi(-|v)P(v) \right)^{-1} \sum_{t \in \Lambda_n} (X_t - \mu(\theta, v))\, 1_{[v_t = v]}$$

We find :

$$\Sigma = \lim_n n \operatorname{Var} \begin{pmatrix} \varepsilon(-2) \\ \varepsilon(0) \\ \varepsilon(2) \end{pmatrix} = \begin{bmatrix} 2 \cdot \frac{1-2p^2(1-p)}{p^2(1-p)} & \frac{2(1-2p)}{p(1-p)} & -\frac{4}{1-p} \\ * & \frac{4}{p} & \frac{2(1-2p)}{p(1-p)} \\ * & * & 2 \cdot \frac{1-2p^2(1-p)}{p^2(1-p)} \end{bmatrix}$$

As : $\lim_{n \to \infty} n \operatorname{Var} \widehat{\theta}_n = (B^T \Sigma^{-1} B)^{-1}$, with : $B = 2 \begin{pmatrix} 1 & -2 \\ 1 & 0 \\ 1 & 2 \end{pmatrix}$ the result for B4) with generalized least squares follows after some calculation.

iii) Looking at identity iii) in theorem 1, we obtain for estimator B5)

$$\Gamma = L^{-1} M L^{-1}, \text{ with } L = E_\theta \left[\sigma^4(\theta, V_t) \begin{pmatrix} 1 & V_t \\ V_t & V_t^2 \end{pmatrix} \right] , \quad M = M_1 + M_2$$

$$M_1 = E_\theta \left[(X_t - \mu(\theta, V_t))^2 \sigma^4(\theta, V_t) \begin{pmatrix} 1 & V_t \\ V_t & V_t^2 \end{pmatrix} \right] = E_\theta \left[\sigma^6(\theta, V_t) \begin{pmatrix} 1 & V_t \\ V_t & V_t^2 \end{pmatrix} \right]$$

$$M_2 = E_\theta \left((X_0 - \mu(\theta, V_0)) (X_1 - \mu(\theta, V_1))\, \sigma^2(\theta, V_0)\, \sigma^2(\theta, V_1) \begin{pmatrix} 2 & V_0 + V_1 \\ V_0 + V_1 & 2 V_0 V_1 \end{pmatrix} \right)$$

When $h = 0$, we can explicitly compute the expectation : we observe that each matrix is diagonal and we obtain our result after a long but straightforward calculation.

A6. Proof of theorem 7

Let us denote $i_k = \pi^{-d} \int_{[-\pi,\pi]^d} [1 - 2\alpha \sum_{i=1}^{d} \cos \lambda_i]^{-k} \, d\lambda_1 \ldots d\lambda_d$; Then we obtain :

$$\operatorname{Var} X = (i_1 .(1 - 2\alpha\rho d))^{-1} \quad \text{and } j = (1 - 2 i_1 + i_2) / (4\alpha^2)$$

Using the inequality : $\quad 1 - \frac{\lambda^2}{2} \le \cos \lambda \le 1 - \frac{\lambda^2}{2}\left(1 - \frac{\pi^2}{12}\right)$

we obtain the bounds : $\delta + a \, \| \lambda \|^2 \le 1 - 2\alpha \sum_{i=1}^{d} \cos \lambda_i \le \delta + b \, \| \lambda \|^2$

where $\| \lambda \|$ is the euclidian norm of λ, $b = \alpha$, $a = \alpha \, [1 - \frac{\pi^2}{12}]$.

Taking the integral in polar coordinates over the domain $r \in [0, 2\pi]$, $\theta_i \in [0, \pi/2]$, $i = 1, d - 1$, we obtain for some finite constant c(d) :

$$c(d) \int_0^{\pi} \frac{r^{d-1} \, dr}{(\delta + br^2)^k} \le i_k \le c(d) \int_0^{\pi} \frac{r^{d-1} \, dr}{(d + ar^2)^k}$$

When $\delta \to 0$, $\delta > 0$, i_k is of the order of j_k

$$j_k = \int_0^{2\pi} \frac{r^{d-1} \, dr}{(\delta + r^2)^k} = \delta^{\frac{d}{2} - k} \int_0^{\frac{2\pi}{\sqrt{\delta}}} \frac{u^{d-1} \, du}{(1 + u^2)^k}$$

and the result of theorem 7 follows.

Références

BESAG J., 1974, Spatial Interaction and the statistical analysis of lattice systems (with discussion), JRSS b 36, p. 192-236.

BESAG J., 1976, Parameter estimation for Markov fields, Tech. report serie n° 2, Dept. of Stat., University of Princeton.

BESAG J., 1977, Efficiency of pseudo-likelihood estimation for simple Gaussian fields, Biometrika, p. 616-618.

BOLTHAUSEN E., 1982, On the Central Limit Theorem for Stationary mixing random fields, The Annals of Proba, vol. 10, n° 4, pp. 1047-1050.

COMETS F., 1989, On Consistency of a class of estimators for exponential family of Markov random fields on the lattice, préprint. Orsay n° 89-30.

COMETS F. and GIDAS B., 1991, Asymptotics of Maximum Likelihood Estimator for the Curie - Weiss model, to appear in Ann. of Stat.

DOBRUSHIN P.L., 1968, The description of a random field by mean of conditional probability and conditions of regularity, Th. Proba. and its Appl., XII, p. 197-227.

GEORGII H.S., 1988, Gibbs measures and phase transition, Studies in Maths. 9, de Gruyter, Berlin.

GLÖTZL E., RAUCHENSCHWANDTNER B., 1981, On the statistics of Gibbsian processes, in the 1rst Pannomian on Math. Stat., Eds REVESZ P., SCHMETTERER L., ZOLOTAREV V.M., Lect. Notes in Stat. 8, Springer-Verlag.

GUYON X., 1987, Estimation d'un champ par pseudo-vraisemblance et application au cas markovien, Spatial Proc. and Sp. Time Series, Proced. 6th Franco-Belgian Meeting 1985, Ed. DROESBEKE, Pub. Fac. Univ. Saint Louis, Bruxelles, p. 15-62.

GUYON X. and HARDOUIN C., 1990, The Chisquare Coding test for nested Markov random Fields hypothesis, same volume.

KINDERMANN R. and SNELL J.L., 1980, Markov Random Fields and their Applications, Contemporary Maths, n° 1, AMS, Providence.

KÜNSCH H., 1982, Decay of correlations under Dobrushin's uniqueness conditions and its applications, Comm. Math. Phys. 84, p. 207-222.

KÜNSCH H., 1989, Jackknife and bootstrap for general stationary observations, Ann. Stat., vol. 17, n° 3, p. 1217-1241.

LEBOWITZ J.L., MARTIN-LÖF A., 1972, On the uniqueness of the equilibrium state for Ising spin system, Comm. Math. Phys. 25, p. 276-282.

LELE S.R., ORD J.K., 1986, Conditional Least Squares estimation for Spatial Processes : some asymptotics results, Tech. Rep. n° 65, Dept. Stat., The Pensylvania State Univ.

MARTIN-LÖF, 1973, Mixing properties, differentiability of the free energy and Central Limit Theorem for a pure phase in the Ising model at low temperature, Comm. Math. Phys, 32, p. 75-92.

PICKARD D., 1987, Inference for discrete Markov field : the simplest non trivial case, JASA 82, p. 90-96.

POSSOLO A., 1986-a, Estimation of binary Markov random fields, Tech. report n° 77, Dept. Stat., Univ. of Washington.

POSSOLO A., 1986-b Sub-sampling a random field, Tech. Report n° 78, Dept. Stat., Univ. of Washington.

WALD A., 1949, Note on consistency of the maximum likelihood estimate, Ann. Math. Stat., 20, p. 595-601.

YOUNES L., 1988, Estimation and Annealing for Gibbsian fields, Ann. Inst. H. Poincaré, 24, p. 269-294.

A REMARK ON THE ERGODICITY OF SYSTEMATIC SWEEP IN STOCHASTIC RELAXATION

CHII-RUEY HWANG, SHUENN-JYI SHEU

Institute of Mathematics
Academia Sinica
Taipei, TAIWAN 11529

ABSTRACT. Let S be a graph with n sites and L be a level set. Let Π be a Gibbs distribution on the configuration space $\Omega = \{x \mid x : S \to L\}$. For any s in S, let P_s denote a transition probability on Ω which is reversible w.r.t Π and changes only the level of a configuration at site s with positive probability. A systematic sweep σ is a bijection from $\{1, \dots, n\}$ to S. We study the ergodicity of the transition matrix $P^\sigma = P_{\sigma(1)} \cdots P_{\sigma(n)}$.

1. INTRODUCTION

Let S be a finite graph with n sites (nodes) and L be a finite level set with l levels. The configuration space Ω is defined by $L^S = \{x \mid x : S \to L\}$.

Consider a given Gibbs distribution Π on Ω. To sample from Π directly or to calculate the phase average (w.r.t.Π) of certain observable defined on Ω is usually not feasible due to the large configuration space Ω. Instead an iterative method by generating an ergodic Markov chain with Π as the equilibrium distribution is used to approximate Π. In this short note we will study the ergodicity of a commonly used generating mechanism.

For any $s \in S$, a transition matrix P_s on Ω is called a local dynamics (updating) at site s if for $y \neq x$,

$$\begin{aligned} P_s(x,y) &> 0 \quad if\ y(t) = x(t)\ for\ all\ t \neq s, \\ P_s(x,y) &= 0 \quad otherwise; \end{aligned}$$

and $\Pi(x)P_s(x,y) = \Pi(y)P_s(y,x)$.

To check Π is invariant under P_s is more difficult than to check the above reversibility condition in practice, since in the latter case one does not have to sum over the whole space but just checks the condition locally.

Note that P_s is self-adjoint w.r.t. the inner product defined by

$$< f,g >= \sum_{x \in \Omega} f(x)g(x)\Pi(x),$$

where $f, g \in R^{|\Omega|}$. Define $|f|^2 =< f, f >$. And for any linear transformation A from $R^{|\Omega|}$ to $R^{|\Omega|}$, $|A|$ denotes the norm of A induced by using $|\cdot|$ on $R^{|\Omega|}$.

Let σ be a bijection from $\{1, \ldots, n\}$ to S. The sites of S are visited according to the order $\sigma(1), \sigma(2), \ldots, \sigma(n)$ for updating. A systematic sweep according to σ has the following transition probability

$$P^\sigma = P_{\sigma(1)} \cdots P_{\sigma(n)}.$$

Note that $\Pi P^\sigma = \Pi$ but P^σ is not reversible w.r.t. Π in general. We will discuss the ergodicity of P^σ.

Related topics and other types of sweep strategy have been discussed in Amit and Grenander (1989), Barone and Frigessi (1990), Frigessi, Hwang , Sheu and Di Stefano (1990), S. Geman and D. Geman (1984), D. Geman (1990), Grenander (1989).

2. Results

Theorem 1. *If L has at least three levels, $l \geq 3$, then P^σ is ergodic for any systematic sweep σ.*

Proof. It is easy to see that for each x in Ω and for each $s \in S$, the set of all configurations which commute with x under P_s constitutes an ergodic class. Hence, P_s can be rearranged into block diagonal form consisting of ergodic matrices. Moreover for P_s^2, the entries in each ergodic class are strictly positive, hence, all the entries in $P^2_{\sigma(1)} \cdots P^2_{\sigma(n)}$ are strictly positive and ergodicity follows immediately.

To prove the ergodicity of P^σ, we first claim

$$\text{(1)} \qquad If\ |P_s f| = |f|,\ then\ P_s f = f.$$

Let λ_x denote the eigenvalues of P_s with the corresponding eigenvectors e_x chosen to form an orthonormal basis with respect to $< , >$. Since P_s can be put into block diagonal form consisting of ergodic matrices, hence, if $|\lambda_x| = 1$ then $\lambda_x = 1$.

$$f = \sum_x < f, e_x > e_x = \sum_{\substack{x \\ < f, e_x > \neq 0}} < f, e_x > e_x.$$

$$P_s f = \sum_{\substack{x \\ < f, e_x > \neq 0}} \lambda_x < f, e_x > e_x.$$

$|f| = |P_s f|$ implies

$$\sum_{\substack{x \\ < f, e_x > \neq 0}} < f, e_x >^2 = \sum_{\substack{x \\ < f, e_x > \neq 0}} \lambda_x^2 < f, e_x >^2.$$

Since $|\lambda_x| \leq 1$, the above equality together with a previous argument implies $\lambda_x = 1$ for $< f, e_x > \neq 0$. Therefore $P_s f = f$.

Now assume $|P^\sigma f| = |f|$. We will show that $P^\sigma f = f$ as follows: Since $|P_s| = 1$,

$$|f| = |P^\sigma f| = |P_{\sigma(1)} \cdots P_{\sigma(n)} f| \leq |P_{\sigma(n)} f| \leq |f|.$$

By (1), $P_{\sigma(n)} f = f$. This implies $|P_{\sigma(1)} \cdots P_{\sigma(n-1)} f| = |f|$. By induction, we have $P_{\sigma(k)} f = f$ for each $1 \leq k \leq n$. Therefore $P^\sigma f = f$.

Now we have established that if λ is an eigenvalue of P^σ with absolute value 1, then $\lambda = 1$. If we can further show that the eigenspace corresponding to 1 is one dimensional then we have proved the ergodicity.

By the same argument in the previous paragraph, $P^\sigma f = f$ implies $P_s f = f$ for all s. Hence,

$$P^2_{\sigma(1)} \cdots P^2_{\sigma(n)} f = f.$$

But $P^2_{\sigma(1)} \cdots P^2_{\sigma(n)}$ is ergodic. This implies the dimension of the eigenspace corresponding to 1 is one. □

For $l = 2$, the situation is more complicated. We will concentrate on the two dimensional stochastic Ising model with periodic boundary: i.e. S is a two-dimensional discrete torus with $n = m^2$ sites. Each site has four neighbors. $L = \{+1, -1\}$,

$$\Pi(x) = \frac{1}{Z_T} exp - \frac{1}{T} U(x),$$

where the energy function $U(x) = -\sum_{\substack{s,t \\ are\ neighbors}} x(s)x(t)$. We will consider the ferromagnetic case $T > 0$ only. For simplicity, let $T = 1$. Moreover we assume that there exists a function F such that for any $x \neq y$ and for any P_s with $0 < P_s(x,y)$, $0 < P_s(x,y) = F(U(y) - U(x))$. I.e. the local flipping rate depends only on the energy difference. The characterization of such dynamics can be found in D. Geman (1990), Frigessi, Hwang, Sheu and Di Stefano (1990).

Note that $0 \le F \le 1$, and $F(\delta) \le e^{-\delta}$ for $\delta \ge 0$, $F(\delta) = e^{-\delta} g(-\delta)$ for $\delta \le 0$. The Metropolis algorithm has $F_M(\delta) = e^{-\delta^+}$ which amounts to the optimal choice of F (D. Geman (1990)).

Consider the following two cases: the Metropolis algorithm and F with $F(\delta) < F_M(\delta)$ for $\delta \in R$, which has the Gibbs sampler as an example.

Theorem2. *If $F < F_M$ then P^σ is ergodic for any σ. For the Metropolis algorithm M there exists a σ such that M^σ is not ergodic.*

Proof. For $F < F_M$, it is easy to see that for each P_s, Ω can be put into ergodic classes w.r.t. P_s, moreover the corresponding entries in each ergodic submatrix are strictly positive. Hence P^σ is ergodic.

Now we discuss the Metropolis case.

For m odd, let σ be defined by the lexicographic order in $\{(i,j) : 1 \le j, k \le m\}$. Then M^σ is not ergodic.

In fact, consider the configurations

$$\begin{aligned} x(j,k) &= +1 \quad if\ j+k\ iseven, \\ &= -1 \quad if\ j+k\ isodd, \\ y &= (-1)x, \end{aligned}$$

$P^\sigma(x,y) = 1 = P^\sigma(y,x)$.

For even $m = 2q$, consider σ defined by

$\sigma(1) = (1,1), if\ \sigma(r) = (j,k)$, define

$$\begin{aligned}\sigma(r+1) =&(j, k+q) && if\ k \leq q,\\ &(j, k-q+1) && if\ 2q > k > q,\\ &(q+j, 1) && if\ 2q = k, j \leq q,\\ &(j-q+1, 1) && if\ 2q = k, 2q > j > q.\end{aligned}$$

Then M^σ is not ergodic. In fact, consider

$$\begin{aligned}x(j,k) = &+1 \quad if && \{j \leq q\ and\ k \leq q\}\ or\\ & && \{j \geq q+1\ and\ k \geq q+1\},\\ &-1 \quad otherwise,\end{aligned}$$

and $y \equiv (-1)x$. $M^\sigma(x,y) = 1 = M^\sigma(y,x)$. □

We still do not know how to characterize the set $\{\sigma \mid M^\sigma\ is\ ergodic\ \}$.

References

1. Y. Amit and U. Grenander, *Compare sweeping strategies for stochastic relaxation*, Div. Appl. Math. Brown Univ. (1989).
2. P. Barone and A. Frigessi, *Improving stochastic relaxation for Gaussian random fields*, Probab. in the Engineering and Inform. Sci. **4** (1990), 369-389.
3. A. Frigessi, C.-R. Hwang, S.-J. Sheu and P. Di Stefano, *Convergence rates of the Gibbs sampler, the Metropolis algorithm, and other single-site updating dynamics*, TR, Inst. Math., Academia Sinica, Taipei (1990).
4. D. Geman, *Random Fields and Inverse Problems in Imaging*, Lecture Notes in Mathematics 1427, Springer-Verlag, 1990.
5. S. Geman and D. Geman, *Stochastic relaxation, Gibbs distribution, and the Baysian restoration of images*, IEEE Trans. PAMI **6** (1984), 721-741.
6. U. Grenander, *Advances in pattern theory*, Ann. Statist. **17** (1989), 1-30.

APPLICATION OF BAYESIAN METHODS TO SEGMENTATION IN MEDICAL IMAGES

N. Karssemeijer

Department of Radiology, University of Nijmegen
P.O.Box 9101, 6500 HB Nijmegen, The Netherlands

Abstract

Two applications of Bayesian image analysis in medicine are discussed, simultaneous segmentation of 3D X-ray Computerized Tomography (CT) scenes and detection of microcalcifications in mammograms. Segmentation by iterative optimization based on Bayesian decision theory appears to be very fruitful if suitable models can be designed to capture prior knowledge of the structure of the image. In practice this requires development of spatial models in which the use of local context is extended from nearest neighbour relations to more complex descriptions.

1. Introduction

The use of digital pictorial information systems in medicine offers important advantages compared to the use of conventional film-based techniques. Until now, applications in clinical routine are mainly concentrated on image acquisition and image display. However, the use of medical workstations for viewing also stimulates interest in automated image analysis as a diagnostic tool. Especially for making quantitative measurements in images the use of fast reproducible computer analysis appears to be advantageous. An essential element of image analysis is segmentation of the image into relevant components. This mostly requires human interaction, in practice, such as manual outlining of regions of interest [1]. The necessity of such laborious procedures for obtaining a segmented image discourages exploration of medical image analysis in many applications. Therefore, segmentation is a key issue in medical imaging.

In medical images, a priori knowledge about the spatial structure of the scene to be segmented is often available. Human anatomy and the use of well defined protocols for data acquisition guarantee a common image structure for each particular type of exam. A variety of models and methods has been developed to use prior knowledge to guide segmentation. However, mostly these can only be applied in a very restricted context. The more recent introduction of stochastic models and Bayesian image analysis has provided a more general framework to model image data and to express prior knowledge [2-4]. Particularly random field models have been shown to be appropriate tools for modelling spatial context. Their use leads to parallel segmentation algorithms which are computationally attractive. These algorithms are basically iterative rules for updating pixel labels in a process of local competition and coöperation. A principal advantage of these methods is that they implement simultaneous exploitation of all information at hand, i.e. both the image data and prior models of the imaging process and scene structure. This is the point where a lot of traditional, sequential methods fail, which aim at finding one region after another, or which are based on a forced separation of image interpretation and low level segmentation. Segmentation cannot be optimized independent of an interpretation of the labels, and interpretation of a segmented image is often impossible without considering the image data and a noise model. Moreover, the notion that putting an optimal edge in between two neighbouring regions needs a model of both area's on each side of the boundary disfavours a sequential approach. To illustrate the possibilities of Bayesian image analysis this paper discusses two applications, segmentation of three dimensional X-ray CT-images and automated detection of microcalcification in mammograms.

2. Statistical methods for segmentation

In the Bayesian approach a segmentation X of a scene S is estimated by searching for pixel labels x_i with maximum a posteriori probability given the data Y. Two alternative choices are maximum a posteriori (MAP) estimation in which $p(X|Y)$ is maximized and maximization of posterior marginals (MPM) which aims at maximizing $p(x_i|Y)$ at each individual site [5]. Given a prior distribution $p(X)$ the posterior distribution of X given Y is

$$p(X|Y) \propto p(Y|X)p(X). \tag{1}$$

In practice suitable models for both the likelihood function $p(Y|X)$ and the prior $p(X)$ have to be adopted in relation to existing procedures for maximizing the posterior probability of X.

Contextual relations are introduced in the prior model $p(X)$. A framework for modelling local context has been developed describing X as a Markov Random Field (MRF). These models are defined in terms of local dependence of individual pixel labels x_i on the labels in their vicinity ∂i. The probability distribution corresponding to a MRF has the following form:

$$p(X) = \frac{1}{Z}\exp[-U(X)], \tag{2}$$

where $U(X)$ is called the energy function and Z is a constant for normalization. We will restrict ourselves to the class of pairwise interaction models. In that case the energy function is given by

$$U(X) = \sum_{i=1}^{N} A_i(x_i) + \sum_{<ij>} B_{ij}(x_i, x_j) \tag{3}$$

in which $< ij >$ denotes summation over all neighbouring pairs of sites in S and N is the total number of sites. A neighbour pair $< i,j >$ is by definition a pair for which $j \in \partial i$.

The interaction function B_{ij} will be assumed to be independent of the positions of the sites i and j in S. For the conditional probabilities at individual sites it can be easily derived that

$$p(x_i = k|x_{\partial i}) \propto \exp\left[-A_i(k) - \sum_{n=1}^{K} B_i(k,n)g_i(n)\right], \tag{4}$$

by comparing the energy functions of segmentations only differing at site i. Here $g_i(n)$ denotes the number of neighbours labelled as n and K is the number of classes.

The conditional probabilities in (4) are used in iterative updating schemes for estimation of the 'true' segmentation X^*. Using 'Iterative Conditional Modes' (ICM) suggested by Besag [3] this is done by each time choosing the most probable label x_i' at a selected site i given the current estimate of the rest of the labeling $\hat{X}_{S\setminus i}$ and the data Y:

$$x_i' = \arg \max_{x_i=1,\dots,K} \left[p(x_i|Y, \hat{X}_{S\setminus i})\right]. \tag{5}$$

Visiting each site in turn or at random the process will converge to a local maximum of the posterior. Using Bayes' relation and the Markov property the probability to be maximized can be written as

$$p(x_i|Y, \hat{X}_{S\setminus i}) \propto p(Y|x_i, \hat{X}_{S\setminus i})p(x_i|\hat{x}_{\partial i}). \tag{6}$$

In practice the logarithm of the expression is evaluated, which requires less computation. The method can only be applied if a good initial estimate X^* is available. This can often be obtained by maximum likelihood (ML), which maximizes the data term in (6) without considering neighbouring pixel labels.

Other methods suggested are based on sampling. If the posterior distribution can be described by a MRF, samples can be obtained by simulation of a Markov chain. Usually simulation is performed by iterative updating of pixel labels, choosing a new label as a random sample from its conditional distribution given by (6). In MPM estimation the labels at each site are chosen as the ones that occur most often during the simulation [6]. Simulated annealing [7] aims at finding the MAP estimate by generating a chain of samples from $p(X|Y)^{1/T}$, while slowly decreasing the parameter T. At small values of T the chain will converge to a global maximum of the posterior. In contrary to ICM these algorithms require no initial estimate. However, as many iterations are needed they are time consuming.

3. Segmentation of tissues in 3D images

Recognition of organs in 3D medical images is an important topic. For instance, in radio-therapy planning organ outlines are used to optimize dose distributions [8]. Also application of three-dimensional display techniques requires segmentation of the volumes to be displayed first [9]. Because the amounts of image data are usually very large, having to rely on human interaction means a severe limitation. Therefore there is an urgent need for fully automatic procedures. Until now, only relative simple cases can be handled, such as recognition of bone in X-ray imagery. Differentiation of soft tissues is much more

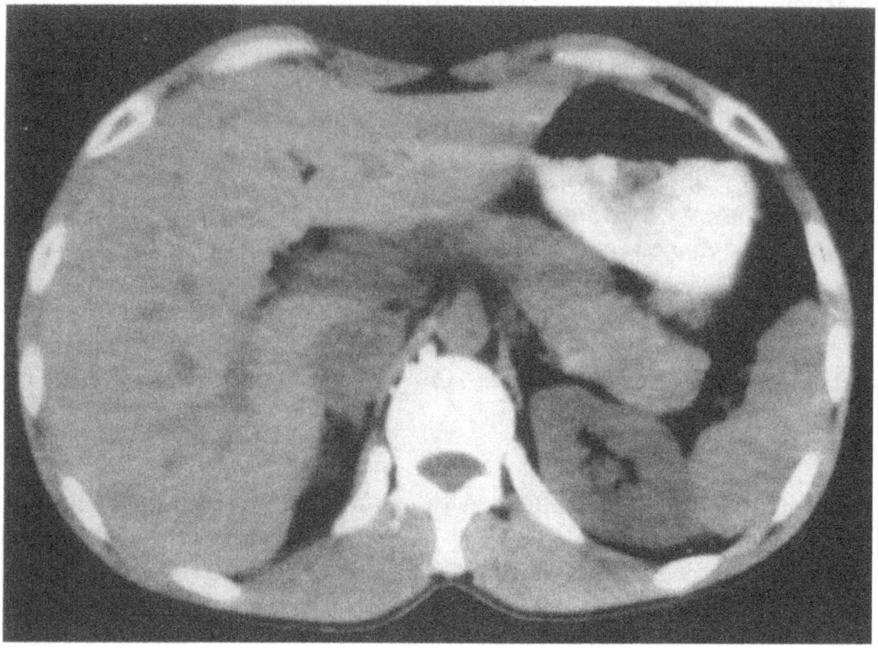

Figure 1. An example of an X-ray CT scan representing an abdominal cross section.

difficult. Magnetic resonance (MR) images, in which pixels are multi-valued variables, add a new dimension to segmentation and tissue characterization.

The problem is the complex spatial structure of the imaged scenes. To find a general scheme to represent prior knowledge about spatial context has turned out to be extremely difficult. Bayesian methods provide a way to make some progress. Combining their use with MRF models based on local contextual relations results in an elegant way to model degrees of spatial continuity. These models are not rich enough, however, to capture more subtle aspects of global shape. Models described by Grenander [10,11] do allow modelling global shape variations, but did not lead, so far, to algorithms applicable in medical practice. A more pragmatic approach is followed here by modelling the spatial structure of a particular class of images by means of a space dependent external field $A_i(k)$ of a MRF model as defined in (4). The strength of the field reflects the prior probability of a label to occur at a given site. Of course, this only makes sense if the imaged scenes have some static common structure. Also, they need to be positioned and scaled in the image matrix in the same way.

The method was applied to simultaneous recognition of twelve different tissue types in abdominal X-ray CT scans. An example of such a scan is shown in figure 1. A labelled set of 18 scan sequences was used. No selection was made with respect to age or sex of the patients. 'True' regions of the spleen, kidneys, pancreas, ribs, and spinal column as well as the outline of the ribcage were designated in each scan, based on a radiologist's

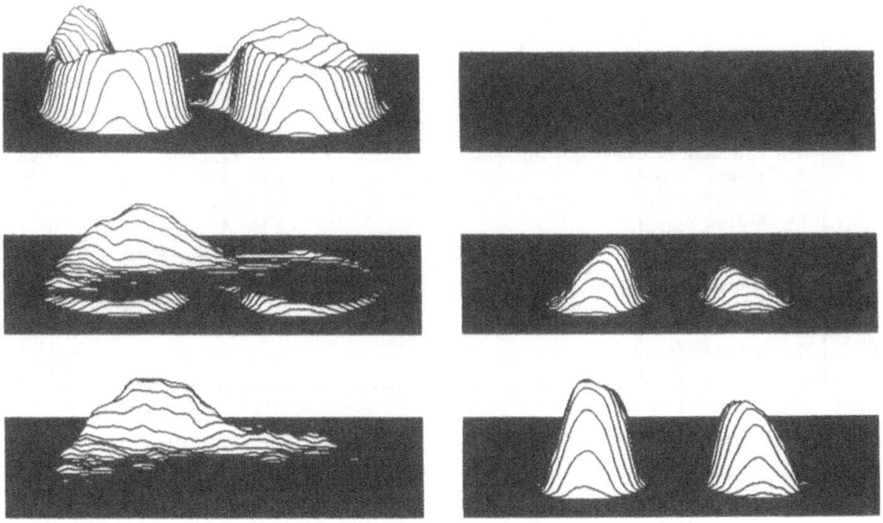

Figure 2. Estimated prior probability functions $p_i(k)$ for the kidneys (right) and air inside the body (left), representing upper, middle and lower levels of the abdomen.

interpretation. By thresholding inside and outside the rib cage the rest of the pixels were labelled as fat, air, contrast fluid, or other tissue. Reference points to normalize the image data, by translating and stretching to the same position and dimensions, were defined on the bone structure and could be extracted automatically.

The prior probability functions $p_i(k)$, representing abdominal geometry, were estimated from the given examples for each of the tissue classes k. Figure 2 shows two examples of these functions. In case of no interaction, by using (4), the external field $A_i(k)$ can be calculated straightforwardly from these functions with $A_i(k) = -\ln p_i(k)$. When spatial interaction is introduced by means of $B(k,n)$ such a simple relation does not exist. The results presented here were obtained by using the negative logarithm of the prior probability functions also after introduction of interaction. Mean densities $\bar{p}(y_i|x_i)$ too were calculated from the labelled image set. These were used to obtain the initial segmentation of a scan sequence using maximum likelihood. In a few cycles of ICM these estimates were updated on basis of the current labeling and the image data at hand.

To define local interaction ten neighbours were considered at each site i. A second order neighbourhood was used inside the scan plane through i and one neighbour was defined in each of the the two adjacent scan planes. Suitable values of the interaction parameters $B(k,n)$ were determined experimentally, minimizing the overall segmentation error by trial and error. They are listed in table 1. The more distant interaction between neighbour pairs in different scan planes was taken twice as small.

Both ICM and simulated annealing were applied to segment the images. An example is shown in figure 3. Error fractions were calculated for each scene, regarding the radiologist's contours as being correct. Mean error fractions, obtained by averaging over all 18 scenes, are listed in table 2. The same set was used for learning and testing, which

		1	2	3	4	5	6	7	8	9	10	11	12
1	spine	0.	1.	∞	∞	∞	∞	1.	1.5	1.	.5	1.	∞
2	ribs		−.5	2.	2.	∞	2.	1.	1.5	1.	.5	1.	∞
3	contrast			0.	2.	1.5	2.	1.	.5	1.	1.	1.	∞
4	spleen				−.3	4.	4.	1.5	1.5	2.	∞	2.	∞
5	pancreas					−.5	4.	∞	1.5	1.5	∞	1.5	∞
6	kidney						−.3	1.5	1.5	2.	∞	2.	∞
7	rest outside							0.	1.	1.5	1.	1.	1.
8	rest inside								0.	1.	1.5	1.	∞
9	fat inside									0.	1.5	1.	∞
10	fat outside										0.	1.	1.
11	air inside											0.	∞
12	air outside												0.

Table 1. Interaction parameter values.

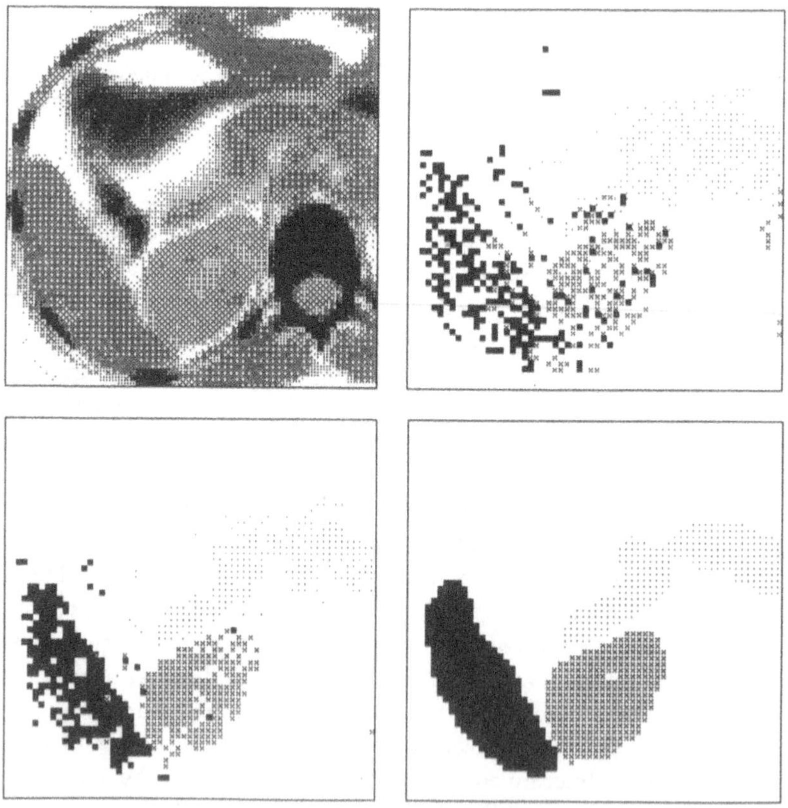

Figure 3. Segmentation by simulated annealing. Only three classes are shown: spleen (•), kidney (×) and pancreas (·). Samples at $T = 20$ and $T = 7$ are shown. The lower right figure shows the resulting segmentation at $T = 1$.

may have induced a slight positive bias of the results. Annealing results were obtained by running 80 iteration cycles, decreasing T by $T = 1/cn$, with n the number of the iteration cycle and c=.02. The experimental results show that the initial maximum likelihood labeling already is rather good, especially as far as the separation of contrast, spine, and ribs and the distinction between in- and outside of the abdomen is concerned. Improvement of the labeling during relaxation concentrates mainly on the internal organ classes four five and six and the rest class eight. Simulated annealing results for spleen and kidney appear to be somewhat better than those obtained by ICM, while for the other classes results are about the same. No accurate segmentation of the pancreas could be achieved, which is not surprising as in a number of cases also visually this organ could hardly be localized. For the other tissue classes the accuracy of the segmentation results is high enough for many clinical applications.

For the given application it was found that ICM is rather insensitive for the exact choice of the external field after the initial step. Satisfactory results were obtained also when using $A_i(k)$ only to restrict a label's possible occurrence to the support of its prior probability function. More details about this study can be found in [12].

class	1	2	3	4	5	6	7	8	9	10	11	12
	Initial Segmentation								E = 7.21 (*sd* 1.81)			
$n-$	2.5	5.9	0.7	11.5	57.2	9.6	5.3	11.1	1.6	7.7	0.0	0.7
$n+$	1.9	4.7	1.4	13.0	33.2	25.9	9.4	12.4	3.7	2.9	0.8	0.1
	ICM					8 cycles			E = 5.63 (*sd* 1.84)			
$n-$	2.4	5.2	0.5	6.4	50.5	4.8	3.7	8.9	1.9	6.5	0.0	0.5
$n+$	1.3	3.8	0.9	7.8	22.4	9.4	10.0	10.7	3.0	3.8	0.4	0.0
	Simulated Annealing					80 cycles			E = 5.47 (*sd* 1.56)			
$n-$	2.7	5.2	0.7	7.0	38.3	5.9	3.2	9.8	2.4	5.4	0.0	0.5
$n+$	1.1	4.2	1.0	5.2	32.2	6.9	10.8	9.3	2.6	4.7	0.3	0.0

Table 2. Mean percentages of incorrectly labelled pixels, taken relative to the 'true' number of pixels in each of the segmented classes, and averaged over all 18 scan sequences (112 scans). Missed pixels are denoted by $n-$ and false positives by $n+$. Values of E are the mean total error percentages. See table 1 for correspondence between class numbers and tissue types.

4. Detection of microcalcifications in mammography

Application of digital image analysis in mammography is investigated to increase the quality and efficiency of breast cancer screening programs. One of the most important radiological signs for early detection of breast cancer is the presence and appearance of microcalcifications. These are small bright spots against the varying background density of the projected X-ray image. On conventional screen/film recorded mammograms microcalcifications as small as 0.1 mm may be imaged. Much smaller spots can be identified histologically. Only clusters of calcifications are suspicious. Especially in a screening program, where more than 99% of the images are normal, automated detection of microcalcifications may improve the radiologist's performance to pick out abnormalities.

Procedures based on local thresholding to detect calcifications have been described by various authors [13-15]. Using these algorithms accurate segmentation is possible to some extent. The performance reduces when the threshold is lowered in order not to miss faint calcification clusters. The false positive rate increases strongly then, due to background structure and noise. Especially background structure is difficult to deal with. An example is shown in figure 4, which is a detail of a about 5 cm in cross section of a mammogram digitized with 12 bits CCD camera and a 0.1 mm pixel size. The line shaped microstructure, which represents connective tissue partitioning the breast in different compartments, is labelled incorrectly. The segmentation result was obtained by applying a fixed threshold on the image of relative pixel values y_i' defined by

$$y_i' = y_i - \sum_{j \in \partial i} y_j, \tag{7}$$

with y_i denoting the original image data at site i. The neighbourhood ∂i was chosen as a square area of 9 pixels in diameter. To optimize the performance of the algorithm the pixel values were rescaled first using an iso-precision criterion [16]. On this scale the variance of the pixel values due to film noise and digitization noise is constant over the whole range of pixel values. This is of great importance in algorithms in which the same detection criteria are applied at each grey level.

Automated calcification detection procedures mostly consist of a second stage, in which it is attempted to improve the result of thresholding. True calcification clusters are classified on basis of various criteria like shape, size, or cluster size. A disadvantage such a stepwise approach is that the rigorous reduction of the image data that takes place in the first step is not justified by a model concerning the nature of the structures to be detected. Difficulties in the second step do arise because image interpretation is based on unreliably segmented areas. The aim here is to avoid such sequential mechanisms by using iterative optimization and a statistical model of the image.

4.1 The image model

To apply statistical methods as described in section 2 both the appearance of microcalcifications and background have to be modeled. It is assumed that the background can be described by a global intensity pattern with independent additive noise superimposed on it. Thin bright patches of connective tissue like in figure 4 do not fit in such a description. Therefore a third class is introduced to model line-like structures. To isolate calcifications from the noisy background local contrast is important. Therefore relative pixel values y_i' have to be used rather than absolute pixel values. These were calculated using (7), with the same neighbourhood ∂i as chosen above.

A distinction between faint microcalcifications and connective tissue can only be made on the basis of shape, as these two structures do have a similar contrast. The most obvious approach to modelling this shape difference would be through the use of $P(X)$, because it reflects geometric properties of the imaged tissue, independent of the imaging

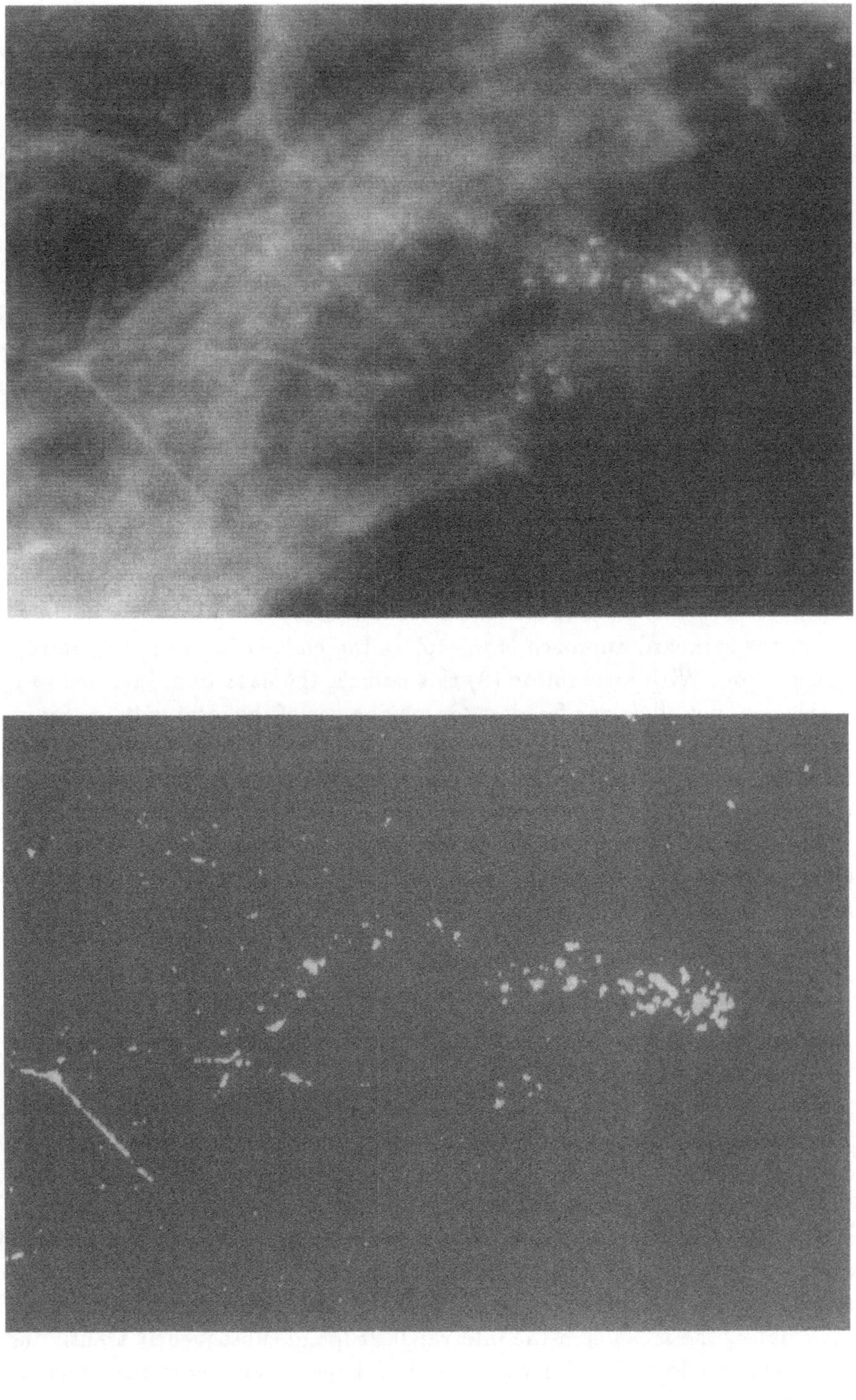

Figure 4. A mammographic detail showing a calcification cluster (top). The segmentation (bottom) was obtained by local thresholding. Thin layers of connective tissue cause significant errors.

process. Having in mind, however, that iterative optimization has to be used to label the scene, this approach does not seem very attractive. As it restricts shape modelling to a description of contextual relations between neighbouring labels, it rules out the possibility to obtain an initial estimate by maximum likelihood. This hampers fast local optimization by ICM, which needs a good initial estimate. Another serious disadvantage is that shape clues modeled in $P(X)$ can only enter the labeling process through slow propagation of labels during iteration. From the point of view of computation this is rather inefficient. Information to decide whether a pixel belongs to a line or dot structure can be extracted much more straightforwardly from the image data. Therefore, in a rule for iterative local updating of pixel labels using all current information at hand, which is the basic idea behind the ICM algorithm, the local data around the site to be updated should not be ignored. For that reason the image data is involved here in modelling local shape characteristics.

Starting point is the use of ICM and relation (6). Conditional independence of the records y_i is assumed, and the imaging process is supposed to be without blur. Consequently, the following relation holds

$$p(Y|X) = \prod_{i=1}^{N} p(y_i|x_i). \tag{8}$$

To proceed, the standard approach is to replace the current labeling $\hat{X}_{S\backslash i}$ in (6) by the true segmentation. With assumption (8) this reduces the data term involved to $p(y_i|x_i)$. Clearly, this does not allow any freedom for making use of the local data $y_{\partial i}$ for updating the label at i. Therefore, another approach is chosen which does permit involvement of the local data. Instead of taking the labels surrounding site i for true they are supposed to be that uncertain that it is better to omit them from the conditioning set of the data term. This restricts influence of the current neighbour labels to the interaction term. Regarding the remaining expression modelling the likelihood of the data the following is assumed:

$$f(Y|x_i) = f(y_i', \theta_i|x_i), \tag{9}$$

where y_i' is the relative pixel value calculated as in (7) and θ_i is a shape parameter extracted from the image data in a small area around i. This shape parameter indicates whether a line or dot feature at site i is likely or not.

Calculation of the shape parameter θ_i from the local data is explained in figure 5. The method resembles the Hough transform, which is used in image analysis for detection of straight lines [17,18]. At each site in a small neighbourhood of i the direction φ and magnitude r of the gradient are calculated. This data is accumulated in a (r, φ) table. The density $f_i(\varphi)$ of the gradient directions in the neighbourhood of i is determined from this table for $r > T$, with T a fixed threshold applied to reduce the influence of noise. After rescaling φ for normalization, by shifting the least occurring gradient direction to $\varphi = 0$, calculating the mean $\bar{\varphi}$ in the interval $[0, 2\pi)$, and subsequently translating again to put the mean gradient direction at $\varphi = \pi$, the shape parameter is calculated by

$$\theta_i^2 = \int_0^{2\pi} (\varphi - \varphi_b)^2 f_i(\varphi) d\varphi, \tag{10}$$

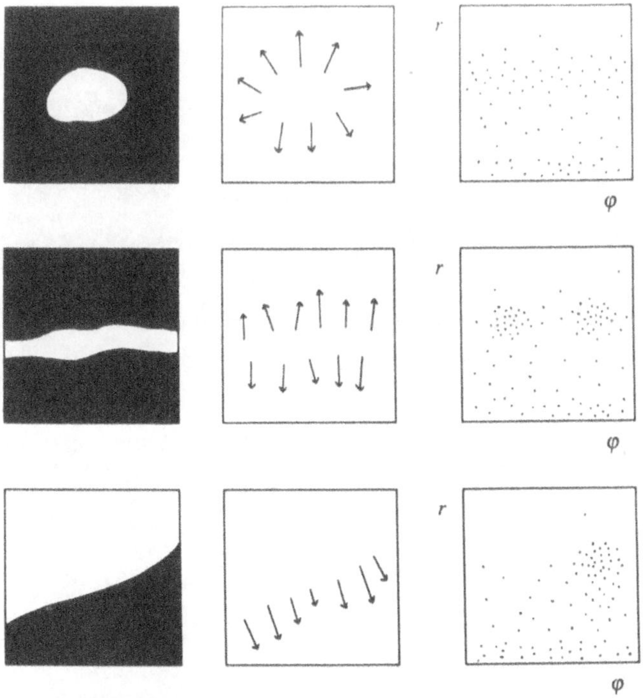

Figure 5. Calculation of the shape parameter θ_i from local data. Column (a) shows schematic diagrams of possible grey level configurations within a small neighbourhood of i. The gradient is represented in (b). In (c) this data is transformed to a (r, φ) space in which classification of local grey level patterns is more easy.

with $\varphi_b = \frac{1}{2}\pi$ in the interval $[0, \pi)$ and $\varphi_b = \frac{3}{2}\pi$ for $\varphi \in [\pi, 2\pi)$. Low values of θ_i indicate that a line structure at i is likely. An example of a shape parameter image is shown in figure 6. A 9×9 pixels neighbourhood was used. The image has been inverted for display purposes. Spatial continuity of the classes to be labelled is assumed to some extent. Due to the small size of the structures to be labelled constraints entering the labeling process through nearest neighbour interaction should be limited. Only the value of the parameter $B(k, n)$ modelling interaction between calcification and line labels has to be chosen high, to disallow patches consisting of a mixture of these two classes. Interaction with the background class has to be chosen small or zero, to allow calcifications consisting of only one or a few pixels to survive if they are bright enough. A second order neighbourhood is used to model nearest neighbour interaction. Apart from this, long range interaction is modeled between calcification and background. This permits bringing in the fact that calcifications occur in clusters. Faint isolated spots should not be interpreted as calcifications. The following model is used

$$p(x_i = k | x_{\partial i}) \propto \exp\left[-\sum_{n=1}^{3} B(k, n) g_i(n) - B'(k, n) h_i(n)\right], \tag{11}$$

in which $h_i(n)$ is the number of pixels labelled as n within a large surrounding of i. In the following all sites within 30 pixels distance are considered, making up a total of

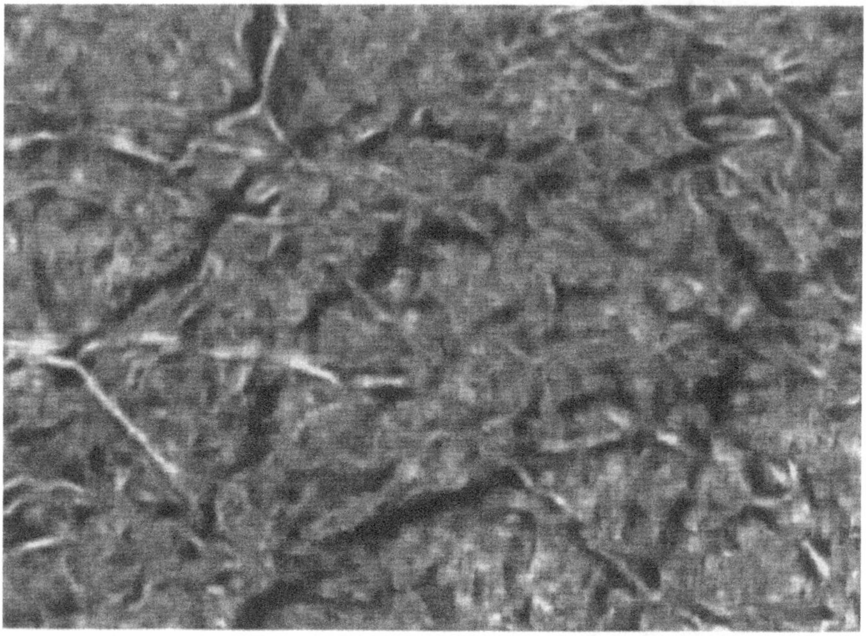

Figure 6. Shape parameter image of the mammographic detail shown in figure 4. Bright patches correspond to areas where connective tissue structures are more likely.

about 2800 distant neighbours. The model may seem to induce an excessive amount of computation to evaluate the long range neighbourhood at each site to be updated. Fast implementation is possible, however, because the number of calcifications is only small and because the only pairs of labels for which $B'(k,n)$ differs from zero are those for which k or n corresponds to calcification. Other pairs of labels do not contribute to the interaction sum. This means that only the number of calcification pixels in the surrounding of i has to be known in order to update x_i. This can be computed efficiently, after each iteration cycle, by a convolution of the current calcification mask with the indicator function of the neighbourhood. A dynamic space dependent field results, which strength relates to the local number of calcifications. Of course, this field can also be updated during labeling instead of at the end of each cycle.

4.2 Results

All digitized mammograms being used were scaled on an iso-precision scale [16]. Digitization was performed using a 0.1 mm pixel size, while the exposure was adjusted in such a way that the maximum output level of the CCD camera was reached at film base density.

To obtain the preliminary result presented here conditional independence of the two parameters y_i' and θ_i describing the local data has been assumed. For the given set of labelled training examples this appeared to be the case in first approximation. Both the densities $f(y_i'|x_i)$ and $f(\theta_i|x_i)$ were estimated from these examples. The following values for the interaction parameters parameters were determined experimentally: $B(1,2) = 0.25$, $B(1,3) = 1.5$, $B(2,3) = 5.0$, and $B'(2,1) = B'(2,3) = 3/N_l$, $B'(2,2) = -120/N_l$, with the labels 1 corresponding to background, 2 to calcification, and 3 to line shaped structures. The parameter N_l denotes the number of pixels in the long range neighbourhood. Other values follow from symmetry conditions, or else were put to zero. To prevent unrestrained expansion of calcifications in dense clusters it appeared to be necessary to limit the strength of the calcification field. A cut-off at two percent of the maximum value N_l was applied.

An example of the results obtained using the procedure described above is shown in figure 7. Comparison to the result of local thresholding in figure 4 shows a large improvement. Comparing the ML estimate and the result after relaxation the effect of interaction can be judged. Inside the cluster calcification pixels are added, whereas faint isolated spots disappear. Some of the bright single spots are film errors. Calcificaltions are adequately distinguished from connective tissue, although the labeling of the latter class itself is not very good. This may be explained from the learning stage, which has not yet been optimized. Only one rather poor example of line shaped structures was used.

5. Discussion

Bayesian image analysis provides a general approach to segmentation of images using prior knowledge. The use of MRF's to model spatial context makes the approach effective. In medical applications, nearest neighbour interaction models generally appear to be too poor to express existing prior beliefs. Specification of spatial context in a more global sense is necessary. In this respect MRF models including an additional interaction with a space dependent external field was shown to be useful. Given the class of images to be segmented has a common geometric structure such a field can be used to indicate regions in which tissues are more likely to occur than others. This idea resulted in a method for automatic recognition of organs in 3D-medical images. Also when modelling long range interaction between pixel labels for detection of calcification clusters the idea of using an additional field term was applied. Here it appeared as a fast implementation of the maximization procedure. The field strength indicates whether there are other calcifications nearby.

In mammographic images, which are recorded as planar projections, each pixel represents in fact a mixture of tissue classes. Therefore, individual labeling of sites based on the absolute pixel values is impossible. Only pixel values taken relative to the local grey levels are informative. It has been shown that a slight modification of ICM enables direct exploitation of the local data in the updating algorithm. Except for using relative grey levels this also offers possibilities to model local shape characteristics by means of parameter images. This provided a basis for an efficient algorithm to distinguish long linear connective tissue structures from faint calcification spots. Using this approach the idea of separate modelling of scene characteristics in $p(X)$ is abandoned. This makes a theoretical justification more difficult. Representation of the image data by parameter images is not new. In vision research it has been suggested as a fundamental step preceding

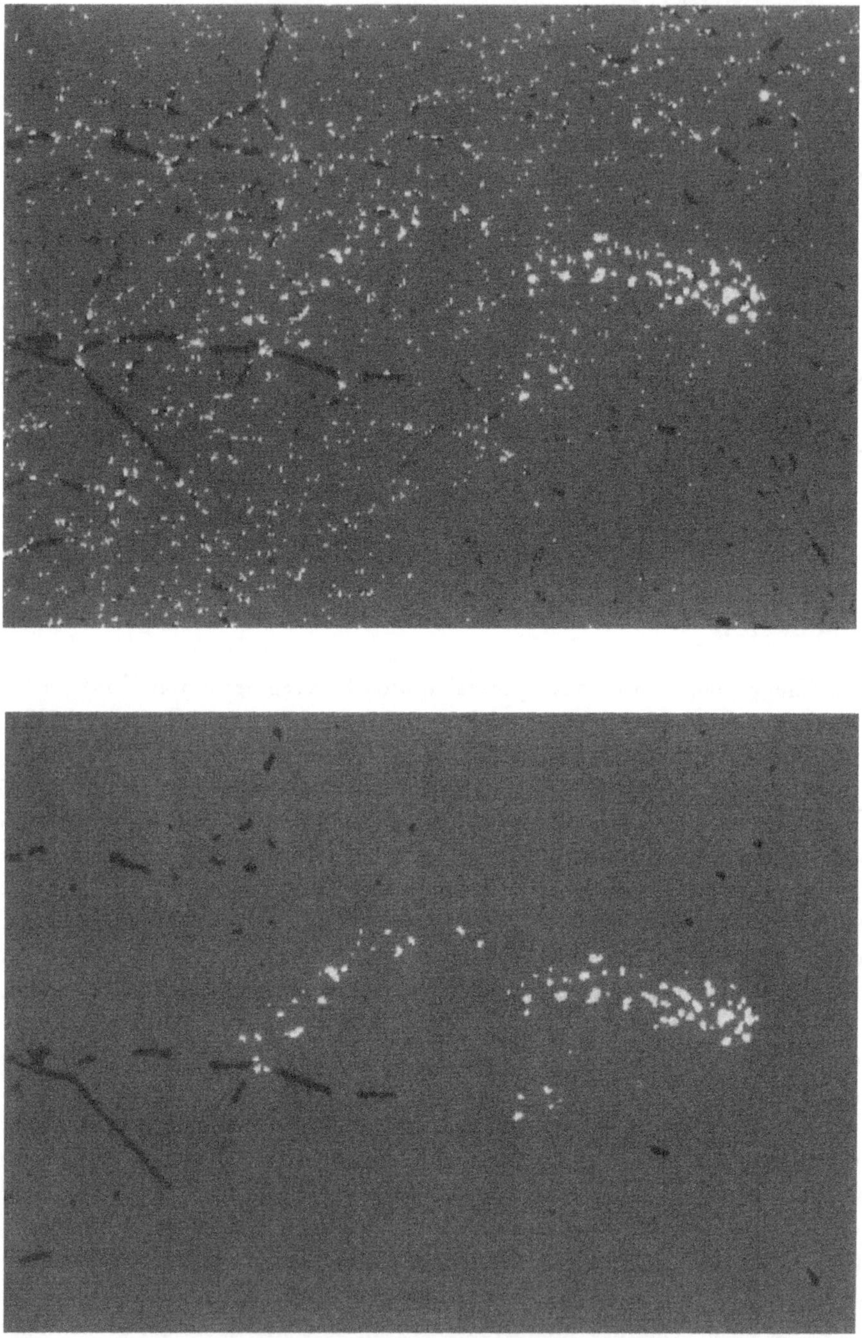

Figure 7. Calcification detection by iterative optimization applied to the image shown in figure 1. In the upper figure the initial ML estimate is shown. White represents calcifications while black stands for line-like structures. Convergence of the labeling to the result presented in lower figure was reached after 5 cycles.

more complex recognition tasks [19], supported experimental evidence. This gives an interesting parallel.

In the future more realistic models of long range interaction will be implemented for calcification detection. For instance, interaction strength could decrease more gradually instead of being constant within a certain range. Also the use of proportionality to the number of calcification pixels nearby probably isn't the best choice for definition of interaction. Large calcifications seem to have too much influence in that way. The number of calcifications could be used instead.

The work on calcification detection was financially supported by a grant of the 'Praeventie-fonds'.

References

1. Coggins J M (1988) Computer aided object definition from digital images. Proc. Joint US Scand. symp. on future directions of computer aided radiotherapy, San Antonio, Texas.

2. Hansen F R, Elliot H (1982) Image segmentation using simple Markov random fields. Comput. Graph. Image. Process 20:101-132

3. Besag J E (1986) On the statistical analysis of dirty pictures. J. Royal. Statist. Soc., Ser. B 48:259-302

4. Geman S, Geman D (1984) Stochastic relaxation, Gibbs distributions and the Bayesian restoration of images. IEEE Trans. Pattern. Anal. Machine. Intell. 6:721-741

5. Dubes R C , Jain A K (1989) Random field models in image analysis Journal of Applied Statistics 16 (2):131-164

6. Marroquin J, Mitter S, Poggio T (1987) Probabilistic solution of ill-posed problems in computational vision. Journal of the American Statistical Association 82(397):76-89

7. Kirkpatrick S, Gelatt C D, Vecchi M P (1983) Optimization by simulated annealing. Science 220:671-680

8. Schlegel W (1987) Progress and trends in computer assisted radiation therapy. In: Proc. Computer Assisted Radiology '87, Springer-Verlag, Berlin

9. 3-D Imaging in Medicine (1990) Höhne K H, Fuchs H and Pizer S M (Ed.). NATO ASI Series F Vol. 60, Springer-Verlag, Berlin.

10. Grenander U and Keenan D M (1989) Towards Automated Image Analysis, Journal of Applied Statistics 16 (2):207-221

11. Amit Y, Grenander U and Piccioni M (1989) Structural image restoration through deformable templates. (preprint) Division of Applied Mathematics, Brown University.

12. Karssemeijer N (1989) A statistical method for automatic labeling of tissues in medical images. Machine Vision and Applications 3:75-86

13. Fam B H, Olson S L, Winter P F and Scholz F J (1988) Algorithm for the detection of fine clustered calcifications on film mammograms. Radiology 169 (2):333-337

14. Chan H P, Doi K, Vyborny C J, Lam K L and Schmidt R A (1988) Computer aided detection of microcalcifications in mammograms. Investigative Radiology 23(9):664-671

15. Davies D H, Dance D R and Jones C H (1990) Automatic detection of clusters of calcifications in digital mammograms. SPIE Vol. 1233 Medical Imaging IV, Image Processing:185-189

16. Karssemeijer N and van Erning L J Th O (1991) Iso-precision scaling of digitized mammograms to facilitate image analysis. SPIE Vol. 1445 Medical Imaging V, Image Processing: 166-177

17. Ballard D H (1984) Parameter nets. Artificial Intelligence 22:235-267

18. Illingworth J and Kittler J (1988) A survey of the Hough Transform. Comput. Graph. Image. Process 44:87-116

19. Ballard D H (1986) Cortical connections and parallel processing: Structure and function. The Behavioral and Brain Sciences, 9:67-120

SOME SUGGESTIONS FOR TRANSMISSION TOMOGRAPHY BASED ON THE EM ALGORITHM

by

John T. Kent and Christopher Wright
Department of Statistics
University of Leeds
Leeds LS2 9JT
England

Summary

The standard reconstruction method in transmission tomography is convolution back-projection. The purpose of this work is to investigate the extent to which the quality of the reconstruction can be improved by taking into account the stochastic nature of the measurement process and the underlying regularity of the image. Some interim results are presented based on the EM algorithm.

1. Introduction

The research in this paper was motivated in some work done by Lewis Burkinshaw and his colleagues at the University of Leeds (Horsman et al., 1987). They were interested in measuring the trabecular density of the two bones in a cross-section of the human wrist using cone-beam tomography with low dose X-rays. As a first approximation the image can be regarded as two thin annuli. The low dosage of X-rays makes it difficult for the standard method of convolution back-projection (CBP) to discriminate clearly between trabecular bone, cortical bone and tissue. They compare CBP and maximum entropy (Burch et al., 1983) and come to the conclusion that for estimation of the density of trabecular bone, the maximum entropy method gives better results when the number of views is small (because CBP reconstructions exhibit ring artefacts) but that when the number of views is large there is little to choose between the two methods.

The purpose of this research is to see if the quality of the reconstruction can be improved by taking the stochastic nature of X-rays explicitly into account.

The outline of the paper is as follows. In Section 2 we review the Radon transform and its approximate inversion via convolution back-projection. In Section 3 we set up a stochastic formulation of the problem and describe how the EM algorithm can be used to implement maximum likelihood estimation. Several miscellaneous points including smoothing of the image are discussed in Sections 4-5. Some examples are given in Section 6.

2. The Radon transform

For any angle θ let

$$\ell_\theta = \begin{bmatrix} \cos\theta \\ \sin\theta \end{bmatrix}, \quad n_\theta = \begin{bmatrix} -\sin\theta \\ \cos\theta \end{bmatrix},$$

so that ℓ_θ is a unit vector in the direction θ and n_θ is a unit normal vector. Given θ, any point t in $\mathbb{R}^2$ can be written in the form $t = a\ell_\theta + bn_\theta$ with $a,b\in\mathbb{R}$.

Next let $f(t)$, $t\in\mathbb{R}^2$ be a well-behaved function, e.g. with compact support. The Radon transform of f is defined by the line integrals

$$P_f(b;\theta) = \int f(a\ell_\theta + bn_\theta)da.$$

The Radon transform is appropriate in transmission tomography because one observes the total attenuation along a ray through the image, which one can equate with this integral.

Next, consider the Fourier transform of the Radon transform

$$\tilde{P}_f(\beta;\theta) = \int e^{-ib\beta}\, P_f(b;\theta)db.$$

If we set $\omega = \beta n_\theta \in \mathbb{R}^2$ then it is easily checked that

$$\tilde{P}_f(\beta;\theta) = \iint e^{-i\omega t} f(t)dt$$

$$= \tilde{f}(\omega), \text{ say,}$$

the bivariate Fourier transform of $f(t)$. Thus, the Radon transform, observed for all b and θ, determines the Fourier transform of $f(t)$, and hence by the Fourier inversion formula, determines $f(t)$ itself.

Convolution back-projection (CBP) is an algorithm for reconstructing f based on an approximate inversion of the Fourier Transform; see e.g. Shepp and Kruskal (1978). Using polar co-ordinates, we can write

$$f(t) = \frac{1}{(2\pi)^2} \iint e^{it\omega}\, \tilde{f}(\omega)d\omega$$

$$= \frac{1}{(2\pi)^2} \int_0^{\pi} \int_{-\infty}^{\infty} e^{ib\omega}\, \tilde{P}_f(\beta;\theta)\ |\beta| d\beta d\theta$$

where $\omega = bn_{\theta}$. The fact that the function $|\beta|$ blows up as $\beta \to \infty$ causes numerical problems so we replace $|\beta|$ by

$$\tilde{\varphi}(\beta) = \begin{cases} \frac{2}{h}|\sin(h\beta/2)|, & |\beta| \le 2\pi/h \\ 0, & |\beta| > 2\pi/h. \end{cases}$$

Note that $\tilde{\varphi}(\beta) \cong |\beta|$ for small $|\beta|$. The parameter $h > 0$ represents the "resolution" and as $h \to 0$ the quality of the approximation improves.

Let $\varphi(b)$ denote the inverse Fourier transform of $\tilde{\varphi}(\beta)$. If we replace $|\beta|$ by $\tilde{\varphi}(\beta)$ in the above integral and use Parseval's relation, we obtain

$$f(t) \cong \frac{1}{(2\pi)^2} \int_0^{\pi}\int_{-\infty}^{\infty} e^{ib\beta}\, \tilde{P}_f(\beta;\theta)\tilde{\varphi}(\beta)d\beta d\theta$$

$$= \frac{1}{2\pi}\int_0^{\pi}\int_{-\infty}^{\infty} P_f(b';\theta)\varphi(b-b')db'd\theta\ .$$

The inner integral is a convolution of the Radon transform with $\varphi(b)$. The outer integral is the back projection of this result through the image.

Shepp and Kruskal (1978) describe how to implement a discretized version of this integral in practice. If the Radon transform is observed on a discrete grid, then typically the resolution h is chosen to be equal to the pixel width.

3. A stochastic formulation

A more realistic model for transmission tomography takes into account the fact that X-rays consist of a beam of photons forming a Poisson process. Further, the probability that a photon will be absorbed by an object is related to the attenuation coefficient of that object.

More specifically, consider a rectangular object of size $h_1 \times h_2$ with constant attenuation coefficient f and suppose a beam of X-rays forming a Poisson process of intensity λ strikes the box in a direction normal to the edge of width h_2. Then the number of photons entering the box, X say, has a Poisson distribution

$$X \sim P(\lambda h_2).$$

Similarly, the number of photons leaving the box, Y say, also has a Poisson distribution

$$Y \sim P(\lambda h_2 e^{-fh_1}).$$

Note that the orientation of the box relative to the beam of X-rays plays a role here.

Next suppose the object consists of a linear array of n square pixels each with edge length 1. Let f_j be the constant attenuation coefficient of the jth pixel. Consider a beam of X-rays passing through this line of pixels, and let

$$X_j = \text{number of photons entering pixel } j,$$
$$Z_j = X_{j+1} = \text{number of photons leaving pixel } j,$$
$$U_j = X_j - Z_j = \text{number of photons absorbed in pixel } j.$$

Also, suppose a detector is placed at the end of the array, and let $Y = X_{n+1} = Z_n$ denote the number of photons received at the detector. Note that it is only Y which is observed in this experiment, though the underlying intensity λ is also assumed known.

The distributions of these random variables can be easily written down:

$$X_j \sim P(\lambda\ e^{-F_{j-1}}), \quad j = 1,\ldots,n+1$$

$$U_j \sim P(\lambda(1-e^{-f_j})\ e^{-F_{j-1}}), \quad j = 1,\ldots,n,$$

where $F_0 = 0$ and $F_j = f_1+\ldots+f_j$. Further, U_j is independent of $X_{j+1},\ldots,X_{n+1}$. In particular, U_j is independent of Y, for $j = 1,\ldots,n$.

The object of tomography is to estimate the attenuation coefficients $\{f_j\}$ given the detector count Y. Natural estimates to use in this setting are maximum likelihood estimates, which can be calculated using a natural iterative algorithm, the EM algorithm. Of course in this section the estimation problem is grossly underdetermined with n parameters to estimate and only one observation. However, it is convenient to explain the methodology in this simplified setting in preparation for more realistic applications in later sections.

The essence of the EM algorithm in the tomography problem is to iteratively update estimates of the attenuation coefficients $\{f_j\}$. Given current estimates $\{\hat{f}_j^{old}\}$, we calculate the expected value of the unobserved counts given Y (the E or Expectation step). Next, treating the expected counts as if they were observed, we compute the MLEs of the attenuation coefficients to get updated estimates $\{\hat{f}_j^{new}\}$ (the M-step). The process is then repeated.

Writing $X_j = U_j+\ldots.+ U_n + Y$ and recalling that the U_j's are independent of Y, we obtain for the E-step

$$\begin{aligned} E[X_j|Y] &= Y + \lambda\ (e^{-F_{j-1}} - e^{-F_n}) \\ &= \hat{X}_j \text{ say.} \end{aligned}$$

Next for the M-step note that the distribution of Z_j given X_j follows a binomial distribution $B(X_j,\ e^{-f_j})$, where $Z_j = X_{j+1}$. Hence the updated estimate of f_j is

$$\hat{f}_j^{new} = \log(\hat{X}_j|\hat{Z}_j).$$

If $Y = \lambda\ e^{-\hat{F}_n}$ then it is easily checked that $\hat{f}_j^{new} = \hat{f}_j^{old}$, reflecting the fact that only $f_1+\ldots+f_n = F_n$ is identifiable in this simple setting.

4. Several beam directions

Next consider the case of a square $n \times n$ array of pixels indexed by $j = (j_1, j_2)$ and with 2 X-ray beams, in the horizontal (H) and vertical (V) directions respectively. A beam can be viewed as a collection of n parallel sub-beams, with each sub-beam passing through a linear array of n pixels. The E-step of the EM algorithm proceeds as before, treating each of the n horizontal and n vertical X-ray sub-beams separately. For the M-step we must combine the information from the horizontal and vertical beams. The estimated total number of photons entering pixel j is, in an obvious notation, $\hat{X}_j^{(H)} + \hat{X}_j^{(V)}$. The estimated total number leaving is $\hat{Z}_j^{(H)} + \hat{Z}_j^{(V)}$. Since each photon has probability e^{-f_j} of being unabsorbed in pixel j we get the updated estimate

$$\hat{f}_j^{(new)} = \log\{(\hat{X}_j^{(H)} + \hat{X}_j^{(V)})/(\hat{Z}_j^{(H)} + \hat{Z}_j^{(V)})\}.$$

In this setting, the notation "$j + 1$" does not make sense, therefore it is better to use the notation X_j and Z_j to denote the number of photons entering and leaving a pixel (in a particular direction).

However, complications arise when we move to more than two beam directions. The problem is that most of the X-ray beams will cross the pixel grid at an oblique angle. Lange and Carson (1984) attempted to deal with the problem by looking at the geometry of each X-ray beam and computing the length of intersection with each pixel in its path. This approach is time-consuming and leads to complications at the M-step of the algorithm.

In contrast, our approach is to introduce several different grids which are then mapped into one another at appropriate stages of the analysis. First consider an $n \times n$ reconstruction grid (or r-grid) on which the discrete reconstruction is to be presented. Next, for each direction θ for an X-ray beam, consider a detector grid (or d-grid) with axes parallel and perpendicular to the X-ray beam. A set of n photon counts (each corresponding to Y of the last section) is made for each direction.

To make our approach work we need a way to map a function from one grid to another. We have used the following simple rule. Let $i = (i_1, i_2)$ label pixels in the r-grid, and let $j = (j_1, j_2)$ label pixels in one of the d-grids. Suppose $\{g_j\}$ is some function defined on the d-grid. Consider a pixel i in the r-grid. Its centre lies in one of the pixels of the d-grid which we denote by $J(i)$, say. If we define a new function $\{h_i\}$ on the r-grid by

$$h_i = g_{J(i)},$$

then $\{h_i\}$ can be viewed as the function $\{g_j\}$, mapped to the new grid. For simplicity we arrange our co-ordinate systems so that $J(i)$ is unique. That is, a pixel centre in one grid never lies on the boundary between pixels in the other grid.

We can now describe an approximate EM algorithm for this setting. Given current estimated attenuation coefficients on the d-grid for each θ, carry out the E-step, as in Section 3, to get estimated numbers of photons entering and leaving each d-grid pixel. Next for each θ, use grid matching to convert these estimated counts to the r-grid. The M-step is carried out in the r-grid and yields the following updating formula for pixel i in the r-grid:

$$\hat{f}_i^{new} = \log \frac{\text{total estimated number of photons entering pixel } i}{\text{total estimated number of photons leaving pixel } i.}$$

The "total" here is obtained by combining information from each direction θ of X-rays. Finally, the updated attenuation coefficients are mapped back to each d-grid for the next iteration of the EM algorithm.

5. Directionality of the EM algorithm

Consider again the one-dimensional situation of section 3. The distribution of Y, the total number of unabsorbed photons, is the same whether the beam goes from left to right, or right to left. However, the resulting EM algorithm differs in the two cases.

More generally in a two-dimensional experiment involving beams from K different directions, say, there are 2^K different ways to set up the EM algorithm. In particular, a beam moving in a direction θ can equally be viewed as a beam in the direction $\theta + \pi$ for the purposes of the EM algorithm. These choices will have no effect on the final MLE if it is unique. However, if the set of MLE's is not unique but forms a ridge of maximum likelihood, then the choice of beam directions may influence the point on the ridge to which the EM algorithm converges.

To illustrate this last point consider a 3×3 array of attenuation coefficients given by the top entry in Table 1. Observations are made in the horizontal and vertical directions with no noise (the case $\lambda \to \infty$). If the detector counts are assumed taken in the left and top parts of the image then the EM algorithm (with a uniform initial solution) yields the reconstruction shown in the second row of Table 1. The directionality effects are evident.

For our practical work we have adopted a "two-sided" approach. Each detector count is assumed to have been observed twice, once in one direction and once in the other direction. This "doubling" of the likelihood has no effect on the MLE's but may improve convergence properties of the EM algorithm. The results of this two-sided approach for an artificial example are given in the bottom row of Table 1.

With more than two beam directions, the directionality effects appear to be minimal. This may be because the grid matching steps involve a small degree of smoothing which makes the MLE unique.

6. Applications

When the number of X-ray beams K is much less than the number of detectors n in each beam, the reconstruction problem is inherently ill-posed (with Kn observations to estimate n^2 attenuation coefficients). Therefore we also included some smoothing steps in the EM algorithm along the lines of the EMS algorithm (Silverman *et al.*, 1990).

The EMS algorithm consists of the two steps of the EM algorithm plus a smoothing step, called the S-step, in which the attenuation coefficient in a pixel is replaced by a weighted average of itself and its eight horizontal, vertical and diagonal neighbours. The weighted average depends on a smoothing parameter of $\sigma > 1$. The pixel of interest is weighted by $1/\sigma$. The remaining $1-1/\sigma$ is shared out between the neighbouring pixels so that diagonal neighbours receive less weight than horizontal or vertical neighbours, by a factor of $1/\sqrt{2}$. In this paper we have taken a small amount of smoothing, $\sigma = 1.1$.

Another possible approach to smoothing uses the penalized likelihood of Green (1990), which is iteratively maximized using his so-called One-Step Late (OSL) algorithm. The results which this method achieves are dependent upon the choice of penalty function, and further investigation is required to find a penalty function which gives the best results.

In general, smoothing had a modest but not dramatic effect on reconstructions from simulated data. The reason is perhaps that the grid matching steps already incorporate a degree of smoothing, but more study is needed.

To illustrate the scope of our methods we present a simple simulated example. Figure 1 represents a simulated phantom for the two bones of the human wrist. The attenuation coefficients are taken as 0.1mm^{-1} for the cortical bone (the annuli), $5 \times 10^{-3}\text{mm}^{-1}$ for the trabecular bone (inside the annuli) and $5 \times 10^{-3}\text{mm}^{-1}$ for flesh, in an image of size 128^2mm^2. For reconstruction purposes, $n = 64$ and there are 12 beam angles, equally spaced between 0° and 180°.

For the case of no noise $(\lambda \to \infty)$ reconstructions based on CBP, EM and EMS are given in Figures 2-4. (For a suitable penalty function the OSL algorithm gave similar results to EMS.) Both CBP and EM are subject to some artefacts, though EMS successfully smooths them away. Of course, some caution is needed when interpreting such pictures because the choice of contrast can dramatically affect one's interpretation.

Next consider the case $\lambda = 200$, a substantial level of noise. Figures 5-7 give the CBP, EM and EMS reconstructions. Here, EM and EMS look very similar to before. However, CBP is entirely swamped by artefacts now. As a reconstruction procedure it seems excessively sensitive to noise.

In the example described in the introduction, the objective is to measure the average trabecular bone density in the interior of the wrist bones. The medical motivation is to develop a screening procedure to identify women at risk from osteoporosis (bone loss) as they get older. More work is needed to study the accuracy and bias of the reconstruction methods used here. In particular care is needed when smoothing to avoid smearing the trabecular bone with other parts of the image.

References

DEANS, S.R., (1983) *"The Radon Transform and some of its applications"*, Wiley-Interscience, New York.

GREEN, P.J., (1990) "On the use of the EM algorithm for penalized likelihood estimation", *J. Roy. Statist. Soc. Ser. B,* **52**, pp. 443-452.

HORSMAN, A., SUTCLIFFE, J., BURKINSHAW, L., WILD, P., SKILLING, J., WEBB, S., (1987) "Isotope computed tomography using cone beam geometry: a comparison of two reconstruction algorithms", *Phys. Med. Biol.,* **32** pp 1221-1235.

LANGE, K. and CARSON, R., (1984) "EM reconstruction algorithms for emission and transmission tomography", *J. Comput. Assist. Tomog.*, **8**, pp. 306-316.

SHEPP, L.A. and KRUSKAL, J.B., (1978) "Computerized tomography: the new medical X-ray technology, *Amer. Math. Monthly*, **85**, pp. 420-439.

SILVERMAN, B.W., JONES, M.C., WILSON, J.D. and NYCHKA, D.W., (1990) "A smoothed EM approach to indirect estimation problems with particular reference to stereology and emission tomography", *J. Roy. Statist. Soc. Ser. B*, **52**, pp. 271-324.

0.005 0.0079 0.0083	0.005 0.0037 0.0034	0.005 0.0079 0.0083
0.005 0.0034 0.0034	0.100 0.0933 0.0935	0.005 0.0034 0.0034
0.005 0.0088 0.0083	0.005 0.0028 0.0034	0.005 0.0088 0.0083

Table 1 Reconstruction using the EM algorithm for noiseless data on a 3 × 3 grid of pixels using 2 X-ray beams. Top entry in each pixel gives the true attenuation coefficients. Middle entry gives one-sided EM reconstruction. Bottom entry gives two-sided EM reconstruction.

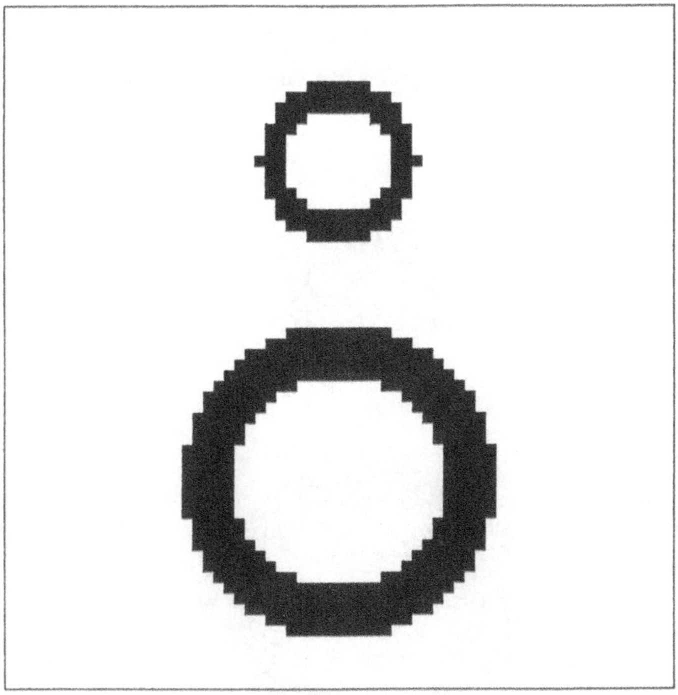

Figure 1. Simulated phantom for two bones in the human wrist.

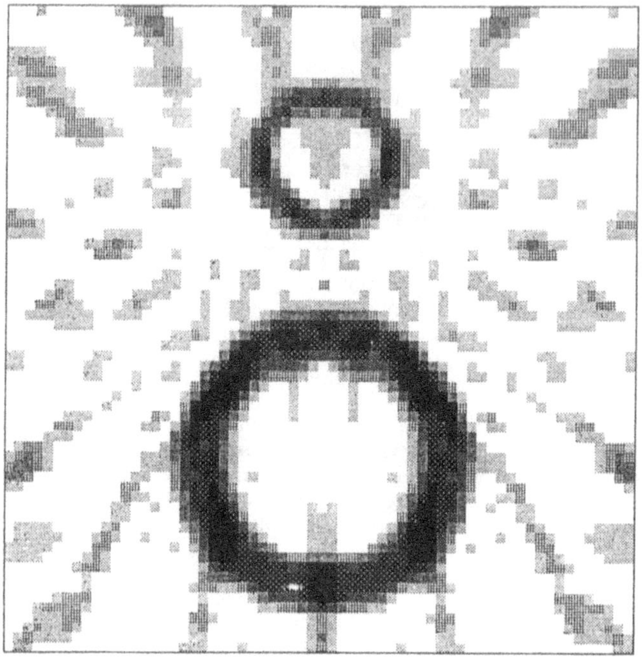

Figure 2. Reconstruction via CBP, no noise $(\lambda \to \infty)$.

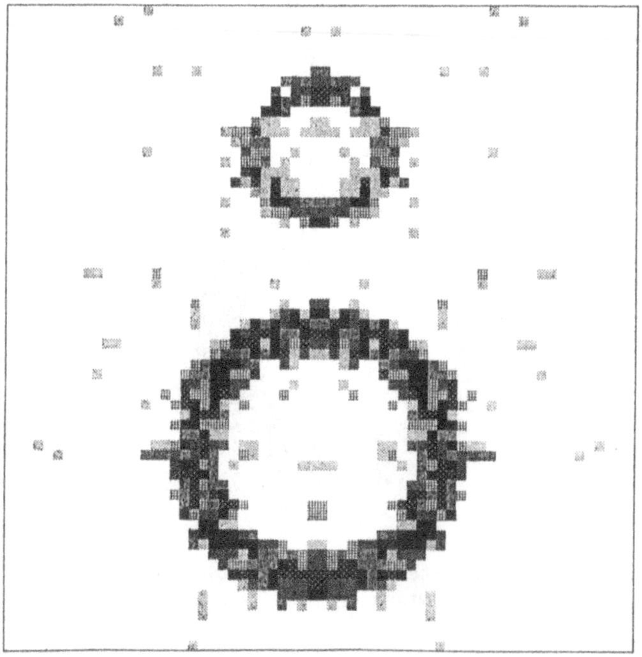

Figure 3. Reconstruction via EM, 500 iterations, no noise $(\lambda \to \infty)$.

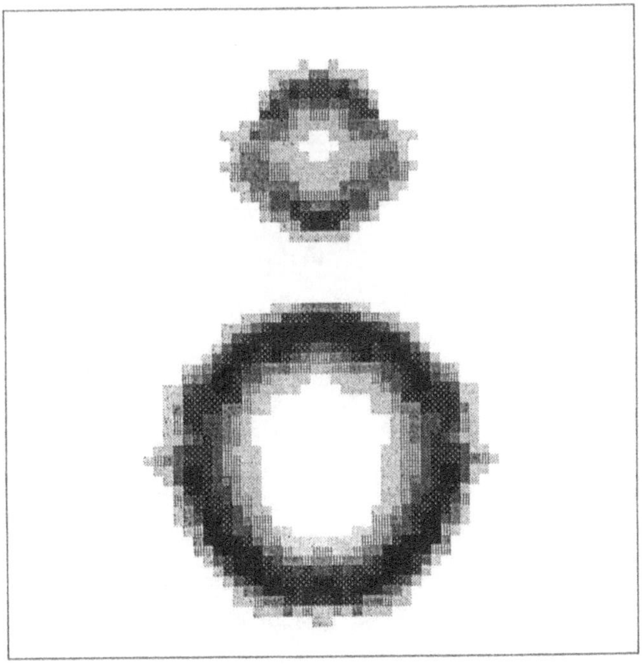

Figure 4. Reconstruction via EMS, 100 iterations, no noise ($\lambda \to \infty$).

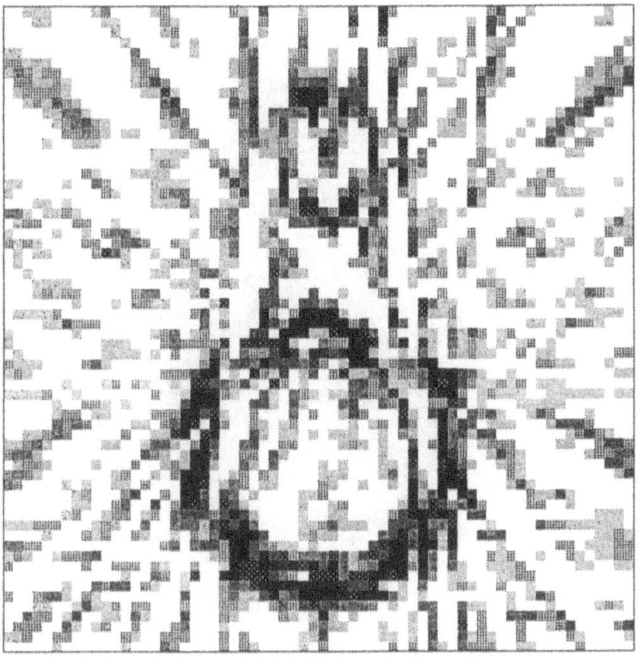

Figure 5. Reconstruction via CBP, λ = 200.

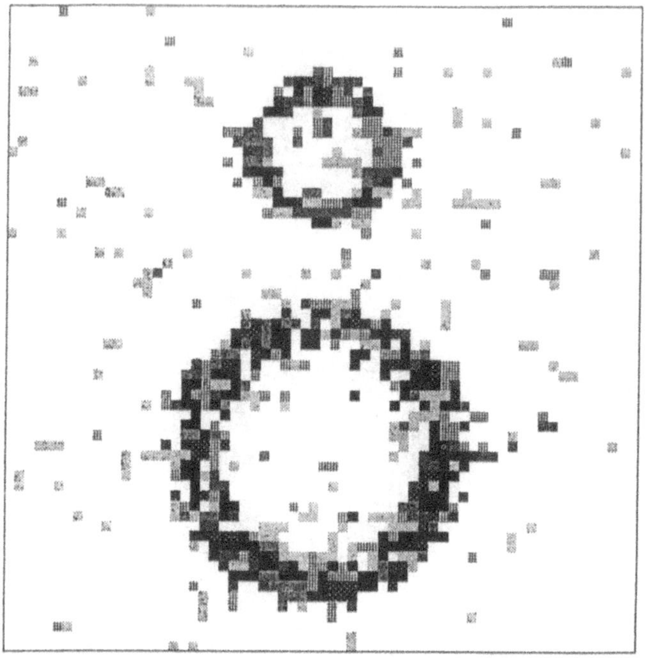

Figure 6. Reconstruction via EM, 500 iterations, $\lambda = 200$.

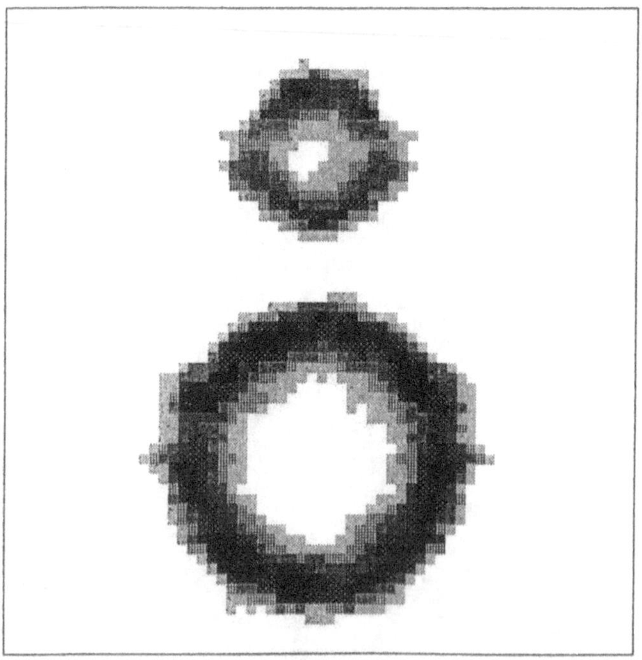

Figure 7. Reconstruction via EMS, 100 iterations, $\lambda = 200$.

Deconvolution in Optical Astronomy. A Bayesian Approach.

R. Molina
Departamento de Ciencias de la Computación e I.A.,
Universidad de Granada, 18071 Granada, España.

B. D. Ripley
Department of Statistics, University of Oxford,
Oxford OX1 3TG, United Kingdom.

Abstract
We describe in this work how the Bayesian paradigm can be applied to a deconvolution problem in optical astronomy. The use of robust statistics in this process is also discussed.

0 Background.

The aim of this paper is to summarize the work developed by the authors in the field of deconvolution in astronomy. The problem was posed to one of the authors (BDR) by members of the Instituto de Astrofísica de Andalucía in 1986 and the results we describe were developed in the period 1987–1990. The computer programs implementing the methods were written by BDR and developed by A. I. Sutherland while funded by the UK Science and Engineering Research Council. Reports of our collaborative works are [9](§5.3), [5] and [6]. Another paper [7] has been written for an astronomical audience.

1 Introduction.

Optical astronomers now normally collect digital images by means of charge coupled device (CCD) detectors. Those images are blurred by atmospheric motion and distorted by physical noise in the recording process. These distortions can change the shape of the objects. For this reason it is very important to take into account the global distortion process when determining physical observational quantities.

The deconvolution process aims at reconstructing the image S which would have been observed under perfect conditions.

In this paper we describe how the Bayesian approach to image analysis can be applied to the deconvolution of smooth varying objects. In section 2 we describe the framework, the Bayesian paradigm. The two main ingredients of this Bayesian paradigm, the prior model and the observation distribution, are studied in sections 3 and 4 respectively. The next step is then to calculate the estimate of the real underlying image; the algorithm implementing this task is described in section 5. Finally, a test example is shown in section 6.

2 Bayesian Paradigm.

Under this paradigm we distinguish between S, the image which would have been observed under ideal conditions, and Z, the observed image. The aim is then to reconstruct S from Z. Both are viewed as arrays of values corresponding to the luminosities of pixels in the CCD array. We use a single index i for these pixels, so that a typical observed value at pixel i is Z_i.

Bayesian methods start with a prior distribution, a probability distribution $P(S)$ over images S. This prior distribution assigns preference amongst the images S. It is also necessary to specify $P(Z|S)$ which models the observational process. Inference about S is based on the posterior distribution, which is given by

$$P(S|Z) = P(Z|S)P(S)/P(Z) \propto P(Z|S)P(S) \tag{1}$$

To show just one restoration it is common to choose the mode of $P(S|Z)$, that is to display the image $\hat{S}$ which satisfies

$$\hat{S} \text{ maximizes } P(Z|S)P(S) \tag{2}$$

This is known as the MAP (maximum a posteriori) estimate of S. This will be the estimate we use in this work.

3 Prior Distribution.

The prior distribution is based on our knowledge about the image; we are interested in images of galaxies, which have a smoothly varying luminosity.

The luminosity distribution of elliptical galaxies has been investigated by many astronomers, and several analytic functions have been proposed to model the distribution. The most commonly used is *"the $r^{1/4}$-law"* proposed by de Vaucoulers (see [11] for references).

$$I(r) = I(0)\exp[-(ar)^{1/4}] \tag{3}$$

where r is the distance from the galactic center and $I(0)$ and a are parameters which differ from galaxy to galaxy. Furthermore, the luminosity distribution of pure disk in a galaxy can be modelled by *"the exponential law"* written as

$$I(r) = I(0)\exp[-br] \tag{4}$$

We noticed that our astronomer colleagues tended to view the observed images on log-scale, and this is supported by the theories cited in the previous paragraph. Thus we decided to specify a smooth surface for $Y = \ln(S + p)$ where p allows for a linear scale at very small counts, ($p = 100$ in our examples). Our prior knowledge about the luminosity distribution makes it possible to model the distribution of Y by a conditional autoregression (see [8]). We have

$$P(Y) \propto \exp -\frac{1}{2\kappa_s} Y^T (I - \phi N) Y \tag{5}$$

where $N_{ij} = 1$ if cells i and j are (horizontal or vertical) neighbours, zero otherwise and ϕ just less than 0.25. The parameters can be interpreted by the following expressions describing the conditional distribution:

$$E(Y_i|Y_j,\ j \neq i) = \phi \sum_{j \text{ nhbr of } i} Y_j \tag{6}$$

$$\text{var}(Y_i|Y_j,\ j \neq i) = \kappa_s \tag{7}$$

4 Observational Distribution.

Atmospheric motion causes random refraction of the incoming light, so that photons which under ideal conditions would arrive at one pixel will be spread over nearby pixels. Although an exact theoretical expression describing the shape of the point spread function (*psf*) is not known, previous studies ([4], [1], [5]) have suggested that the *psf* can be modelled as

$$h(r) = \frac{\beta}{\pi R^2}\left(1 + \frac{r^2}{R^2}\right)^{-\beta} \tag{8}$$

where r is the distance from the source to the receiving pixel and β is about 3. In our studies $\beta \approx 3$ and $R \approx 2.7 - 6$ pixels, depending on the atmospheric conditions of the night when the images were recorded.

The noise in the observational process is due mainly to the discrete nature of photons and electronic noise in the CCD detector. Physical considerations suggest that both sources will give noise which is independent from pixel to pixel with variance which may depend on the (average) number of photons arriving at the CCD cell.

Let μ denote the effect of blurring the true S (assumed known in this section) by the point spread function h. Then the photon count should be Poisson with mean μ_i and hence variance μ_i. Adding electron noise suggests

$$\text{var}(\mu_i) = a + \mu_i \tag{9}$$

but since not all photons are recorded, we might expect

$$\text{var}(\mu_i) = a + b\mu_i \tag{10}$$

with $b < 1$. Let Z_i^N be the value at pixel i minus the average of the value of its four neighbours. Then the mean of Z_i^N is close to zero, but the variance is $(5/4)\,\sigma^2(\mu_i)$. The idea of our estimation process is to divide the values of μ_i into small intervals, and to estimate $(5/4)\,\sigma^2(\mu_i)$ from the variance of the (Z_i^N) with μ_i in the interval. There are two problems with this idea. First we do not know μ_i, only the observed value Z_i. Fortunately, on the scale we are using, Z_i has a mean of at least a few hundred and a variance of about equal to its mean. Thus the relative error in replacing μ_i by Z_i is only a few percent at most. The second problem is the occasional "gross errors" in the observations Z_i caused by detector malfunctions, cosmic-ray strikes and so on. These can cause an observation to be out by thousands, and result in values of Z_i which lack credence given the observed values Z_j in the neighbourhood of pixel i. We must use a robust estimator of the variance ([2], [3]). Since we have so much data, we can afford to use a simple and slightly inefficient method.

For observations $x_1, \ldots, x_n$ from a normal distribution with mean known to be zero but unknown variance σ^2 we would estimate σ^2 by

$$s^2 = \frac{1}{n}\sum x_i^2 \tag{11}$$

Since the observations are squared, this is extremely susceptible to outliers. Consider instead an estimator proportional to $\sum |x_i|$ or median $|x_i|$. Each of these would be expected to be proportional to σ, and much less susceptible to "gross errors", especially the second. We use

$$\hat{\sigma}^2(\mu_i) = \left[0.6745 \times \text{median } |Z_i^N|\right]^2 /(5/4) \tag{12}$$

where the median is over pixels with Z_i in the range for μ_i. The constant 0.6745 is that needed to obtain the correct value if there is a normal distribution, and so no gross errors.

Under the noise model just described, we have

$$-2 \ln P(Z|S) = \text{const} + \sum_{\text{pixels } i} \left\{[Z_i - (HS)_i]^2/\sigma^2((HS)_i) - \ln \sigma^2((HS)_i)\right\} \tag{13}$$

where $\sigma^2((HS)_i) = a + b(HS)_i$. The effect of $\ln \sigma^2((HS)_i)$ is negligible (see [6] for details). However, even the expression

$$-2 \ln P(Z|S) = \text{const} + \sum_{\text{pixels } i} \left\{[Z_i - (HS)_i]^2/\sigma^2((HS)_i)\right\} \tag{14}$$

is unrealistic near anomalous pixels where either detector errors. Robust statistics can come in our help again. The idea is to downweight observations which are far away from their mean. Such values are given too much weight in (14). The squared term in $D_i = (Z_i - (HS)_i)/\sigma((HS)_i)$ represents the number of standard deviations that Z_i is away from its mean. In robust statistics we replace $\sum D_i^2$ by $\sum \rho(D_i)$ for a function ρ which penalizes extreme values less severely. We used Huber's 'proposal 2' function for ρ, defined by

$$\rho(x) = \begin{cases} x^2 & \text{for } |x| \le c \\ 2c|x| - c^2 & \text{for } |x| > c \end{cases}$$

This is quadratic in the centre, but penalizes large deviations linearly rather than quadratically. Equivalently, observations D_i are downweighted if $|D_i|$ exceeds c. In practice we choose c about 2, thus downweighting only those observations more than two standard deviations from their means.

5 Calculating the MAP

The MAP estimator is that 'true image' S which minimizes

$$-2 \ln P(S|Z) = \text{const} - 2 \ln P(Z|S) - 2 \ln P(S) \tag{15}$$

We work not with S but with the transformation $Y = \ln(S + 100)$ for each pixel. Thus we choose to find the MAP estimate of Y, not of S. (It would be desirable if $\hat{S}_i = \exp(\hat{Y}_i) - p$ gave the MAP estimate of S, but this is not the case. Since we must choose, we give Y higher priority). Thus we choose Y to minimize

$$-2 \ln P(Z|S) - 2 \ln P(Y) \tag{16}$$

since $P(Z|S) = P(Z|Y)$.

Let ψ denote the derivative of ρ, defined in the previous section, then $\psi(x) = x$ for $|x| \leq c$, and $\psi(x) = c\ \mathrm{sgn}(x)$ for $|x| > c$. Differentiating (16) and ignoring the dependence of $\sigma^2((HS)_i)$ on Y (see [6] for details), the robust MAP estimator of Y satisfies

$$Y = \max\left(\ln(p)\, I, \phi N Y + \kappa_s \,\mathrm{diag}(e^Y) H^T W^{-\frac{1}{2}} \psi(W^{\frac{1}{2}}(Z - HS))\right) \quad (17)$$

$$S = \exp(Y) - p \quad (18)$$

where $W = \mathrm{diag}[\sigma^2(HS_i)]$ is a diagonal matrix of variances.

These equations are solved by iteration. Suppose that $Y^{(t)}$ is the current image. Forming $S^{(t)} = \exp(Y^{(t)}) - p$ and substituting $Y^{(t)}$ and $S^{(t)}$ on the right hand side of (17) gives Y', say. To stabilize the convergence procedure we only move part of the way from $Y^{(t)}$ to Y'. For each pixel i we have

$$Y^{(t+1)} = \max\left(Y_i^{(t)} - \ln(2),\ \min\left(Y_i^{(t)} + \ln(2), cY' + (1-c)Y^{(t)}\right)\right) \quad (19)$$

where c is a small positive constant. This shows that the value at each pixel moves only a small proportion of the way to Y', and the new value of S_i differs from the old by at most a factor of 2.

As can be seen, the iterative method we have just described relies heavily on the calculation of convolutions, which we do *via* Fourier transforms. So an efficient FFT are essential to the efficient implementation of our method. After some experimentation, we settled on the base 2 and 4 routines of Singleton ([10]).

We comment on two further problems. The parameter κ_s represents the degree of smoothness we assume the true image to have, and has to be chosen. So does the starting point of the iterative scheme. Although κ_s can be estimated on a trial-error basis, it is possible to calculate an approximately unbiased estimator of κ_s (see [6] for details). One obvious choice of the starting point is given by

$$S_i^{(0)} = \max(0, Z_i) \quad (20)$$

However, this will usually be a long way from the solution. It is possible to chose a starting point closer to the expected solution (see [6] for details).

6 Test Example

The image we use to describe the method contains two galaxies NGC450 and UGC807. The size of the image is 512 × 512. Some contour levels are shown in figure 1. We applied the deconvolution process we have described to this image. The parameters of the *psf* were estimated from stars in the image as $\beta = 5$, $R = 3$. The other parameters were estimated as $a = 200$, $b = 0.7$ and $\kappa_s = 0.07$. Finally, the relaxation constant c was set to 0.1 and the iterative scheme was run for 50 sweeps over the image.

Figure 2 shows the contours at photon level 500 for the original and deconvolved images. We have plotted just one level for clarity. There are two interesting points to note. The shape of the small galaxy is much clearer in the deconvolution. Second, there are small dots in the deconvolved image. It might be thought that the deconvolution was very noisy and that was the reason why the dots appeared. However, they correspond to HII regions whose existence have been confirmed by comparing them with a redshifted Hα image of the same objects (see [7] for details).

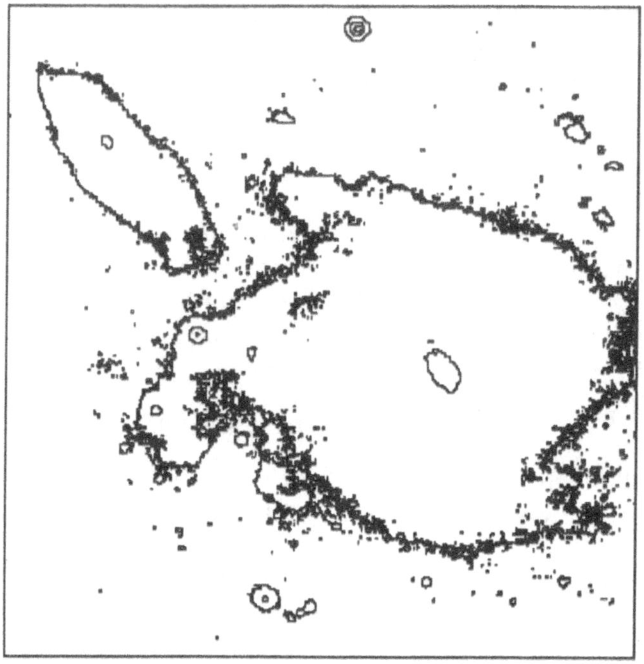

Figure 1: Contours at levels 200, 2000, 20,000 of the original image.

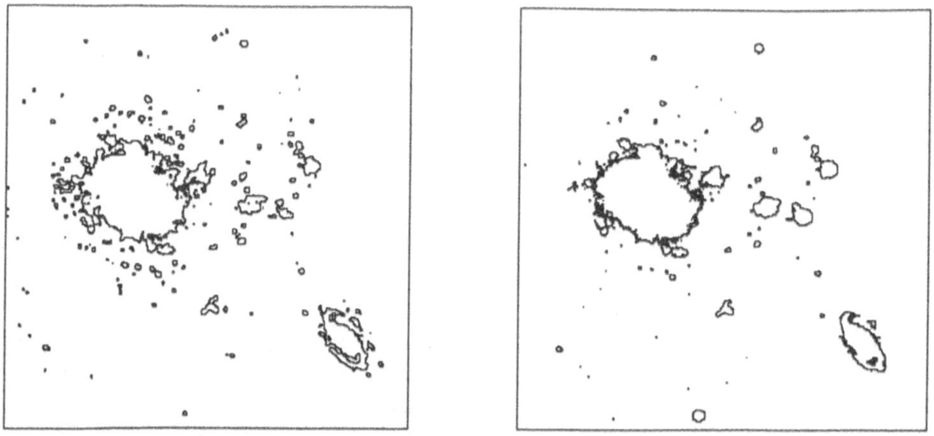

Figure 2: Contour level 500 of the original image **left** and deconvolved image **right**.

Our implementations were on a VAX 6320 and on Sun 3 and 4 workstations. One iteration took 72 seconds on a VAX 6320 and 27 seconds on a Sun 4/370, for a 512×512 image.

References

[1] Buonanno, R., Buscema, G., Corsi, C.E., Ferraro, I. and Iannicola, G. 1983. Automated photographic photometry of stars in globular clusters. *Astron. Astrophys.*, **126**, 126–278.

[2] Hampel, F.R., Ronchetti, E.M., Rousseeuw, P.J. and Stahel, W.A. 1986. *Robust Statistics: The Approach Based on Influence Functions.* Wiley.

[3] Huber, P.J. 1981. *Robust Statistics.* Wiley.

[4] Moffat, A.F.J. 1969. A theoretical investigation of focal stellar images in the photographic emulsion and application to photographic photometry. *Astrom. Astrophys.*, **3**, 455-462.

[5] Molina, R., Pérez de la Blanca, N. and Ripley, B.D. 1989. Statistical restoration of astronomical images. In Di Gesu, V.D., Scarsi, L., Crane, P. and Friedman, J.H. Eds. *Data Analysis in Astronomy III*, Plenum Publishing Corporation,. 75–82.

[6] Molina, R. and Ripley, B.D. 1989. Using spatial models as priors in image analysis. *J. Appl. Statist*, **16**, 193–206.

[7] Molina, R., del Olmo, A., Perea, J. and Ripley, B.D. 1990. Bayesian deconvolution in optical astronomy. *Submitted to Astrom. J.*

[8] Ripley, B.D. 1981. *Spatial Statistics.* Wiley.

[9] Ripley, B.D. 1988. *Statistical Inference for Spatial Processes.* Cambridge University Press.

[10] Singleton, R.C. 1969. An algorithm for computing the mixed radii fast Fourier. *IEEE Transactions on Audio and Electroacoustics, AU–17.* 93–103

[11] Takase, B., Kodaira, K. and Okamura, S. 1984. *An Atlas of Selected Galaxies.* University of Tokyo Press.

Parameter Estimation For Imperfectly Observed Gibbs Fields and Some Comments on Chalmond's EM Gibbsian Algorithm.

Laurent Younes

28 January 1991

Abstract

We make a short review of existing algorithms for parameter estimation from imperfectly observed Gibbs field. We then focus on one of these methods: the EM Gibbsian algorithm, making new comments and simulations.

1 Introduction.

In this paper, we are concerned with general parameter estimation methods for partially observed Markov random fields (MRF) (by general methods, we mean methods that are valid for a reasonably wide class of models). This issue is of main importance for numerous problems that involve MRF (we particularly have in mind the Bayesian framework for image analysis developped in [15], which represents one of the most powerful and one of the most popular methods for low level processing of pictures).

Unfortunately, this problem is quite awkward, and there does not exist, as far as we are informed, totally satisfactory methods designed for it. The point is that, due to the computational restrictions that are inherent to MRF modeling, standard algorithms of parameter estimation for hidden models (such as EM) are almost useless in this context. Beside various clever solutions to this problem for very particular models, (cf [13], [5], and other references given in [17]), we are aware of only three general algorithms for it : stochastic gradient (SG) algorithm for solution of maximum likelihood equations ([47], [48]) , modification of variational estimator (VE) for partially observed data ([1]), and the so-called EM Gibbsian (EMG) algorithm ([6]). In the following, we shall briefly present the first two of them, and then, focus on the third making some remarks concerning its properties that did not appear in the original paper of Chalmond. This will be done in sections 3 and 4. Before that, in section 2, we fix some notations and present the model that we use.

2 Model.

We consider MRF on the regular lattice $\mathbf{Z}^d$: let F be a set, the single site state space, and denote by Ω the set of all configurations $x = (x_s, s \in \mathbf{Z}^d)$. If $D \subset \mathbf{Z}^d$, we denote Ω_D the set of

[0]Université Paris Sud. Laboratoire de statistique appliquée. Bat 425. 91405 ORSAY Cedex (France) and DMI, Ecole Normale Supérieure.

all $x_D = (x_s, s \in D)$. We assume that F is provided with a measure μ_1 and denote by μ (resp. μ_D) the corresponding product measure on Ω (resp. Ω_D).

Let $\Theta \subset \mathbf{R}^\nu$ be the set of parameters, and consider a family of functions: $(h_C, C \subset \mathbf{Z}^d, C$ finite), where h_C is defined on Ω and takes values in $\mathbf{R}^\nu$, and $h_C(x)$ only depends of coordinates x_s, $s \in C$. In all the sequel, we shall assume that h_C vanishes for all C of diameter larger than a given non-negative number γ (bounded range). This assumption simplifies notations and proofs and is absolutely necessary for any of the preceding methods to be feasible.

For $\theta \in \Theta$ we define the potential:

$$\lambda_\theta = (\langle \theta, h_c \rangle,\ C \subset \mathbf{Z}^d).$$

($\langle\ ,\ \rangle$ being the usual euclidian scalar product).

Let $\mathcal{G}(\theta)$ be the set of Gibbs fields associated to this potential; this means that, if D is a finite subset of $\mathbf{Z}^d$, and $x' \in \Omega_{D^c}$ is given ($D^c = \mathbf{Z}^d \setminus D$), a regular version of the conditional law for $\pi \in \mathcal{G}(\theta)$ on Ω_D given x' outside D has a density with respect to μ_D which is given by:

$$\pi(x|x') = exp(-\langle \theta, H(x|x') \rangle)/Z_\theta(x') \tag{1}$$

where $H(x|x') = \sum_{C \subset \mathbf{Z}^d} h_c(x.x')$, $x.x'$ being the element of Ω for which the coordinate at site s is x_s if $s \in D$ and x'_s if not (and it is assumed that the normalizing constant Z_θ is well defined, ie. that the exponential is μ_D integrable).

Although it has no importance in practice, we shall need, for rigorous results, to assume that h_C is space homogeneous: let's denote T_s the translation operator in the direction s, ie. $(T_s x)_t = x_{s+t}$, we assume that $h_{C+s} = h_c \circ T_s$.

In the case of perfect observation, the inference problem is the following: estimate θ_* from the restriction to Ω_D of some realization $x \in \Omega$ of one law $\pi_{\theta_*} \in \mathcal{G}(\theta_*)$. We assume that there exists at least one law in $\mathcal{G}(\theta_*)$, (this is a consequence of our assumptions when F is finite), but note that there may exists several ones. Identification is possible provided $\mathcal{G}(\theta)$ and $\mathcal{G}(\theta')$ are disjoints for distincts $\theta, \theta' \in \Theta$ (note however that it is in general impossible to estimate π_{θ_*} itself on the basis of a single realization of the field). In this context, several estimation methods exist, the most popular being the maximum pseudo likelihood (MPL) estimator (see section 3.3.2), other are maximum likelihood using stochastic gradient ([46]) which involves more computation while remaining feasible and has the advantage to be efficient, variational estimators ([1]), logit estimator for binary fields ([39]), various estimators based on approximations of the likelihood in the Gaussian case ([26], [23]), etc... All these estimators are proven to be consistent (this means that if D tends to $\mathbf{Z}^d$ in a reasonable way, the corresponding estimators converge to the true θ_*). See for example [18], [21], [22], [9] for various results concerning minimum of contrast estimation with this context.

In the case of partial observations, the situation is the following. Let F^h and F^o be some sets, and μ_1^h, μ_1^o be measures on these sets. Denote Ω^h (resp. Ω^o) the set of configurations on $\mathbf{Z}^d$ with coordinates in F^h (resp. F^o), and let $F = F^h \times F^o$, $\mu_1 = \mu_1^h \otimes \mu_1^o$ be the one site state spaces and measures for the joint process. We are therefore given a Gibbs field on the corresponding set $\Omega = \Omega^o \times \Omega^h$. Inference has to be made on the basis of the observation of the restriction to Ω_D^o of the second coordinate, $y \in \Omega^o$ of some realization $(x^h, y) \in \Omega$ of a law $\pi_{\theta_*} \in \mathcal{G}(\theta_*)$ (we shall use letter y as well as x^o to denote the observed data).

Let's remark that for the VE, F^h should be $\mathbf{R}^k$, whereas for the EMG estimator, F^h has to be a finite set with a small number of elements. Concerning the SG method, all kind of state spaces may be used, provided some good Monte Carlo simulation algorithm is available, but all rigorous results related to it were proven only for finite (or compact) F^h. Consistency for maximum of likelihood estimator in this context is proven in [8] and [48].

In both cases of complete and incomplete observations, it is of course possible to assume that, in the preceding model, part of the coordinates of the true parameter are known. For example, for noisy data, the noise mechanism can have been estimated before. Most of the time, we shall not consider this, because this would introduce extra notations. But, since it is straightforward to modify all the equations for this case, we shall make references to this situation in the remarks.

If π is some field over Ω, we shall denote by π^o (or ψ) and π^h its marginals on Ω^o and Ω^h.

In almost all cases, the marginal on Ω_D of some $\pi_\theta \in \mathcal{G}(\theta)$ is impossible to compute. The first reason is that, if we are only given θ, a precise description of $\mathcal{G}(\theta)$ is seldom available, so that π_θ itself cannot be described. The second reason is that, even if we know exactly what π_θ we are dealing with (for example in the absence of phase transition), there is no close form for the expression of probability on Ω_D in term of the potential (which is the only thing provided with the model). Therefore, in practice one is forced to use some approximate expression of the likelihood, which is most of the time of the kind (1) with some fixed boundary condition x' (variants are free boundary condition, periodic boundary condition, differing only in terms involving coordinates near D^c). Consistency of maximum of likelihood estimation, for any of the preceding variants is not affected. It is unfortunately not the case for asymptotic normality and efficiency (see [22]), unless some more stringent conditions are imposed, such as Dobrushin's ones for unicity of the field, and suitable modifications are made in the energy $H(x|x')$ to eliminate edge effects (see [49]).

In the following, we assume that we have chosen some $x' \in \Omega_D$ for each D to approximate π_θ by $\pi_\theta(\ |x')$, and we shall not, unless necessary, let this appear in the notations, keeping abusively the expression π_θ, or π_θ^D. For each D, we are therefore using as raw materials for estimation a probability measure on Ω_D of the kind:

$$\pi_\theta^D(x^h, x^o) = \exp(-\langle\ \theta, H(x^h, x^o)\ \rangle)/Z_\theta.$$

We shall denote by E_θ expectations with respect to this law, by ψ_θ its marginal law over the observed space Ω_D^o, and for a given y, by π_θ^y the conditional law on Ω_D^h given that the field on Ω_D^o is y. We shall also use (X^h, Y) to designate a field on Ω.

Implementation of the algorithms requires some methods for simulating Markov random fields. They generally consist in a sequence of local updatings of the configurations ("local" means that only one or a small number of sites are changed), using transition kernels that are in detailed balance with the law to simulate. There also exist non local methods, such as multi-grid Monte Carlo algorithms, Swendsen-Wang algorithms, that are more efficient for some classes of models. See [44] for a precise description of these algorithms, as well as for a bibliography. All these methods are dynamic, which means that they provide a Markovian sequence of configurations that converges in distribution.

3 General methods for inference on partial observations

3.1 Maximum of likelihood.

Let y be the observation. A straightforward computation shows that the maximum of likelihood estimator is a solution of the equation

$$E_\theta(H) - E_\theta(H|y) = 0. \tag{2}$$

(In the case of perfect observation, this equation simplifies in: $E_\theta(H) - H(y) = 0$).

Computation of expectations with respect to some Gibbs field can only be done by simulation and such an operation is time consuming. Solving this equation, even for perfect observations is therefore a non trivial problem. If one is sure to use the same model a large number of times, and if the dimension of θ is small (1 or 2), it is possible to draw once for all the function $f(\theta) = E_\theta(H)$ and use this for subsequent estimation (see [18] for such an attempt). However, this cannot be done in general, in particular in the case of partial observation. We thus are generally reduced to using a gradient like algorithm to solve (2). An exact gradient descent algorithm would be

$$\theta_{n+1} = \theta_n + a[E_{\theta_n}(H) - E_{\theta_n}(H|y)].$$

Evaluating with a reasonable precision these expectations at each step requires a lot of computation time, but such an approach was succesfully used in [30] for a maximum of entropy estimation that led to similar equations. It is however possible to significantly improve on this by applying ideas from stochastic approximation techniques. Basically, the method is to mix the gradient descent with Monte Carlo simulation, without necessarily waiting for stabilization of the estimated expectations. A short formulation of this kind of algorithms is the following. Let $P_\theta(x,x')$ (resp. $P_\theta^y(x,x')$) be a transition kernel corresponding to some elementary part of the chosen Monte Carlo simulation algorithm for the distribution π_θ (resp. π_θ^y), for example, one site updating or one sweep of a Gibbs sampler. Then define the sequence (θ_n, X_n^1, X_n^2) by fixing $\theta_0 \in \Theta$, X_0^1, $X_0^2 \in \Omega_D^h$, and by the following stochastic algorithm:

$$\begin{aligned} P(X_{n+1}^1 = x'|X_n^1 = x) &= P_{\theta_n}(x,x') \\ P(X_{n+1}^2 = x'|X_n^2 = x) &= P_{\theta_n}^y(x,x') \\ \theta_{n+1} &= \theta_n + \frac{a}{n+b}[H(X_{n+1}^1) - H(X_{n+1}^2)] \end{aligned} \tag{3}$$

In [48] and [49] details are given on practical remarks about the implementation of the algorithm, and improvements that can be done. Applying results and methods from the stochastic approximation literature, (see [2]), it is possible to prove some rigorous facts concerning convergence. In particular, when observation is complete (in which case the second process X_n^2 disappears), θ_n converges almost surely, provided the constant a is small enough. In the general case, one can prove a local result of quasi convergence, which is much weaker. See [46], [47] for details. Recently, some stochastic algorithms have been studied for global optimization of functions for a context that inclдes ours ([24], [27], [14]); they are in some way random perturbations of (3) reminiscent of annealing algorithms for discrete or continuous optimization ([15], [20]). They might give interesting results in this context although we are not aware of any related numerical experiments.

Finally, we point out that, since the conditional law given the observations is simulated during (3), it is possible, in the context of a Bayesian analysis of images, to perform some reconstruction on line, using the MPM criterium (cf. [30]). This possibility is also available for both algorithms that follow.

3.2 Variational Estimator.

This estimator, which is preferably designed for complete observations, is mainly valid for $F = \mathbf{R}^n$ and μ_1 = lebesgue measure on F. We give of it a summarized version, refering to [1] for details and variants. In the case of perfect observations, the principles are as follows. The first step is to choose statistics

$$(W_s^\alpha,\ \alpha = 1, \ldots, \nu,\ s \in D)$$

which are local (ie. $W_s^\alpha(x)$ depends on x_t for t in a restricted neighborhood of s), and such that the divergence theorem holds:

$$\int_{\Omega_D}\{\nabla.[W_D^\alpha(x)\pi_\theta(x)]\}dx = 0$$

where W_D^α is the $|D|$-dimensional vector with components equal to W_s^α.

Developing these identities and replacing expectations with respect to π_θ with empirical ones yields a set of linear equations in θ, and the variational estimator is defined as a solution of these equations. One possibility for W_s^α is $\partial H_s^\alpha/\partial x_s$, where $H_s = \sum_{C/\ s\in C} h_C$ and H_s^α is the α-th component of H_s. The obtained equations, that are very simple to solve, are proven to provide consistent estimators.

In the case of partial observations, the W's are chosen so that

$$\int_{\Omega_D}\{\nabla.[W_D^\alpha(y)\psi_\theta(y)]\}dy = 0$$

(the state space for y still being $\mathbf{R}^n$).

This leads to the equations :

$$\sum_{i=1}^{\nu}\theta_\beta[W^\alpha(y).E_\theta(\nabla H^\beta|Y=y)] = \nabla.W^\alpha(y). \tag{4}$$

It is not obvious to obtain some natural way of finding W in this context. There is no proof of consistency for estimators obtained from (4), but authors in [1] report satisfying numerical resuts, at least for small noise.

Note that for solving (4), it is necessary to use a (stochastic) Newton algorithm, since conditional expectations given Y generally cannot be put into close form. This algorithm, however, is less complex than the one used for maximum of likelihood estimation. This is balanced by the fact that (4) does not come from the gradient of some function to maximize, and this can yield more risk of instability. Note also that in both cases, it is likely that several solutions of the equations exist.

3.3 EM algorithm algorithm and EM Gibbsian algorithm.

3.3.1 EM algorithm.

The EM algorithm is a way for obtaining local maxima of the likelihood for partial observations. For several applications, it is a very efficient tool, widely used, especially for one dimensional hidden Markov chains.

As said before, the maximum likelihood estimator solves the equation:

$$E_\theta[\frac{\mathrm{d}}{\mathrm{d}\theta}l(\theta)|Y=y]$$

where $l(\theta)$ is the logarithm of the likelihood. The principle of the EM is as follows : start with some parameter θ_0, and construct θ_{n+1} from θ_n as a maximizer of:

$$g_n(\theta) = E_{\theta_n}[\frac{\mathrm{d}}{\mathrm{d}\theta}l(\theta)|Y=y] \tag{5}$$

For exponential models, under some identifiability condition, g_n is strictly concave and this provides a well defined algorithm. One can prove that the sequence θ_n is such that $\psi_{\theta_n}(y)$

is increasing. Thus, if θ_0 is suitably chosen, the algorithm will converge to the maximum of likelihood, while wrong choices of θ_0 can lead to a non global maximum, or even divergence. In the absence of *a priori* knowledge on the true value of the parameter, it is generally recommended to try several θ_0.

We now specialize to the case of partially observed MRF. The difficulty is of course to maximize g_n, which is equivalent to solve the equation:

$$E_\theta(H) = E_{\theta_n}(H|y). \tag{6}$$

Computation of $E_{\theta_n}(H|y)$ requires Monte Carlo simulation, and the solution of (6) can hardly be obtained without a stochastic gradient algorithm. We propose a procedure for that purpose, in the vein of what has been done in section 3.1.

Let θ_0 be a starting point, and define the doubly indexed sequences $(\theta_{n,k}\ n \geq 1,\ 0 \leq k \leq K(n))$, $(X^1_{n,k})$, $(X^2_{n,k})$ by :

i) $\theta_{n,0} = \theta_n = \theta_{n-1,K(n-1)}$.

ii) $(X^2_{n,k},\ k = 1, \ldots, K(n))$ is an ergodic sequence of configurations that converge in distribution to $\pi_{\theta_n}(\ |y)$.

iii) $\theta_{n,k+1} = \theta_{n,k} + \gamma_k[H(X^1_{n,k+1}) - \hat{E}_{\theta_n}]$
where $(X^1_{n,k+1})$ is obtained from $(X^1_{n,k})$ by making one step of some Monte Carlo simulation algorithm for $\pi_{\theta_{n,k}}(\ )$, γ_k is a decreasing sequence of numbers, and $\hat{E}_{\theta_n}$ is an empirical estimation of $E_{\theta_n}(H|y)$ on the basis of the sequence $(X^2_{n,k},\ k = 1, \ldots, K(n))$.

We believe that some rigorous study of this algorithm, using results concerning stochastic approximations (cf [2]) might be feasible, maybe requiring some technicalities. We shall however not get into that direction, mainly because the full application of the procedure would be quite unrealistic in practice, because of the large number of computations it involves ($K(n)$ should be large enough so that $\theta_{n,K(n)}$ should be close to the maximizer of g_n). On the other hand, this algorithm is surely more stable than the one in section 3.1, since it mimics the EM algorithm rather than a gradient algorithm. In the case of high noise, when the latter performs poorly, it might be fruitful to make a few iterations of the former as a preliminar analysis. This would have the effect to place the initial point of the SG algorithm deeper into some attraction basin of the log-likelihood, increasing its trend to convergence (the risk of terminating in a non global maximum is of course still the same).

In [6], Chalmond proposed a modified procedure for the case of finite F^h with a small number of elements, in which the maximization step is simplified. He proposed to replace the likelihood in the expression of g_n (cf (5)) by the pseudo likelihood, of which we now recall the definition.

3.3.2 Pseudo likelihood.

Pseudo likelihood has been introduced in the case of complete observations as an alternative to maximum of likelihood estimation. It requires much less computation at the risk of losing some efficiency. It has been inspired by the coding technique that Besag ([3]) developed for fields with bounded range. For such fields, it is possible to extract some maximal subset of $\mathbf{Z}^d$ such that the variables at sites from this sublattice are mutually independent given variables from the rest of the sites. Let's denote this sublattice by L, the coding estimator associated to L is a maximizer of the logarithm of the conditional probability on $L \cap D$, given the data at $D \setminus (L \cap D)$:

$$p_L(\theta) = \sum_{s \in L \cap D} \log \pi_\theta(x_s | x_t\ t \neq s)$$

where $\pi_\theta(x_s|x_t, t \neq s)$ is the conditional law at site s given the rest of the sites.

Noting that efficiency can be expected to be improved if a sum of p_L's for several L's is maximized, it is natural to maximize the function (pseudo likelihood):

$$p(\theta) = p_D(\theta) = \sum_{s\in D} \log \pi_\theta(x_s|x_t,\ t \neq s)$$

Since conditional probabilities for bounded range models are easy to deal with, this function can be maximized with the help of some deterministic gradient descent algorithm. This method does not naturally generalize to the case of incomplete observations.

3.3.3 EM Gibbsian algorithm.

As said before, this algorithm essentially consists in replacing $l(\theta)$ by $p(\theta)$ in (5). We now describe it, with a parametric formulation a little bit different from the one given in [6].

The function g_n is now

$$g_n(\theta) = E_{\theta_n}[\frac{\mathrm{d}}{\mathrm{d}\theta}p(\theta)|Y = y] \tag{7}$$

Let's put:

$$H_s(x) = \sum_{c/\ s\in C} h_c(x)$$

and

$$K_s(x) = \sum_{c/\ s\in C} \frac{1}{|C|} h_c(x)$$

We have: $H(x) = \sum_{s\in D} K_s(x.x')$. Since the boundary condition x' will have negligible influence on what will follow, we shall make the abuse of notation: $H(x) = \sum_{s\in D} K_s(x)$.

We also have: $\frac{\mathrm{d}}{\mathrm{d}\theta}p(\theta) = \sum_{s\in D} \hat{H}_s(\theta) - H_s$, where

$$\hat{H}_s(\theta, x) = E(H_s|x_t,\ t \neq s)$$

(if $f(\theta, x)$ is a function on $\mathbf{R}^\nu \times \Omega$, we denote by $f(\theta)$ the function on Ω that associates $f(\theta, x)$ to x).

The maximization step is therefore done by solving:

$$E_{\theta_n}[\sum_{s\in D} \hat{H}_s(\theta) - H_s|Y = y] = 0$$

This is not completely trivial because $\hat{H}_s$ can depend in a non trivial way on θ. This however can be simplified when $|F^h|$ is small as well as the range of the field. This implies that $\hat{H}_s(\theta, x)$ will only depend on a few coordinates x_t for t in a neighborhood $V(s)$, of s, and will therefore take a reasonable number of different values. It is thus possible, by considering the different configurations $(x_t^h, t \in V(s))$, to divide $\Omega^y = \Omega^h \times \{y\}$ into K subsets $\Omega_l^y(s)$, $l = 1, \ldots, K$, independent of θ such that $\hat{H}_s(\theta, x) - H_s$ takes the constant value $f_{l,s}(\theta, y)$ on $\Omega_l^y(s)$. Since it depends on conditional probabilities for π, $f_{l,s}(\theta, y)$ can be explicitely expressed in terms of θ, y. Because of the bounded range, the property for a configuration to be element of $\Omega_l^y(s)$ will depend on its coordinates in a small neighborhood of s. Moreover, because of stationarity, one can arrange these sets so that:

$$\Omega_l^y(s + t) = T_s\Omega_l^y(t)$$

and

$$f_{l,s+t}(\theta, y) = f_{l,t}(\theta, T_s y).$$

It is indeed always possible to take as many $\Omega_l^y(s)$ as configurations $(x_t^h, t \in V(s))$, but symmetry properties for particular models can allow to choose larger $\Omega_l^y(s)$, and thus smaller K.

With these notations, the equation to be solved becomes:

$$\sum_{s\in D}\sum_{l=1}^{K} f_{l,s}(\theta, y)\pi_{\theta_n}[\Omega_l^y(s)|y] = 0 \tag{8}$$

The $\pi_{\theta_n}[\Omega_l^y(s)|y]$ have to be computed by Monte Carlo simulation. The final equation can the be solved by using standard deterministic procedures. We illustrate this with an example at the end of section 4.

In [6], the treatment is somewhat different. The conditional law of the observed field given the original one, ie. $P(Y = y|X^h = x^h)$ is assumed to be of the kind:

$$\prod_{s\in D} p(y_s|x_s). \tag{9}$$

The model is overparametrized by the family $p(y|x)$, for $y \in F^o$, $x \in F^h$ (both these sets being assumed finite; a different approach, consistent with ours, is used for Gaussian noise), and by the family of all local specifications of the law of the hidden field X^h. This yields explicit formulas for updating, once similar quantities to the $\pi_\theta[\Omega_l^y(s)|y]$ have been computed. At the end of the algorithm, the parameter θ_* is computed on the basis of local specifications by a least square procedure. We refer to [6] for details. Our approach, while requiring some extra computation, has the advantage of being more consistent with the parametric formulation; moreover, it permits more flexibility in the structure of the noise, since there is no need for asumptions like (9).

In the next section, we explore a little more some features of this algorithm.

4 Comments and simulation results on the EM Gibbsian algorithm.

The first remark that can be made, is that, if the algorithm converges, it provides a solution of the equation

$$\begin{aligned} E_\theta[\frac{\mathrm{d}}{\mathrm{d}\theta}p_\theta|y] &= 0 \\ \text{ie. } E_\theta[\sum_{s\in D}(\hat{H}_s(\theta) - H_s)] &= 0 \end{aligned} \tag{10}$$

We are therefore dealing with a moment estimator. Remark that there are similarities between the preceding equation, and equation (4) for the variational estimator. A stochastic gradient algorithm could also be used, as an alternative to the EMG for solving (10).

Several questions arise at this point. The first one is whether equation (10) has or not a unique solution. The answer is likely to be false, since this is not the case for the maximum of likelihood equations. The second question is whether the method provides a consistent estimator. In view of the first point, we do not expect to obtain more than a local result of consistency, of the kind: there exists a fixed neighborhood, $\mathcal{V}$, of θ_* such that, for large enough D, there exists only one solution of (10) in $\mathcal{V}$, denoted by $\hat{\theta}_D$, and $\hat{\theta}_D$ converges to θ_* when D tends to $\mathbf{Z}^d$. The last issue is of course: does the EM Gibbsian algorithm converge? there again, we should not expect more than local results, saying, for example that the algorithm converges provided its initial point lies in a small enough neighborhood of a solution of (10). Unfortunately, we were not able to give any positive answer to these questions, except under the quite restrictive hypothesis that are made in propositions 1 and 2. We shall then present a few numerical results.

Let D be fixed for a moment. Let's denote by $\phi(\theta_1, \theta_2)$ the function $E_{\theta_1}[p'(\theta_2)|y]$, where p' is the derivative of the logarithm of the pseudo likelihood. Step n of the EMG algorithm requires therefore the solution of

$$\phi(\theta_n, \theta) = 0.$$

In the following we assume that quantities that have to be estimated with simulation are exactly computed. We just study the deterministic part of the algorithm, which has its own complexity. To take into account perturbations due to imperfect estimation of these quantities would require to introduce concepts related to stochastic approximation. However, it is clear from this theory that, since estimation can never be perfect, the sequence θ_n computed in the algorithm will have some "large deviation" behaviour, preventing it from converging. In good cases, this theoretical limitation will never be seen in practice (because time required for "explosion" is very large), but it can happen (and it indeen happens, see the end of section 4) that this behaviour becomes typical. To remedy to this, one possibility is to replace θ_{n+1} by $\bar{\theta}_{n+1} = \theta_n + \frac{a}{n+b}(\theta_{n+1} - \theta_n)$ (reminiscent of stochastic approximation with decreasing steps), and, if necessary, to determine some suitable compact set in which the parameter must lie, and prevent θ_n to go out of it. [1] At this point, it might even be better to directly use a stochastic gradient algorithm of the kind:

$$\theta_{n+1} = \theta_n + \frac{a}{n+b}(\hat{\phi}(\theta_n, \theta_n))$$

where $\hat{\phi}(\theta_n, \theta_n)$ is an approximation of $\phi(\theta_n, \theta_n)$ computed by Monte Carlo simulation. Following ideas of section 3.1, this approximation can be rough provided a is small enough and simulation and updating of the parameter are mixed.

We first study the behavior of ϕ with regard to the second coordinate. Note that if θ_1 is fixed, $\phi(\theta_1, \theta)$ is the derivative of $E_{\theta_1}[p(\theta)|y]$, which is concave in θ. Indeed, we have, by a straightforward computation:

$$\frac{\partial \phi}{\partial \theta_2} = -E_{\theta_1}\{\sum_{s \in D} \mathrm{var}_{\theta_2}[H_s|X_t,\ t \neq s]|y\}$$

We shall assume that there exists no non zero solution $\tau \in \mathbf{R}^\nu$ of the system of equations:

$${}^t\tau \mathrm{var}_{\theta_2}[H_s|x_t,\ t \neq s]\tau = 0, \text{for } x \in \Omega.$$

Note that the number of equations is finite, because of the bounded range assumption, and that, by stationarity, this set of equations does not depend on s. This is equivalent to the assumption that there is no solution to the system ${}^t\tau H_s(x) = 0, \text{for } x \in \Omega$ (because all conditional probabilities are positive). Since all marginal probabilities on finite domains for the law given y are positive, and since $\mathrm{var}_{\theta_2}[H_s|x_t,\ t \neq s]$ only depends on a finite number of coordinates x_t, this

[1] In fact, this procedure (in the case when it converges) does not exactly provide a solution of (10). If, at each step, a sequence $X = (X_1, \ldots, X_N)$ of configurations is simulated to estimate all probabilities that are involved, the stable point of the algorithm is a solution of $\bar{E}_\theta(\hat{\eta}(\theta, X)) = \theta$, where $\bar{E}$ means expectation for the law of the sequence X when simulation is made with parameter θ, and $\hat{\eta}(\theta, X)$ is the solution τ of the approximated version of equation $\phi(\tau, \theta) = 0$ based on the simulated data X. If N is reasonably large, this limit can be expected to be very close to a fixed point of η (see below for a definition of η).

assumption implies that $E_{\theta_1}[p(\theta)|y]$ is a strictly concave in θ, and one obtains a well defined function, that we denote $\eta(\theta_1) = \eta_y(\theta_1)$ by:

$$\theta_2 = \eta(\theta_1) \iff \phi(\theta_1, \theta_2) = 0.$$

The EMG algorithm is thus an iterative procedure defined by : $\theta_{n+1} = \eta(\theta_n)$.

Let $\hat{\theta}$ be a solution of (10), or equivalently a fixed point of η, a sufficient condition for the existence of a neighborhood $\mathcal{V}$ of $\hat{\theta}$ such that $\theta_0 \in \mathcal{D} \Rightarrow \theta_n \to \hat{\theta}$, is that the largest eigenvalue of $\eta'(\hat{\theta})$ is smaller than one. Here we have

$$\eta'(\hat{\theta}) = -\frac{\partial \phi}{\partial \theta_2}^{-1}|_{(\hat{\theta},\hat{\theta})} \cdot \frac{\partial \phi}{\partial \theta_1}|_{(\hat{\theta},\hat{\theta})}$$

where

$$\frac{\partial \phi}{\partial \theta_1}|_{(\hat{\theta},\hat{\theta})} = \text{cov}_{\hat{\theta}}\{\sum_{s \in D}[H_s - \hat{H}_s(\hat{\theta})], \sum_{t \in D} K_t | \, y\}.$$

Two simple remarks can be made at this point.

1) In a situation of "maximal noise", when the conditional law given y does not depend on y (for example if $\langle \theta, H \rangle$ can be written $\langle \bar{\theta}, H_1(x^h) \rangle + \langle \alpha, H_2(x^h, y) \rangle$ and α is *known* to be 0), we have $\phi(\theta, \theta) = 0$ for all θ, and η' is the identity matrix, whose eigenvalues are 1. So, in this very bad situation, these values are at the boundary of their validity domain. Of course, there is no reason for this situation (which is the worst, for example in the case of maximum of likelihood estimation) to be the worst for the EMG case.

2) The opposite situation is when there is a one to one relation between Y and X. With our modeling, this typically corresponds to cases when some parameters tend to infinity. In that case, we obviously have $\eta' = 0$ (we compute pseudo maximum of likelihood for perfect observations), and there therefore exists some neighborhood of these (maybe infinite) parameters for which EMG algorithm is valid. In other terms, this algorithm can be used for large enough signal to noise ratio.

In this last remark, the validity domain can depend on D. To be able to make some uniform asymptotics, we shall need some additional assumptions.

Let y be given. For $D \subset \mathbf{Z}^d$, and some configuration $x' \in \Omega^h_{D^c}$, we define the law $\pi^y_{\theta,D}$ on Ω^h_D by

$$\pi^y_{\theta,D}(x|x') = \exp\{-\langle \theta, H_D[(x^h, y_D)|(x', y_{D^c})] \rangle)\}/Z_{\theta,D}(y) \tag{11}$$

where H_D has been defined near equation (1), and y_V is the restriction of y to the subset V of $\mathbf{Z}^d$.

This is the "approximate conditional neighborhood" with edge condition x'. The family $\pi^y_{\theta,D}(.|x')$, $D \subset \mathbf{Z}^d$, $x' \in \Omega^h_{D^c}$ forms a consistent family of conditional laws. We shall assume that, *for all* y, the family satisfies Dobrushin's unicity condition (cf [10]) which is based on bounds on the distances between $\pi_{\theta,\{s\}}(.|x')$ and $\pi_{\theta,\{s\}}(.|x'')$, for x' and x'' that differ on only one coordinate, and this for θ in a neighborhood of θ_*.

For example, assume that the energy has the following form (we omit D and x' in the notations):

$$\langle \theta, H(x^h, y) \rangle = \langle \theta_1, H_1(x^h) \rangle + \langle \theta_2, H_2(y) \rangle + \theta_3 \sum_s h_3(x^h_s, y_s).$$

(this is the case when the original field is degraded independently at each site). For this energy, Dobrushin's conditions are true provided θ_1 is small enough, or, in the case when F^o is finite and when for all $y_s \in F^o$, $h_3(.,y_s)$ has a unique minimum, when θ_3 is large enough . These conditions are very convenient to use, and yield explicit bounds (cf [43]). The weaker conditions given in [11] would be sufficient, but they appear to be very difficult to check (even with the help of a computer), because of the non-homogeneity of $\pi^y_{\theta,D}$.

We assumed these conditions for all y. In fact, it would be enough to assume them for ψ_{θ_*} almost all y, where ψ_{θ_*} is the true (unknown) law of the observations.

With these assumptions, there exists a neighborhood $\mathcal{V}$ of θ_* such that for all $\theta \in \mathcal{V}$ the following is true

i) There exists a unique field, denoted π^y_θ associated to this family, that is, for all $D \subset \mathbf{Z}^d$, finite:

$$\pi^y_{\theta,D}(x^h) = \int_{\Omega^h_{D^c}} \pi^y_{\theta,D}(x|x')\pi^y_{\theta,D^c}(dx')$$

where $\pi^y_{\theta,D}$ is the marginal of π^y_θ on Ω^h_D and we made the abuse of notation of writing $\pi^y_{\theta,D}(x^h)$ in spite of $\pi^y_{\theta,D}(x^h_D)$ (and the same for D^c).

ii) Let T_s be the shift operator on configurations (we use the same notation whenever this operator acts on Ω^h, Ω^o or Ω). We have $\pi^{T_s y}_\theta = \pi^y_\theta \circ T_{-s}$.

To see that, put $q = \pi^{T_s y}_\theta \circ T_s$.

We fix a finite D and a configuration $x^h = (x^h_t, t \in D) \in \Omega^h_D$, and denote by $T_{-s}x^h$ the element of Ω^h_{-s+D} for which the coordinate at site $-s+t$ is x^h_t. We omit to write the index θ.

$$\begin{aligned} q_D(x) = \pi^{T_s y}_{-s+D}(T_{-s}x^h) &= \int_{\Omega^h_{-s+D^c}} \pi^{T_s y}_{-s+D}(T_{-s}x^h|x')\pi^{T_s y}_{-s+D^c}(dx') \\ &= \int_{\Omega^h_{D^c}} \pi^{T_s y}_{-s+D}(T_{-s}x^h|T_{-s}x')q_{D^c}(dx') \\ &= \int_{\Omega^h_{D^c}} \pi^y_D(x|x')q_{D^c}(dx') \end{aligned}$$

(Note that these equalities do not rely on the uniqueness assumption; in the general case they imply relations between the sets of laws associated to families $\pi^y_{\theta,D}(.|x')$ and $\pi^{T_s y}_{\theta,D}(.|x')$; in the last one, we used translation invariance of the potential).

iii) Correlations for π^y_θ decay exponentially fast, in the sense that if f_1 and f_2 are respectively measurable functions of x_{D_1} and x_{D_2}, with $D_1 \subset \mathbf{Z}^d$ finite and $D_2 \subset \mathbf{Z}^d$, we have:

$$\mathrm{cov}(f_1, f_2) \le A\|f_1\|_\infty\|f_2\|_\infty|\partial D_1|\exp(-g.\mathrm{dist}(D_1, D_2))$$

where A and g are positive constants that only depend on the potential, and ∂D_1 is the boundary of D_1 in $\mathbf{Z}^d$. We assume that they do not depend on θ (by restricting if needed the set $\mathcal{V}$.

iv) Let $\pi_\theta \in \mathcal{G}(\theta)$; then π^y_θ is a version of the conditional law for π_θ given y.

Assuming this, we can state two propositions of which we shall only sketch the proof without entering into details. The first one is about the asymptotic behavior of the derivatives of ϕ.

Proposition 1 *Assume that π_{θ_*} is ergodic. Then for all $\theta_1 \in \mathcal{V}$, $\theta_2 \in \Theta$, for π_{θ_*} almost all y*

a)

$$\lim_{D\to\mathbf{Z}^d} \frac{1}{|D|} \sum_s E^y_{\theta_1,D}[\mathrm{var}_{\theta_2}(H_s|X_t, t \neq s)] = E_{\theta_*}\{E^y_{\theta_1}[\mathrm{var}_{\theta_2}(H_0|X_t, t \neq 0)]\}$$

The indices θ, D under expectations or covariances mean that they are computed accorded to some given approximate (conditional) marginal on D (the result is independent of the boundary condition). The limit $D \to \mathbf{Z}^d$ should be understood in the sense of Van Hove, or, for simplicity, for any sequence of d-dimensional cubes increasing to $\mathbf{Z}^d$.

b)

$$\lim_{D\to\mathbf{Z}^d} \frac{1}{|D|} \mathrm{cov}^y_{\theta_1,D}[\sum_s H_s - \hat{H}_s(\theta_2), \sum_t K_t] = \sum_t E_{\theta_*}\{\mathrm{cov}_{\theta_1}[H_0 - \hat{H}_0(\theta_2), K_t]\}$$

Boundary condition has an effect of order $\sqrt{|D|}$, which is negligible in the preceding limits.

To prove a), remark that, by i), the difference between $E^y_{\theta_1,D}[\mathrm{var}_{\theta_2}(H_s|X_t, t \neq s)]$ and the exact expectation $E^y_{\theta_1}[\mathrm{var}_{\theta_2}(H_s|X_t, t \neq s)]$ tends to 0; ii) and the ergodic theorem give the conclusion (we do not get into details).

To prove b), remark that, by i) and iii), it suffices to consider the limit of

$$\frac{1}{|D|} \mathrm{cov}_{\theta_1}[\sum_s H_s - \hat{H}_s(\theta_2), \sum_t K_t]$$

which, by ii) has the same limit as

$$\sum_t \frac{1}{|D|} \sum_s \mathrm{cov}^{T_s y}_{\theta_1}[H_0 - \hat{H}_0(\theta_2), K_t]\},$$

and this converges to

$$\sum_t E_{\theta_*}\{\mathrm{cov}_{\theta_1}[H_0 - \hat{H}_0(\theta_2), K_t]\}. \diamond$$

In addition, it can be proven by similar techniques that these functions are continuous in θ_1, θ_2. We shall denote by W_1 and W_2 the expectations given in the right-hand terms of a) and b) in the preceding proposition, computed for $\theta_1 = \theta_2 = \theta_*$. Note that

$$W_2 = W_1 - \sum_t \mathrm{cov}_{\theta_*}\{E^y_{\theta_*}[H_0 - \hat{H}_0(\theta_*)], E^y_{\theta_*}[K_t]\}.$$

We have the following (under the same hypothesis as in proposition 1)

Proposition 2 *For π_{θ_*} almost all y,*

a) If $W_1 - W_2$ is invertible, there exists a neighborhood of θ_* in which for all large enough D, there exists a unique solution $\hat{\theta}_D$ of $E^y_{\theta,D}[p'(\theta)] = 0$ and θ_D converges to θ_*.

b) if the largest eigenvalue (in modulus) of $W_1^{-1}W_2$ is strictly smaller than 1, there exists a neighborhood of θ_* such that any starting point in this neighborhood, the EMG algorithm converges to $\hat{\theta}_D$ defined in a).

a) is a straightforward application of the inverse mapping theorem and of the fact that

$$\frac{1}{|D|} E_{\theta_*,D}[p'(\theta_*)|y]$$

tends to 0 a.s. and b) is obvious from the preceding results. $\diamond$

Unfortunately, a) and b) in proposition 2 appear to be very difficult to check, even for very simple models. They are however true for *weak enough noise* (because of the continuity of W_1, W_2 in the parameters), with the advantage over remark 2 above that this statement is uniform with respect to $D \subset \mathbf{Z}^d$.

We have tried to check b) for a very simple non trivial case of imperfect observations on which some computation could be made. The original field is an isotropic nearest neighbor 2-D Ising field, and observation is made on even sites, ie. sites $(i,j) \in \mathbf{Z}^d$ such that $i+j$ is even. Even in that simple case, we couldn't go all the way down, relying on simulations at some point. Since computations are quite cumbersome, and this example has little practical interest, because we can precisely describe the law of the observations, and use any estimation method for complete observations, we do not give details here, just saying that, in trying to check $|W_2| < W_1$ (in that case, these quantities are real numbers), we had a rigorous proof of $W_2 < W_1$ and used simulation to check the other inequality that appeared to be true.

We conclude this paper by giving some numerical results in some situations that are closer to reality. All simulations (except when specified) are done on a 64×64 lattice.

The first one is the case of an Ising field with Bernouilli noise. In that case, the joint energy is:

$$U(x^h, y) = \sum_{<st>} \beta x_s^h x_t^h + \alpha \sum_s y_s x_s.$$

Where the first sum is over all pairs of neighbor sites, and x_s and y_s take values in $\{-1, 1\}$. X^h is therfore a regular isotropic Ising field and Y is obtained by flipping each site independently with probability $p = e^\alpha/(e^\alpha + e^{-\alpha})$. We assumed that the value of α was known. To illustrate the description of section 3.3.3, we explicitely express the algorithm. Equation (7) becomes:

$$g_n(\theta) = \sum_s E_{\theta_n}[v_s(x_s + \tanh(\beta v_s)) \mid y]$$

where v_s is the sum of x_t for all t nearest neighbors of s.

Since x_s takes values $-1, +1$, and v_s is one of $-4, -2, 0, 2, 4$, we define 5 sets $\Omega_k(s)$, $k = 1, \ldots, 5$, and the associated value $f_k(s)$ of $v_s(x_s + \tanh(\beta v_s))$ on these sets by:

- $\Omega_1(s) = \{(x_s, v_s) = (-1, 4) \text{ or } (x_s, v_s) = (1, -4)\}$, $f_1 = -4(1 - \tanh 4\beta)$.
- $\Omega_2(s) = \{(x_s, v_s) = (-1, 2) \text{ or } (x_s, v_s) = (1, -2)\}$, $f_3 = -2(1 - \tanh 2\beta)$.
- $\Omega_3(s) = \{(x_s, v_s) = (-1, 0) \text{ or } (x_s, v_s) = (1, -0)\}$, $f_3 = 0$.
- $\Omega_4(s) = \{(x_s, v_s) = (-1, -2) \text{ or } (x_s, v_s) = (1, 2)\}$, $f_4 = 2(1 + \tanh 2\beta)$.
- $\Omega_5(s) = \{(x_s, v_s) = (1, 4) \text{ or } (x_s, v_s) = (1, 4)\}$, $f_5 = 4(1 + \tanh 4\beta)$.

Put $\pi_k = \sum_s \pi_{\theta_n}[\Omega_k(s)]$, that can be computed by simulation; the equation to solve is

$$4(\pi_1 + \pi_5)\tanh 4\beta + 2(\pi_2 + \pi_4)\tanh 2\beta + 4(\pi_5 - \pi_1) + 2(\pi_4 - \pi_2) = 0.$$

We computed by simulation the mean values, that we still denote by W_1 and W_2, of the partial derivatives of ϕ at point (β, β), for several values of β and p. Even if propositions 1 and 2 were proven in situations that here correspond to small β or large α, we believe that these expectations still give interesting information in other cases. We present several figures that give the value of W_2/W_1 in function of α for fixed values of β. This value appears to always be between 0 and 1, but gets very close to 1 as soon as the noise becomes significant (note that

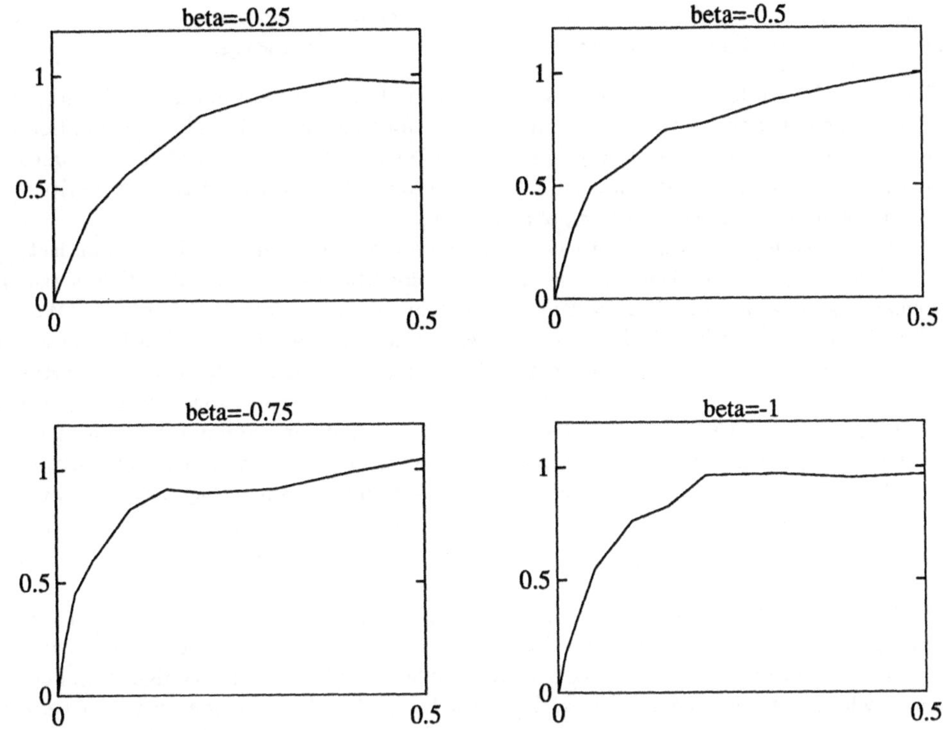

Figure 1: W_2/W_1 in function of α for 4 values of β

these are mean values, and that we observed in the simulations situations when the observed value of η' was larger than 1, in which case the algorithm might diverge.

This case of imperfect observations was studied in [13], where estimators for β and p were proposed. This paper shows the same kind of feature for their method, and the estimation of β for high noise was also very unaccurate. We also could observe this difficulty while experimenting the algorithm of section 3.1 for that situation. Bernouilli noise seems much more critical for estimation than, for example, additive gaussian noise.

The second example concerns precisely Gaussian noise. The hidden field is still an isotropic Ising field, and the observed field Y_s is defined by its conditional law given x^h which is gaussian with mean μ and variance σ^2 if $x_s^h = 1$ and with mean μ' and variance σ'^2 if $x_s^h = 1$. We shall also consider a submodel for which $\sigma = \sigma'$ (this submodel includes additive Gausian noise). We computed, for various values of the parameter, the largest eigenvalue of $W_1^{-1}W_2$. It is worthwhile to remark that for that model, this matrix is not the same for various laws in $\mathcal{G}(\theta_*)$, and we made the simulation with boundary condition equal to 1. This problem did not appear for the preceding model, because we had to estimate expectations of even functions. The mean values W_2 and W_1 are here meaningless in the case of non-ergodic π_{θ_*}, since they are not converging to the empirical expectations they are supposed to approximate. In fact, even when the true law π_{θ_*} is not ergodic, the relevant quantities are values of W_1 and W_2 for ergodic Gibbs fields in $\mathcal{G}(\theta_*)$ of which π_{θ_*} is a mixture.

In all our simulations, this largest eigenvalue was smaller than 1 (see table 4), apparently indicating some stability properties for the algorithm (note however that these values are very close to 1). But, computer experiments with this model show that the EMG algorithm does not converge, whereas it has a completely satisfying behavior for the submodel in which variances are assumed to be equal. To understand this, several remarks can be made. First, propositions 1 and 2 are not valid for all range of the parameter, but EMG behaves poorly even for very weak noise, in which case they should apply. Secondly, proposition 1 and 2 are asymptotic results: large size of the lattice is needed for the estimated parameter $\hat{\theta}_D$ to be close enough of θ_* to ensure that W_1 and W_2 are good approximations for the derivatives of η at $\hat{\theta}_D$. The third reason for non convergence is related to the remark we made at the beginning of this section, concerning the problem of accuracy of the estimation of the $\pi^y(\Omega_k(s))$, and consequence it may have in divergence of the algorithm. This clearly appears in the expressions of the estimated μ' and σ' at step n which are:

$$\mu' = \frac{\sum_s y_s \pi^y(x_s = -1)}{\sum_s \pi^y(x_s = -1)}$$

$$\sigma'^2 = \frac{\sum_s y_s^2 \pi^y(x_s = -1)}{\sum_s \pi^y(x_s = -1)} - \mu'^2$$

The fact that, for Ising fields below phase transition, the number of -1 is very small (edge condition is 1), implying imprecision in the estimation of $\pi^y(x_s = -1)$, yielding wrong updating of μ' and particularly of σ' (in the case of the sub-model, the variance is estimated using all sites and is therefore accurate, improving stability of the algorithm). In addition, we can note that if during the algorithm, it happens that $\mu = \mu'$ and $\sigma = \sigma'$, the field X^h does not depend on Y anymore, and the algorithm is lost and has no chance to find its way back on the right track.

To illustrate this we ran the EMG algorithm, estimating two variances for lattices of size 64×64 and 128×128, and various numbers N of iterations of a Swendsen-Wang procedure to estimate the $\pi^y(\Omega_k(s))$. Here are the results ($\beta = -0.8$, $\mu = 5$, $\mu' = 9$); it seems hard to determine which of the preceding reasons is relevant (probably all of them).

Table 1: Largest eigenvalue, λ, of $W_1^{-1}W_2$ for various parameters

β	μ	μ'	σ^2	σ'^2	λ
-1	-1	1	2	2	0.93
-1	-.5	.5	2	2	0.95
-1	-.05	.05	2	2	0.99
-1	1	5	1	1	0.8
-1	1	5	1	2	0.88
-1	1	5	1	5	0.92
-1	1	5	1	10	0.87
-0.8	5	9	1	1	0.68
-0.8	5	9	5	5	0.75
-0.8	5	9	7.5	7.5	0.97
-0.8	5	9	15	15	0.94

- size: 128×128, $\sigma = \sigma' = 2.5$ $N = 60$. After 21 iterations, $\sigma' = 0$ and the image degenerates in all 1's.
- size: 128×128, $\sigma = \sigma' = 2.5$ $N = 100$. After 24 iterations, $\sigma' = 0$ and the image degenerates in all 1's.
- size: 128×128, $\sigma = \sigma' = 1$, $N = 60$. After 24 iterations, one gets $\mu = \mu'$, $\sigma = \sigma'$, and the rest of the algorithm becomes independent on the observations...
- size: 128×128, $\sigma = \sigma' = 1$, $N = 100$. The algorithm seems stable.
- size: 64×64, $\sigma = \sigma' = 2.5$ $N = 60$. After 24 iterations, the image degenerates in all 1's.
- size: 64×64, $\sigma = \sigma' = 2.5$ $N = 100$. After 16 iterations, the image degenerates in all 1's.
- size: 64×64, $\sigma = \sigma' = 1$, $N = 60$. The algorithm seems to converge, with a very unaccurate value for σ'^2: 0.05.
- size: 64×64, $\sigma = \sigma' = 1$, $N = 100$. $\mu = \mu'$, $\sigma = \sigma'$ after 16 iterations.

When trying smaller values of $-\beta$ that yield a larger number for -1's, performances are improved although the same kind of phenomena can still be observed from time to time.

References.

[1] Almeida, B. Gidas: A variational Method for Estimating the Parameters of MRF from Complete or Incomplete Data. (to appear in *Annals of Applied Probability* (1991).

[2] A. Benveniste, M. Métivier, P.Priouret: *Algorithmes Adaptatifs et Approximations Stochastiques, Théorie et Application. Techniques Stochastiques.* Masson (1987).

[3] J. Besag: Spatial Interaction and the Statistical Analysis of Lattice Systems. *J. of Roy. Stat. Soc.* B-36 pp 192-236 (1974).

[4] J.Besag: On the Statistical Analysis of Dirty Pictures. *J. of Roy. Stat. Soc.* B-48 n^3, pp259-303 (1986) (with discussion).

[5] B. Chalmond: Image Restoration Using an Estimated Markov Model. *Signal Processing* 15 pp. 115-129 (1987)

[6] B. Chalmond: An iterative Gibbsian Technique for reconstruction of m-ary images. *Pattern Recognition.*Vol. 22 No. 6 pp. 747-761.(1989)

[7] T.S. Chiang, Y. Chow and C.R. Hwang : Diffusion for Global Optimization in $\mathbf{R}^n$. *SIAM J. control and optimization*, 25 pp 737-753.

[8] F. Comets and B. Gidas: Parameter estimation for Gibbs distributions, II: Partially Observed Data (to appear in *Annals of Applied Probability* (1991).

[9] F. Comets: On the Consistency of a Class of Estimators for Exponential Families of MRF on the Lattice. preprint univ. Paris X.

[10] R.L. Dobrushin: The Description of a Random Field by Means of Conditional Probabilities and Conditions of its Regularity. *Thry. Prob. Appl.* XIII-2 p. 197-224 (1968).

[11] R.L. Dobrushin: Prescribing a System of Random Variables by Conditional Distributions. *Thry. Prob. Appl.* XV-3 p.458-486 (1970).

[12] R.L. Dobrushin and S. B. Schlosman in *Statistical Mechanics and Dynamical Systems* Ed. Fritz, Jaffe, Szasz. Birkauser- Boston (1985).

[13] A. Frigessi - M. Piccioni: Parameter Estimation for the two-dimensional Ising Fields Corrupted by Noise. *Stoch. Proc. and Appl.* 34 pp. 297-311. (1990)

[14] S.B. Gelfand and S. Mitter: Recursive Stochastic Algorithm for Global Optimization in $\mathbf{R}^d$. preprint (1990).

[15] D. and S. Geman: Stochastic Relaxation, Gibbs Distribution and Bayesian Restoration of Images. *IEEE TPAMI.* Vol PAMI-6 pp 721-741 (1984).

[16] D. Geman: Parameter Estimation for MRF with Hidden Data and Experiments With the EM Algorithm. Preprint. (1984)

[17] D. Geman: Random Fields and Inverse Problems in Imaging. Lecture notes at the Ecole D'Eté de Saint Flour.

[18] D. Geman and D.E. Mc Clure: Statistical Methods for tomographic image reconstruction. In: *Proceedings of the 46-th session of the International Statistical Institute* Bulletin of the ISI, vol 52. (1987).

[19] S. Geman and C. Graffigne : Markov Random Field Image Models and Their Applications to Computer Vision. In *Proceedings of the International Congress of Mathematicians 1986* Ed. A.M. Gleason. American Mathematical Society, Providence (1987).

[20] S. Geman and C-R Hwang :Diffusion for Global Optimization. *SIAM J. control and optimization*, 24 pp 1031-1043.

[21] B. Gidas: Consistency of Maximum Likelihood and Pseudo Maximum Likelihood for Gibbs Distributions. In *Stochastic Differential Systems With Appl. Elec. Comp. Eng., Op. Res.* Springer Verlag.

[22] B. Gidas: Parameter estimation for Gibbs distributions, I: Fully Observed Data (to appear in *Markov Random Fields: Theory and Applications*, Academic Press (1991).

[23] X.Guyon: Pseudo Maximum de Vraisemblance et Champs Markoviens. In *Spatial Processes and Spatial Time Series Analysis. Proc. 6th. Franco-Belgian Meeting of Statisticians.* Droesbeke F. (Ed.). (1987)

[24] C.R. Hwang and C.R. Sheu: On the behavior of a stochastic algorithm with Annealing. preprint (1990)

[25] H.Künsch: Asymptotically Unbiased Inference for Ising Models. *J. of Appl. Prob.* 19 A 345-357 (1982).

[26] H. Künsch et H. Dalhaus: Edge Effect and Efficient Parameter Estimation for Stationary Random Fields. Preprint 49. ETH Zentrum Zürich. (1986).

[27] H.L. Kushner: Asymptotic global behavior for Stochastic Approximations and Diffusion with Slowly Decreasing Noise Effects: Global Minimization via Monte Carlo. *SIAM J. of Appl. Math* 47 pp. 169-185 (1987).

[28] S. Lakshamanan and H. Derin: Simultaneous Parameter Estimation and Segmentation of Gibbs Random Fields. *IEEE trans. Pattern ana. and Machine Intell.* 11 pp 799-813.

[29] O.E. Landford et D. Ruelle: Observable at Infinity and States with Short Range Correlations in Statistical Mechanics. *Comm. Math. Phys.* 13 p.194-215 (1969)

[30] A.Lippman: A Maximum Entropy Method for Expert Systems. Brown University Thesis (1986).

[31] J. L. Marroquin, S. Mitter and T. Poggio: Probabilistic solution of Ill Posed Problems in Computational Vision. *J. Am. Stat. Assoc.* 82 pp 76-89 (1987).

[32] M. Métivier et P. Priouret: Théorèmes de Convergence p.s. pour une Classe d'Algorithmes Stochastiques à Pas Décroissants. *Prob. Th. Rel. Fields* 74, pp. 403-428. (1987).

[33] N. Metropolis, A.W. Rosenbluth, M.N. Rosenbluth, A.H. Teller et E. Teller: Equations of State Calculation for Fast Computing Machines. *J. Chem. Phys.* Vol 21. p.1087-1091. (1953).

[34] Nguyen Xuan Xanh et H. Zessin: Ergodic Theorems for Spatial Processes. *Z. Wahr. Verw. Geb.* 48; p. 133-158 (1979).

[35] D. Pickard: Asymptotic Inference for an Ising Lattice. *J. Appl. Prob.* 13, pp 486-497, 1976.

[36] D. Pickard: Asymptotic Inference for an Ising Lattice II. *Adv Appl. Prob.* 9, pp 479-501 (1977).

[37] D. Pickard: Asymptotic Inference for an Ising Lattice III. *J. Appl. Prob.* 16, pp 12-24 (1979).

[38] D. Pickard: Inference for General Ising Models. *J. Appl. Prob.* 19 A, pp 345-357 (1982).

[39] A. Possolo: Estimation of Binary Markov Random Fields. University of Washington, Technical Report (1986).

[40] C. Preston:*Random Fields.* In Lect. Notes in Math. Vol. 534. Berlin, heidelber, Newyork. Springer (1976).

[41] B. Prum: *Processus sur un réseau et mesures de Gibbs; applications.* Techniques stochastiques. Masson. (1986).

[42] D. Ruelle: *Thermodynamics Formalism.* In Encyclopedia of Mathematics and its applications. Vol. 5. Addissin-Wesley. (1978).

[43] B. Simon: A Remark on Dobrushin Uniqueness Theorem. Comm. Math. Phys. p.183. (1979).

[44] A. Sokal: Monte Carlo Methods in Statistical Mechanics. Lecture notes: Cours de troisième cycle de la physique en Suisse Romande.

[45] L.Younes: Couplage de l'estimation et du recuit pour des champs de Gibbs. *C. R. Acad. Sc. Paris, t. 303*, série I, No. 13 (1986).

[46] L.Younes: Estimation and Annealing for Gibbsian Fields. *Ann. de l'Inst. Henri Poincaré* vol 2 (1988).

[47] L. Younes: Parametric Inference For Imperfectly Observed Gibbsian Fields.*Prob. Thry. Rel. Fields* 82, pp 625-645. (1989)

[48] L. Younes: Maximum of likelihood estimation for Gibbs Fields. To appear in *Proceeding of the 1988 AMS-IMS-SIAM joint conference on Spatial Statistics and Imaging* (1988).

[49] L. Younes: Problèmes d'estimation paramétrique pour des champs de Gibbs Markoviens. Application au traitement d'images. Thesis at Université Paris Sud. (1988).

Lecture Notes in Statistics

Vol. 44: D.L. McLeish, Christopher G. Small, The Theory and Applications of Statistical Inference Functions. 136 pages, 1987.

Vol. 45: J.K. Ghosh, Statistical Information and Likelihood. 384 pages, 1988.

Vol. 46: H.-G. Müller, Nonparametric Regression Analysis of Longitudinal Data. VI, 199 pages, 1988.

Vol. 47: A.J. Getson, F.C. Hsuan, {2}-Inverses and Their Statistical Application. VIII, 110 pages, 1988.

Vol. 48: G.L. Bretthorst, Bayesian Spectrum Analysis and Parameter Estimation. XII, 209 pages, 1988.

Vol. 49: S.L. Lauritzen, Extremal Families and Systems of Sufficient Statistics. XV, 268 pages, 1988.

Vol. 50: O.E. Barndorff-Nielsen, Parametric Statistical Models and Likelihood. VII. 276 pages, 1988.

Vol. 51: J. Hüsler, R.-D. Reiss (Eds.), Extreme Value Theory. Proceedings, 1987. X, 279 pages, 1989.

Vol. 52: P.K. Goel, T. Ramalingam, The Matching Methodology: Some Statistical Properties. VIII, 152 pages, 1989.

Vol. 53: B.C. Arnold, N. Balakrishnan, Relations, Bounds and Approximations for Order Statistics. IX, 173 pages, 1989.

Vol. 54: K.R. Shah, B.K. Sinha, Theory of Optimal Designs. VIII, 171 pages. 1989.

Vol. 55: L. McDonald, B. Manly, J. Lockwood, J. Logan (Eds.), Estimation and Analysis of Insect Populations. Proceedings, 1988. XIV, 492 pages. 1989.

Vol. 56: J.K. Lindsey, The Analysis of Categorical Data Using GLIM. V, 168 pages. 1989.

Vol. 57: A. Decarli, B.J. Francis. R. Gilchrist, G.U.H. Seeber (Eds.), Statistical Modelling. Proceedings, 1989. IX, 343 pages. 1989.

Vol. 58: O.E. Barndorff-Nielsen, P. Blæsild, P.S. Eriksen, Decomposition and Invariance of Measures, and Statistical Transformation Models. V, 147 pages. 1989.

Vol. 59: S. Gupta, R. Mukerjee, A Calculus for Factorial Arrangements. VI, 126 pages. 1989.

Vol. 60: L. Györfi, W. Härdle, P. Sarda, Ph. Vieu, Nonparametric Curve Estimation from Time Series. VIII, 153 pages. 1989.

Vol. 61: J. Breckling, The Analysis of Directional Time Series: Applications to Wind Speed and Direction. VIII, 238 pages. 1989.

Vol. 62: J.C. Akkerboom, Testing Problems with Linear or Angular Inequality Constraints. XII, 291 pages. 1990.

Vol. 63: J. Pfanzagl, Estimation in Semiparametric Models: Some Recent Developments. III, 112 pages. 1990.

Vol. 64: S. Gabler, Minimax Solutions in Sampling from Finite Populations. V, 132 pages. 1990.

Vol. 65: A. Janssen, D.M. Mason. Non-Standard Rank Tests. VI, 252 pages. 1990

Vol. 66: T. Wright, Exact Confidence Bounds when Sampling from Small Finite Universes. XVI. 431 pages. 1991.

Vol. 67: M.A. Tanner, Tools for Statistical Inference: Observed Data and Data Augmentation Methods. VI, 110 pages. 1991.

Vol. 68: M. Taniguchi, Higher Order Asymptotic Theory for Time Series Analysis. VIII, 160 pages. 1991.

Vol. 69: N.J.D. Nagelkerke, Maximum Likelihood Estimation of Functional Relationships. V, 110 pages. 1992.

Vol. 70: K. Iida, Studies on the Optimal Search Plan. VIII, 130 pages. 1992.

Vol. 71: E.M.R.A. Engel, A Road to Randomness in Physical Systems. IX, 155 pages. 1992.

Vol. 72: J.K. Lindsey, The Analysis of Stochastic Processes using GLIM. VI, 294 pages. 1992.

Vol. 73: B.C. Arnold, E. Castillo, J.-M. Sarabia, Conditionally Specified Distributions. XIII, 151 pages. 1992.

Vol. 74: P. Barone, A. Frigessi, M. Piccioni (Eds.), Stochastic Models, Statistical Methods, and Algorithms in Image Analysis. Proceedings, 1990. VI, 258 pages. 1992.

GPSR Compliance

The European Union's (EU) General Product Safety Regulation (GPSR) is a set of rules that requires consumer products to be safe and our obligations to ensure this.

If you have any concerns about our products, you can contact us on ProductSafety@springernature.com

In case Publisher is established outside the EU, the EU authorized representative is:

Springer Nature Customer Service Center GmbH
Europaplatz 3
69115 Heidelberg, Germany

Batch number: 10704771

Printed by Printforce, the Netherlands